Climate Warming in Western North America

Climate Warming in Western North America

Evidence and Environmental Effects

Edited by
FREDERIC H. WAGNER

THE UNIVERSITY OF UTAH PRESS
Salt Lake City

The Defiance House Man colophon is a registered trademark
of the University of Utah Press. It is based upon a four-foot-tall,
Ancient Puebloan pictograph (late PIII) near Glen Canyon, Utah.

13 12 11 10 09   1 2 3 4 5

LIBRARY OF CONGRESS CATALOGING-IN-PUBLICATION DATA

Climate warming in western North America : evidence and environmental
effects / edited by Frederic H. Wagner.
        p. cm.
    Includes index.
    ISBN 978-0-87480-906-0 (pbk. : alk. paper)   1. Climatic changes—
Great Basin.   2. Water-supply—Environmental aspects—Great Basin.
3. Great Basin—Climate.   4. Climatic changes—Rocky Mountains Region.
5. Water-supply—Environmental aspects—Rocky Mountains Region.   6. Rocky
Mountains Region—Climate.   I. Wagner, Frederic H.
    QC903.2.G68C56 2009
    551.679—dc22   2009035396

Contents

Figures

Tables

Preface

As the space committed to global warming in the media grows almost daily, the American public is increasingly persuaded of its reality, but it may also be confused by the skeptical signals coming from some quarters. Part of the confusion arises from failure in thought and discourse to separate two distinct aspects of this complex issue: (1) whether the global climate is in fact warming, what is causing it, and what are its consequences; and (2) what, if anything, should be done about it?

Answers to the first aspect are provided by the natural sciences: atmospheric physics, climatology, oceanography, ecology, hydrology, agronomy. Answers to the second are primarily economic, political, and legal. The confusion is exacerbated by individuals who, opposing any action for ideological, political, or economic reasons, attempt to head off action by raising doubts about the validity of the scientific evidence.

This book presents evidence on the scientific aspects for policy makers who need facts on the issue in order to formulate their courses of action. Western governors and mayors have been especially aggressive in addressing the issue. The evidence is also provided for citizens who wish to form knowledgeable opinions on the nature, cause, and effects of global warming so that they can adopt their own positions and lend appropriate support to their elected officials.

The book presents evidence on the reality of climate warming in western North America during the twentieth century; provides evidence that it is human caused; reviews effects to date on water resources, agriculture, and natural biota; and projects changes in these sectors in the twenty-first century in response to continued warming.

Western North America is unique for assessing the reality and effects of climate warming because the latter impinges on the existing, lengthy latitudinal (~38°) and altitudinal (6,280 m) climatic gradients of the region along which biotic, hydrologic, economic, and social gradients have adapted. General circulation models (GCMs) predict different levels of warming at different latitudes and altitudes, and thus responses are likely to vary along these gradients, making the total effect exceedingly complex. Evidence bearing out those predictions will be presented in the following chapters.

Much of the West is arid to semiarid, and is thus highly sensitive to any trends in precipitation. Warming alone will increase evapotranspiration and, if not compensated by increased precipitation, will have a xerifying effect on the region's ecosystems and land uses.

The region is also unique in its heavy dependence on mountain streams, fed by seasonal melt and runoff from the previous winter's snowpacks, for its water resources. Some 85 percent of water used by humans in the West comes from surface water, three-fourths of which is fed by mountain streams.

Western water resources can be affected in three ways by climate warming. One is the increased evapotranspiration, and consequent loss, associated with rising temperatures. A second is the question of whether precipitation will increase or decrease as climates change.

Both of these effects will influence the amount of water available for human use in the region. Water resources in the West are already oversubscribed, and yet the human population is predicted to double in 30 to 40 years—the fastest growth in the U.S. At present, 80 percent of the region's water is allocated to agriculture, which is already losing its access to the growing urban demands. Perhaps ironically, two-thirds of the urban use is spent on watering lawns and gardens. Per capita culinary use in the region is a third higher than the national average according to US Geological Survey statistics (Utah Division of Water Resources 1997).

Climate warming also influences water resources by affecting the seasonal pattern of streamflow. Warming is changing snow to rain, which reduces the size of snowpacks and alters the seasonal timing and magnitude of spring thaw, runoff, and streamflow. These changes will have yet-to-be determined effects on the uses of western streams that are keyed to runoff seasonality: diversion for urban use, agricultural irrigation, power generation, and spawning runs of anadromous fish.

Another distinctive feature of the West is the high percentage—perhaps as much as 80 to 90 percent—of its area in natural vegetation. The biota is extremely diverse, with varying assemblages of plants and animals adapted to different points along the latitudinally and topographically imposed climatic gradients. Climate change is likely to produce complex changes in the biota. Some of these are already being observed.

The chapters that follow are based on (1) information presented in an assessment of the potential effects of climate change on the nine-state Rocky Mountain/Great Basin region (Wagner 2003), part of the national assessment ordered by Congress in the Global Change Research Act of 1991; (2) on updated papers presented in a symposium entitled "Future Climate Change: Implications for Western Environments," convened on June 15, 2004, as a part of the annual meeting of the American Association for the Advancement of Science, Pacific Division, held on the Utah State University campus; and (3) other research conducted by the many scientists in the region who have coauthored the chapters of this book. Drawing on these sources, this volume summarizes climate-change projections for the West during the twenty-first century; evidence of climate change from weather records and two proxies for change during the twentieth century; evidence of environmental effects during the twentieth century; and projections of potential effects during the twenty-first and subsequent centuries.

The first chapter, by F. H. Wagner, summarizes temperature and precipitation projections for the twenty-first century over the nine-state Rocky Mountain/Great Basin (RMGB) region by two general circulation models programmed with a projected two-fold increase in carbon dioxide. It also presents a detailed analysis of twentieth-century weather records for the region as a partial test of the twenty-first-century projections, given that atmospheric carbon dioxide rose by a third during the twentieth century.

The next two chapters report studies on two different temperature proxies. In Chapter 2, D. A. Chapman, R. N. Harris, and M. Bartlett infer trends in surface air temperatures over time from temperature profiles in boreholes in the earth. In Chapter 3, D. L. Naftz, L. Oswald, P. F. Schuster, and K. Miller deduce air-temperature

trends from western North American glacial ice cores and streamflow measurements obtained at high altitudes in the Wind River Mountains of Wyoming.

In Chapter 4, P. W. Mote describes the twentieth-century decline in spring snow-water equivalents in western, montane snowpacks, even in cases of increased precipitation. In Chapter 5, J. D. Lundquist, M. D. Dettinger, I. T. Stewart, and D. R. Cayan present changing hydrographs for California streams associated with shrinking snowpacks and precipitation changing from snow to rain.

In Chapter 6, M. A. White, G. V. Jones, and N. H. Diffenbaugh present evidence that the production of high-quality wine grapes—wine grapes are the most economically valuable agricultural crop in California—is affected by climate, and project that climate warming will shift the optimum climate for the crop from California northward to Oregon, Washington, and British Columbia.

Three chapters examine ecological changes stimulated by climate change. In Chapter 7, J. H. Matthews and C. Parmesan discuss shifts in animal species distributions, abundance, and seasonal timing that are altering natural community composition and ecosystem function. Their descriptions are prefaced, however, with a discussion of the scientific methodologies required to correctly infer cause and effect between changes in climatic parameters and ecological changes.

In Chapter 8, J. A. Logan and J. A. Powell review research on geographic and altitudinal shifts in mountain pine beetle distribution in the western United States and Canada that have resulted in widespread mortality of pine forests. They also point out the potential of pine beetle range extension in pines across the entire North American continent.

In a lengthy chapter (9), V. A. Barber and 12 coauthors summarize far-reaching changes in Alaskan ecosystems, where temperatures have risen more than at lower latitudes, and where the natural biota and human cultures have adapted for millennia to extreme cold.

A final chapter (10) by F. H. Wagner synthesizes results from these studies and relevant literature, summarizing evidence of the reality of change, of human causation, and of the twentieth-century environmental effects; he also projects environmental and socioeconomic consequences associated with continued warming.

References Cited

Utah Division of Water Resources. 1997. *The Utah Water Data Book.* Salt Lake City.

Wagner, F. H. (ed.). 2003. *Preparing for a Changing Climate: The*

Potential Consequences of Climate Variability and Change, Rocky Mountain/Great Basin. A Report for the U.S. Global Change Research Program, Utah State Univ., Logan.

PART I

Evidence and Projections
of Climate Warming

Climate Change Projected for the Twenty-First Century and Measured for the Twentieth in the Rocky Mountain/Great Basin Region

Frederic H. Wagner

Abstract

The nine-state Rocky Mountain/Great Basin (RMGB) climate-change assessment, part of the national assessment ordered by Congress with the Global Change Research Act of 1991, was carried out between 1998 and 2003. Its strategy was to develop climate-change scenarios for 2100 in the region, based in part on the projections of two general circulation models (GCMs) adapted for the national assessment at its outset. The models were programmed with a projected twofold increase in CO_2 by that date.

This chapter reports the projections of those models adapted for the RMGB, more-recent global projections by the 2007 report of the Intergovernmental Panel on Climate Change (IPCC), and current model simulations that project latitudinal shifts in the entire global climate system, including western North America. All sets of models make qualitatively similar projections for the end of the twenty-first century.

Since atmospheric CO_2 increased ~32 percent during the twentieth century, the RMGB assessment conducted a detailed analysis of the twentieth-century weather records for the nine-state region to explore whether the climate changes projected by the models for the twenty-first century had begun to materialize during the twentieth, in direction if not in degree.

The RMGB models have projected twenty-first-century temperature increases for the entire region, and Karl et al. (1996), analyzing National Climatic Data Center (NCDC) records, inferred increases throughout the West during the twentieth century. National Weather Service records also show temperature increases through the century averaged over all weather stations in the state of Utah.

But the RMGB analysis, using data from the Vegetation/Ecosystem Modeling and Analysis Project (VEMAP), found statistically significant temperature increases in five northwesterly and northern subregions, but not in three subregions, two of them eastern and one southeastern.

Similarly, Kittel et al. (2002), analyzing U.S. Historical Climatology Network (HCN) data, found significant twentieth-century temperature increase in the northern and central Rockies, but none in the southern Rockies. These results coincide with model projections of northward increase in the degree of warming.

The models also project greater increase in night-time minimum temperatures than in day-time maxima. This proved to be the case during the twentieth century as shown by the VEMAP and HCN data. The VEMAP data also show declining interannual variability of temperatures during the 1900s.

Although the RMGB models projected twenty-first-century precipitation increase for the nine-state region, the most recent models are projecting an increase for the northern portion of the region, but a decline for the southern. The RMGB twentieth-century weather record analysis shows significant increase in annual precipitation in five westerly and northern subregions, but no significance in three southeasterly and southern subregions.

Similarly, Kittel et al. (2002) found twentieth-century precipitation increase in the northern and central Rockies, but not in the southern. And the RMGB analysis of twentieth-century annual streamflows found increase in four northern streams in subregions where precipitation increased, but not in a stream at the southeastern edge of the region.

The analysis also found no increase in twentieth-century flow of five major tributaries of the Colorado

River, or in the Colorado itself, from 1926 to 2000. Moreover, these analyses showed increasing interannual variability both in northern annual precipitation and streamflow.

Introduction

In 1991, the U.S. Congress passed the Global Change Research Act directing the government's executive branch to assess the potential effects of predicted climate change and variability on the nation. The U.S. Global Change Research Program (GCRP) was established in response to this directive. Under coordination of the GCRP, the nation was divided into nineteen regions and five socioeconomic sectors that cut across the regions: agriculture, coastal and marine systems, forests, human health, and water. Each region and sector was then assigned the task of assessing the potential effects of climate change on likely-to-be influenced socioeconomic sectors within its borders or within the national sector.

The Rocky Mountain/Great Basin region, one of the nineteen identified nationally, encompasses parts of nine states (Fig. 1.1). The regional assessment was begun in 1998 and completed in 2003 (Wagner 2003). The strategy used to assess the potential effects of climate change was to select a set of five socioeconomic sectors particularly important to the region and most likely to be affected by climate change: those selected were water resources, cultivated agriculture, livestock ranching, outdoor recreation and tourism, and natural ecosystems.

The next step was to formulate a set of climate-change scenarios for the twenty-first century based on (1) deliberations of western North American climatologists at a 1998 workshop (Stohlgren 2003); (2) projections of two major general circulation models (GCMs) parameterized with a projected doubling of atmospheric CO_2 by 2100, and adapted for the national assessment at that time (Mearns 2003); and (3) a detailed analysis of twentieth-century weather and streamflow records for the RMGB region to ascertain whether the changes projected by the GCMs for the twenty-first century had begun to appear in the twentieth, during which atmospheric CO_2 increased ~32 percent (Baldwin 2003, Baldwin et al. 2003). The climate-change scenarios were then interfaced with the five sectors to project how the latter would be affected—whether hydrologically, socioeconomically, or ecologically—by the end of the twenty-first century.

Since completion of the RMGB assessment, climate scientists around the world have worked to develop and refine numerous general circulation models, and they have recently developed models incorporating new insights into major shifts in the earth's climate system. Hence there is now a basis for comparing twenty-first-century climate-change projections of models at varying stages of development and sophistication that incorporate an ever-increasing understanding of the climate system as developed by different scientists.

The primary purpose of this chapter is to report the climate changes projected for 2100 in the RMGB by the two models used in the assessment (the British Hadley Circulation Model 2 [HadCM2] and the Canadian Coupled General Circulation Model 1 [CGCM1]) and the broader predictions of the newer models, which, while global in scope, embrace the RMGB region. The second purpose is to present the results of the detailed analysis of RMGB weather and streamflow trends during the twentieth century, which provide tentative evidence of the validity of the model projections.

Model Projections of Climate Change in the Rocky Mountain/Great Basin Region in the Twenty-First Century

Hadley and Canadian Model Projections in 2003

General circulation models are approximations of real-world climate systems structured on the basis of the major physical forces shaping climate behavior. At the time of the regional climate assessment, they were structured to simulate climate behavior at global scales. Thus their development was not yet at sufficient resolution to simulate changes for every localized area or, very satisfactorily, at subregional levels.

Climate variability is greatest at the local level—particularly in the RMGB region, with its complex topography. Lesser variability is found at subregional scales, and the least at regional scales. Hence model outputs must be considered simulations of the averages of local climates over large spatial scales. Since the models are of necessity simplifications of real-world systems, their outputs must be considered hypotheses of the behavior of those systems.

Table 1.1 summarizes the seasonal projections for the period 2080–2100 by two atmospheric general circulation models that were considered state-of-the-art and selected for use at the beginning of the U.S. regional assessments; the British Hadley and the Canadian General Circulation Model were adapted for western North America with a projected doubling of CO_2 (Mearns 2003).

Table 1.1. Projected seasonal changes in temperature and precipitation for the Rocky Mountain/Great Basin region by the HadCM2 and CGCM1 models for the period 2080–2100

| Season | Temperature Change | | | |
| | HadCM2 | | CGCM1 | |
	degrees C	degrees F	degrees C	degrees F
Winter	+4.5	+8.1	+8.0	+14.4
Spring	+2.5	+4.5	+6.0S., 7.0N	+10.8S, 12.6N
Summer	+4.0	+7.2	+5.0	+9.0
Fall	+3.5	+6.3	+6.0	+10.8
Mean Annual	+3.6	+6.5	+6.3–6.5	11.3–11.7

| Season | Precipitation Change | | | |
	mm/day	in./mo	mm/day	in./mo
Winter	+2.0 to 3.0	+2.4 to 3.5	+2.0 to 4.0[1]	+2.4 to 4.7[1]
Spring	0.0 to +0.5	0.0 to +0.6	+1.0	+1.2
Summer	0.0 to +0.5	0.0 + 0.6	-0.75 to +0.5	-0.9 to +0.6
Fall	+0.5 to 1.5	+0.6 to 1.8	+0.5 to 3.0	+0.6 to 3.5
Annual	+225.0–495.0	+9.0–19.5	+247.5–765.0	+9.9–30.0

[1]Higher in the west.

Both models projected temperature increases over the RMGB region in all seasons by the end of the twenty-first century. The greatest changes are projected for winter: 4.5°C (8.1°F) by HadCM2 and 8.0°C (14.4°F) by CGCM1. The higher CGCM1 projections were considered relatively extreme scenarios, while the Hadley values were considered moderate. The projected increases for the mean annual temperatures (Table 1.1), obtained by averaging the four seasonal values, are 3.6°C (6.5°F) for HadCM2, and 6.3–6.5°C (11.3–11.7°F) for CGCM1.

The precipitation projections, while again generally similar, differed slightly between the two models (Table 1.1). Both project some increase in all seasons, but the Canadian model is somewhat equivocal on summer change. Both indicate the greatest increase in winter. As with temperature, the Canadian model projects more increase than the Hadley.

The per annum GCM-projected increases by 2080–2100 were calculated by multiplying each seasonal mm/day value in Table 1.1 by 90 and adding all four; and multiplying each inch/month value by 3 and adding all four. These become 225–495 mm and 9.0–19.5 inches in 2100 above 2000 levels, according to the Hadley projections; and 247.5–765.0 mm and 9.9–30.0 inches according to the Canadian model.

These numbers, when transformed into percentage increases over statistically derived Year 2000 precipitation, become:

HadCM2: +59–127 percent
CGCM1: +65–196 percent

The methodology for deriving these latter values will be explained below after calculating the Year 2000 precipitation values.

More-Recent Model Projections

The Intergovernmental Panel on Climate Change published its fourth five-year assessment report in 2007 based on the work of hundreds of scientists around the world. Its Summary for Policy Makers (Albritton et al. 2007), coauthored by fifty-nine investigators internationally in Working Group 1, is a useful, up-to-date summary of the scientific aspects of climate change and the recent research on climate models.

Based on more than a decade of research using numerous models, the report's climate-change projections based on the more advanced models can be compared among themselves, and also collectively with the projections of the Hadley and Canadian models used earlier for the

RMGB assessment. Although these IPCC models project global change, it is instructive to compare them with the projections adapted for the West at the time of the national assessment.

The summary report (Albritton et al. 2007:13) states that "The globally averaged [mean annual] surface temperature is projected to increase by 1.4 to 5.8°C [2.5–10.4°F]…over the period 1990 to 2100." As stated above, the Hadley and Canadian models at the time of the assessment projected 3.6°C and 6.3–6.5°C (6.5°F and 11.3–11.7°F). While the two groups of models have geographically different perspectives, they still project roughly comparable order-of-magnitude temperature change despite a decade of additional research.

The summary report also projects "reduced diurnal temperature range over most land areas." This pattern had already appeared in RMGB temperatures during the twentieth century, as will be shown below, as a consequence of the more pronounced rise in night-time minimum temperatures than in daytime maxima.

The report also projects "for a wide range of [model] scenarios" that:

> …global average water vapour concentration and precipitation are projected to increase during the twenty-first century. By the second half of the twenty-first century, it is likely that precipitation will have increased *over northern mid- to high latitudes* [emphasis added] in winter. At low latitudes there are both regional increases and decreases over land areas. Larger year to year variations in precipitation are likely over most areas where an increase in mean precipitation is projected.

Thus the newer models, like the Hadley and Canadian, are projecting increased precipitation in the mid and northern latitudes.

Most recently, climate scientists have been measuring, and projecting with their models, massive shifts in the earth's climate system (Seager 2007). Between the equator and 30° north latitude, there is a massive cell of atmospheric circulation, with rising air and heavy precipitation at the equator producing the equatorial low-pressure area with its high rainfall. The air then flows northward at the top of the atmosphere, descending at ~30° north to form the subtropical high zone, then returning toward the equator at the surface. The descending air at 30° warms, absorbs moisture, and dries the surface. This feature is termed "the Hadley Cell," and a similar pattern is found in the Southern Hemisphere.

While there is some regional variation associated with air currents off the oceans, the subtropical high zones encompass the arid zones of the world. The 30° north latitude line passes through the Sahara, the arid zone of the Middle East, and western India. In North America it passes through the arid region of northern Mexico and is a partial determinant of the southwestern U.S. "hot" (Mojave, Sonoran, and Chihuahuan) deserts.

In an important summary publication, Richard Seager (2007), of the Lamont-Doherty Earth Observatory at Columbia University, states:

> Projections of anthropogenic climate change conducted by nineteen different climate modeling groups around the world, using different climate models, show widespread agreement that Southwestern North America…[is] on a trajectory to a climate even more arid than now…. As the planet warms, the Hadley Cell…expands poleward… [and] expands the subtropical dry zones. At the same time…the rain-bearing mid-latitude storm tracks shift poleward…. In the Southwest the levels of aridity seen in the 1950s multiyear drought, or the 1930s Dust Bowl, become the new climatology by mid-century: a perpetual drought…. According to the models the drying should already be underway….

In short, the models project intensified aridity and expansion of the subtropical highs, and they portend a northward shift and increased precipitation of the temperate-latitude rainfall zones. Australia and the Sahelian region of Africa already have been suffering prolonged drought. Limited evidence presented in this chapter shows that this process may have begun in the twentieth century in the RMGB, and more evidence for the West as a whole is discussed in Chapter 10.

Seager (2007) includes a map of North America with a line running from southern Florida diagonally northwest through Louisiana to western Kansas, then westward through central Colorado and Utah, then northwestward through Nevada to the California-Oregon border. Precipitation "averaged over 19 different climate models" is projected to decrease below this line and to increase above it by 2021–2040.

It is important to bear in mind that this line is the average for several models that vary to some degree in positioning, and it is thus uncertain precisely where this demarcation will occur. Nevertheless, evidence presented below, and in the final chapter, shows drying in the southern portion of the western United States and increased precipitation in the north during the twentieth century.

Twentieth-Century Rocky Mountain/ Great Basin Trends Disclosed by Weather-Record Analyses

Analytic Procedures

Since atmospheric CO_2 rose ~32 percent during the twentieth century (Backland et al. 1999), the RMGB weather record for the twentieth century was analyzed to determine whether the climate changed in the same directions—qualitatively, though not quantitatively—as those projected by the general circulation models for the twenty-first century with a twofold increase in CO_2. If so, it would be taken as partial, qualitative support for the validity of the models and their projections. Baldwin (2003) analyzed the twentieth-century trends using data from the Vegetation/Ecosystem Modeling and Analysis Project (VEMAP). These data are spatial averages of climate variables for weather stations on 0.5° quadrangles, with corrections for altitude to account for sampling biases produced by weather stations at different locations.

Because of its size and topographic complexity, and the possibility that climate trends might vary across its extent, Baldwin (2003) subdivided the RMGB region into eight subregions (Fig.1.1): Northern Great Basin (NGB), Central Great Basin (CGB), Southern Great Basin (SGB), Northern Rockies (NR), North Central Rockies (NCR), West Central Rockies (WCR), East Central Rockies (ECR), and Southern Rockies (SR). The numbers of VEMAP quadrangles in these eight subregions are, respectively, 66, 66, 63, 29, 19, 9, 37, and 19, for a total of 308 for the region.

Analyzing trends in the individual quadrangles would have emphasized local variability, which the general circulation models could not yet simulate, and might have been so variable as to obscure regional or subregional trends. Averaging over the entire region, however, would have masked variations in trend within this large, nine-state region. Thus Baldwin chose subregions of moderate size both to disclose variations within the region and to provide replications to test any inferences of trend.

Baldwin analyzed the VEMAP records to determine whether there were any statistically significant secular trends in the eight subregions in annual means of daily minimum temperatures (Tmin), daily maximum temperatures (Tmax), mean annual temperatures, and total annual precipitation (PPT) during the twentieth century. He also calculated standard deviations around 30-year

moving averages of Tmin and PPT to determine whether interannual variability changed as the means changed over time.

The basic analytic units were the monthly values for each year: mean monthly Tmin and Tmax, and monthly PPT. The seasonal Kendall test was used because it is nonparametric and does not require the data to conform to any particular distribution (Gilbert 1987). The test used time series for each of the monthly values throughout the century. The slope was calculated for each monthly time series, as was the probability that it is significantly different from zero, using the Mann-Kendall test. The test then aggregated the monthly values to produce slope and probability values for the years.

The mean annual Tmin and Tmax values, and annual PPT were arrayed in time series, and the trends tested with ordinary least squares (OLS) and Sen slopes (Gilbert 1987) along with p-values to determine whether any slopes differed significantly from zero. Standard deviations around 30-year moving averages of these variables were calculated to determine whether interannual variability changed over time.

To analyze the trends in mean annual temperatures, the Tmin and Tmax values for each subregion were combined, and the resulting time series were fit with OLS and Sen slopes and tested for significance.

"Natural" Climatic Variability

In order to ascertain whether any "natural" (that is, not human-induced) climatic variability influenced trends shown by the weather analyses, and out of which any anthropogenic forcings would have to emerge in order to be detectable, Baldwin analyzed data on the El Niño/Southern Oscillation (ENSO), the Pacific Decadal Oscillation (PDO), and the Arctic Oscillation (AO). The NIÑO3 sea-surface anomaly index was used as a measure of ENSO, the Mantua et al. (1997 and updates on the Internet) PDO index, and Thompson and Wallace (1998) AO index values.

Four, three-month seasonal averages were calculated for each oscillation index for each year (e.g., winter (DJF) av. NIÑO3 for 1900, 1901, etc.). Each seasonal average was then regressed on each month's climate variable (Tmin, Tmax, PPT) in each subregion in which a variable showed a significant positive trend over time.

To determine whether trends, measurable over century time scales, might coincide with the twentieth-century trends under analysis here, Baldwin (2003) examined temperature proxies based on tree-ring analyses for the period 1600 to the middle 1900s. These data were taken

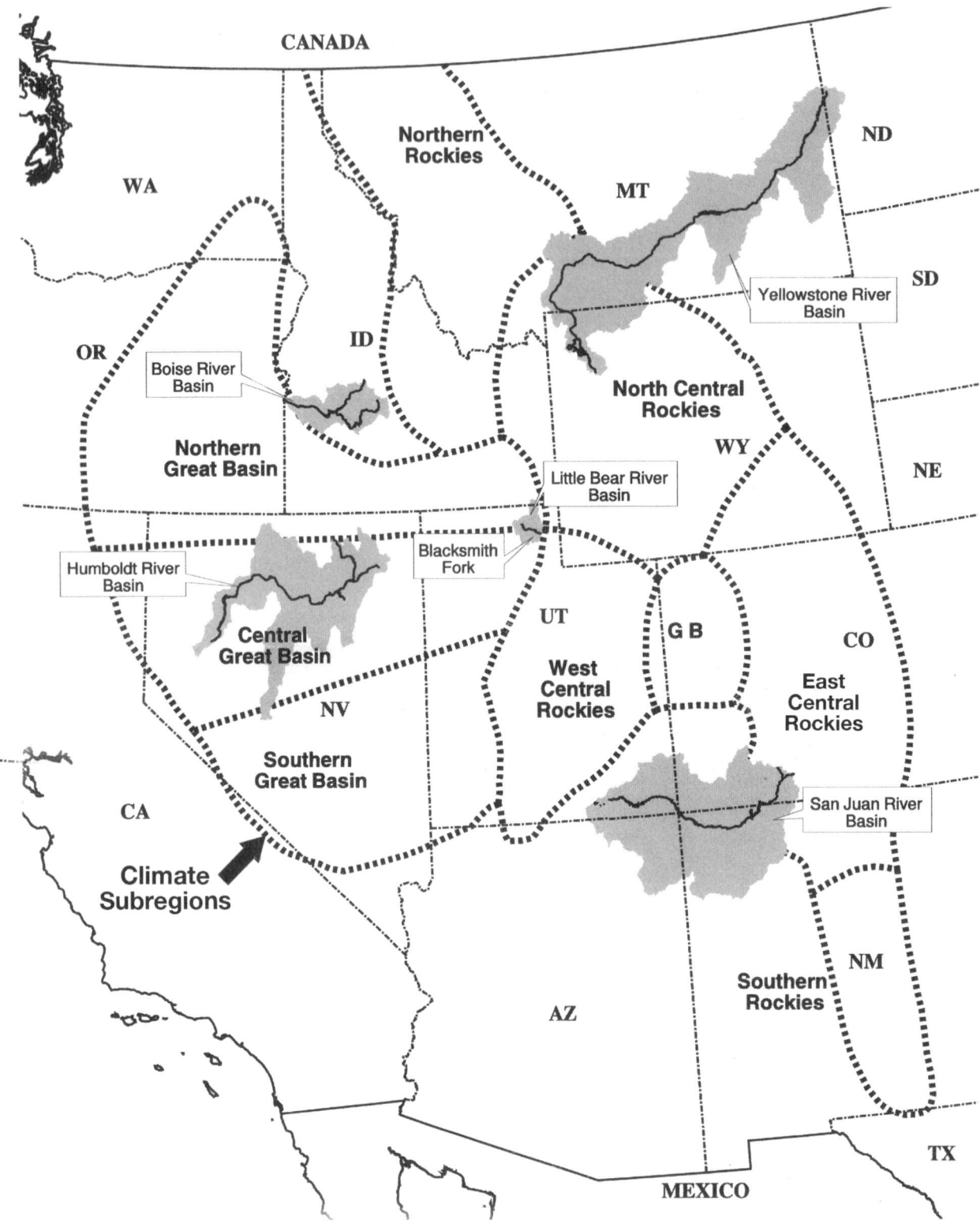

Figure 1.1. Subregional divisions used for analyzing twentieth-century weather records for the Rocky Mountain/Great Basin Regional Climate-Change Assessment. Shaded areas are watersheds of five streams for which twentieth-century annual flows were compared with precipitation records. (Map prepared by Susan Selby, Las Vegas Valley Water District, and presented in Baldwin et al. 2003.)

from the National Oceanographic and Atmospheric Administration (NOAA) Web site.

Three natural (not human-induced) geophysical forces have been shown to influence western North American climate patterns, and any anthropogenic climate change must emerge out of the variability imposed by these forces in order to be detectable. Consequently, the effects of these on twentieth-century RMGB weather patterns were explored to determine whether the observed patterns were correlated with the natural forcings.

One such forcing is the El Niño/Southern Oscillation. The latitudinal positioning of winter storm tracks in western North America varies with the minimum values of the low-pressure cell off the coast of the Aleutian Islands and with associated phases of ENSO (Mote 1999). Very low pressures in the Aleutian cell coincide with the warm-water El Niño phase of ENSO and a split in the mid-latitude winter storm tracks. One track moves northward, bringing storms to Alaska, while a southern track carries storms to southern California and the Southwest. The Pacific Northwest experiences warmer and dryer-than-normal winters, with reduced montane snowpacks and streamflows. During the cold-water La Niña phase, the mid-latitude storm tracks remain intact, carrying winter storms and precipitation across the northern part of the western United States. The ENSO periodicity varies between two and seven years, and averages three to four.

There is evidence that an ENSO effect extends southward through the RMGB region, but because of the region's latitudinal extent, the northern effects are those of the Pacific Northwest, while the southern effects are those of the Southwest, and therefore inverse. Baldwin (1998) has shown that winter precipitation and montane snowpacks are inversely correlated with the ENSO NIÑO3 index in the northern third of Utah, but positively correlated in the southern two-thirds. Precipitation and snowpack in the north tend to be below normal in the El Niño phase, but above average in the south. The reverse tends to be the case in the La Niña phase.

Moreover, the influence is spatially variable, not only in terms of the north-south oscillation but, according to Cayan and Webb (1992), affecting patches over Idaho, Montana, and Wyoming. As an indication of the ENSO effect on precipitation generally over the West, Miller (1997) stated "that only 25 percent of the interannual variation in streamflow can be explained by ENSO in those patches where the impact is most significant."

Baldwin (1998), however, found almost no correlation between time series of the NIÑO3 index and the twentieth-century individual monthly temperature time series generated by the seasonal Kendall test for the eight RMGB subregions. Only September minimum temperatures in a single subregion (Southern Great Basin) were weakly correlated ($R^2 = 0.10$) with the winter ENSO index. Annual ENSO variability may be altered by increases in greenhouse gases (Karl et al. 1996, Miller 1997, Timmerman et al. 1999), but there is no evidence of a secular change in the mean trend of the oscillation over time that would correlate with the century-long trends discussed below.

Mote (1999) detected an influence of the Pacific Decadal Oscillation on the Pacific-Northwest climate. Mote's coworkers observed four alternating phases, averaging about 24 years each during the twentieth century, in the region's temperature and precipitation time series: two of low temperatures and high precipitation, and two of high temperatures and low precipitation.

Dettinger (see Stohlgren 2003) observed that the PDO effect is most marked in the northern part of the western United States and fades toward the south. The reverse is true of ENSO. The ENSO effect has not been sufficiently strong to prevent expression of the PDO in the data analyzed by Mote (1999) for the Pacific Northwest. And in Baldwin's analyses of the 12 monthly time series for the eight RMGB subregions, a PDO correlation could only be detected with March Tmin in the Northern Rockies and October Tmax in the Southern Rockies.

Thompson and Wallace (1998) have described the Arctic Oscillation as the dominant mode of sea-level pressure (SLP) variability in the Northern Hemisphere north of 20° north. It is expressed as the difference in SLP between 45° north and the North Pole. It affects the positioning of both the polar and subtropical jet streams. Its periodicity coincides roughly with that of the PDO. In the present analysis, a statistically significant correlation was detected between AO in only four of the 96 subregional time series.

ENSO, PDO, and AO, as their names imply, are quasi-cyclic changes that oscillate over shorter phases than a century: ENSO at three to four years, PDO and AO at approximately 24 years. No rising or declining portion of these periodicities could covary with a century-long trend.

Moreover, ENSO undergoes several oscillations with each PDO and AO phase. As Mote (1999) comments, in years when La Niña patterns in the Pacific Northwest (high precipitation) coincide with the cool-wet phase of the PDO, the two accentuate each other's effects. But when the El Niño dry phase coincides with the PDO cool-wet phase, there is some dampening of both influences.

Table 1.2 Twentieth-century trends in mean annual minimum and maximum temperatures in eight subregions

Subregion	Trend (per century) Min. Temp. (°C)	Test p-value
NGB	0.38	0.02
CGB	0.21	0.16
SGB	0.38	0.02
NR	0.62	< 0.01
NCR	0.40	< 0.01
WCR	0.54	< 0.01
ECR	0.19	0.16
SR	0.01	0.94
Max. Temp. (°C)		
NGB	-0.06	0.80
CGB	-0.01	0.96
SGB	0.33	0.12
NR	0.67	< 0.01
NCR	0.18	0.42
WCR	0.20	0.26
ECR	0.05	0.76
SR	0.13	0.44

Thus, the effects of these influences, functioning at different frequencies, alternately dampen, amplify, or have no effect on each other's signals to the climate variables. In areas where one index is stronger than the others, its influence may be measurable, as Mote discerned with the effects of PDO on the Pacific-Northwest temperatures and precipitation, and, as discussed above, an ENSO effect on winter snowpacks in Utah. But whether imparting these oscillatory variations into the climate-variable time series, or whether the sources of natural variation cancel each other to the extent of imparting an essentially random signal, secular twentieth-century climate trends emerge as Mote (1999) showed for the Pacific Northwest and as will be shown next for the RMGB region.

Twentieth-Century Temperature Trends

Calculations of the seasonal Kendall tests for trends in Tmin and Tmax during the twentieth century produce Tmin increases significant at $p < 0.05$, in five of the eight subregions (Table 1.2) and ranging from +0.38–0.62°C (+0.68–1.12°F) for the century. Two subregions showed lesser increases (+0.19–0.21°C, +0.34–0.38°F) that were significant at only $p < 0.20$. An eighth region, the South-

ern Rockies, showed no increase. Hence the increases in Tmin occurred throughout the RMGB except in the Southern Rockies, and were more pronounced in the region's northerly portion.

The seasonal Kendall test showed significant increase ($p < 0.20$) in Tmax during the twentieth century in only two subregions: the Northern Rockies (+0.67°C, +1.21°F) and the Southern Great Basin (+.33°C, +0.59°F). The high Tmax increase in the Northern Rockies coincides with the high Tmin increase in the same subregion (Table 1.2).

With alternative statistical methodologies (ordinary least squares and Sen-slope estimates), the twentieth-century trends in mean annual temperatures in the eight subregions were tested and the results summarized in Fig. 1.2. Mean annual temperatures increased significantly ($p \leq 0.05$) in both tests in three of the subregions—the Southern Great Basin, Northern Rockies, and West Central Rockies—with the largest increase in NR at +0.6°C per 100 years. The Sen tests for NGB and NCR were significant at $p < 0.10$, but the OLS were not. Neither test was significant for CGB, ECR, or SR.

To determine whether or not interannual variability in temperatures changed during the twentieth century, Baldwin calculated standard deviations around 30-year moving averages of mean annual Tmin and Tmax, and plotted these as time series. Following high values in the 1920s and 1930s, these tended to decline during the remainder of the century. The one exception was the Southern Rockies, one of the subregions that showed no significant temperature increase during the century. Thus, interannual variability in temperatures has declined over most of the RMGB region since the 1920s or 1930s while temperatures were rising.

Twentieth-Century Precipitation Trends

Baldwin analyzed the VEMAP precipitation records for the RMGB region with the seasonal Kendall test and calculated OLS and Sen slopes for the time series of annual precipitation expressed as mean precipitation (PPT) per month (Fig. 1.3). All showed positive trends over the 1900s, but only those for the Northern, Central, and Southern Great Basin, and the Northern and North Central Rockies showed statistical significance. In these five subregions, nine of the ten tests were significant at $p \leq 0.10$, the OLS test for the Northern Rockies being the single exception. None of the six tests for the West Central, East Central, and Southern Rockies was significant at $p < 0.20$.

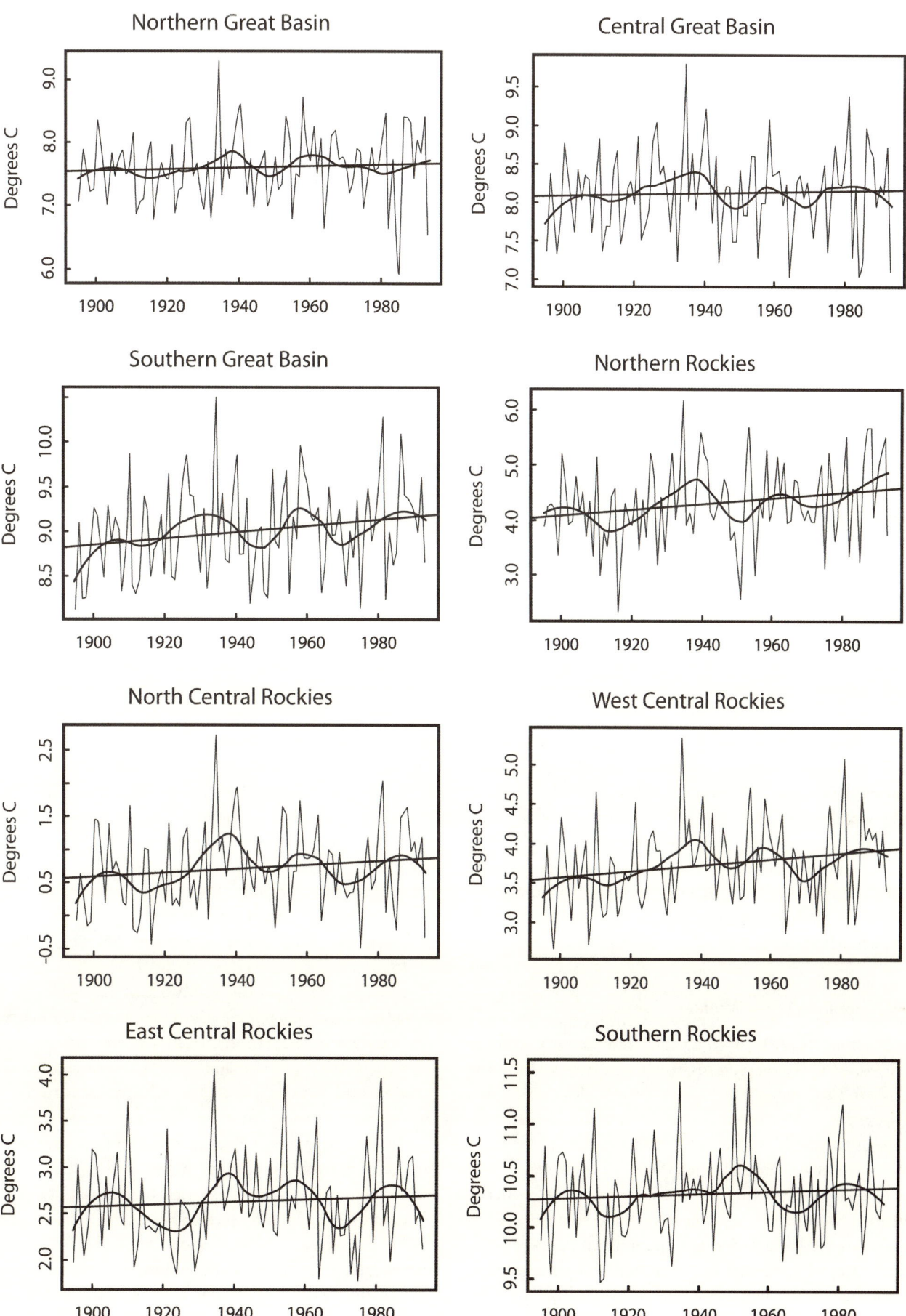

Figure 1.2. Time series of mean annual temperatures in eight subregions (Fig. 1.1). Sinuous lines are Lowess smooths of the annual values; straight lines are slopes produced by least-squares regressions. The latter are significant at $p < 0.05$ for SGB, NR, WCR; $p < 0.10$ for NGB, NCR. P-values for the remaining three are > 0.10. (Redrawn from Fig. 3.20 in Baldwin 2003.)

The percentage increase in annual precipitation over the twentieth century for the five subregions with significant tests, as projected by the OLS slopes, are:

NGB   +10 percent
CGB   +16 percent
SGB   +16 percent
NR   +6 percent
NCR   +13 percent

As with the temperature analyses, Baldwin calculated trends in interannual variability in precipitation by calculating 30-year moving averages of precipitation time series (Fig. 1.3) and standard deviations around these means. The latter are plotted as time series in Figure 1.4.

Trends for the five subregions that showed statistically significant precipitation increases (Fig. 1.3) showed progressive standard-deviation increases beginning in the 1920s or 1930s. No such trends were evident for the East Central and Southern Rockies subregions, neither of which had significant precipitation increase (Fig. 1.3). The trend for the West Central Rockies, also a subregion with no significant precipitation increase, was ambiguous.

In order to convert the projected 2080–2100 model precipitation trends (Table 1.1) to the percentage-increase values shown above, the Year 2000 intercepts in the eight subregions in Figure 1.3 were converted to annual rates in inches per year. The mean annual precipitation for the RMGB region was then converted by multiplying these subregional values by the respective number of VEMAP quadrangles listed above; the sum of these products was then divided by 308, the total number of quadrangles. The result was an area-weighted, average annual precipitation for the region in 2000. That average was then compared with the model-projected 2080–2100 annual increases shown in Table 1.1.

Evidence from Other Weather Sources

Karl et al. (1996) analyzed twentieth-century climate trends using NCDC records. Their results show uniform temperature increases throughout the RMGB region except for a single point in southern Colorado. The increases ranged largely between 1 and 2°C (1.8–3.6°F).

Kittel et al. (2002) divided the Rockies into northern, central, and southern segments and analyzed twentieth-century weather trends in each of these based on HCN station data. Annual average minimum temperatures rose 0.7°F and 0.9°F in the northern and central segments, both statistically significant, but not significantly in the

southern Rockies. The slopes for maximum temperatures were short of significance.

Most recently, Brian McInerny (pers. comm., 2006) of the National Weather Service in Salt Lake City has calculated mean annual temperatures, averaged over all Utah weather stations, for 1895–2004. A positive trend (Fig. 1.5) shows an increase of ~3°F (1.9°C) over the period.

These sources produced similar precipitation results. Karl et al. (1996) found uniform precipitation increases over the Great Basin, but mixed results for the Rockies. Kittel et al. (2002) found increases in twentieth-century summer precipitation of 29.5 percent and 32.8 percent in the northern and central Rockies, but no significant increase in the southern Rockies.

McInerny's (pers. comm., 2006) analysis of Utah data showed an increase in mean annual precipitation, averaged over all state weather stations, of 13 percent for the period 1895–2004 (Fig. 1.6).

Streamflow Evidence

Since analyses of twentieth-century weather records for the RMGB region have indicated precipitation increase in six of the VEMAP subregions, several data sets on the flow of streams in the region were analyzed to determine whether increased precipitation was reflected in the region's hydrology. The purpose of the first analysis by Baldwin et al. (2003) was to develop a regionwide overview of streamflow behavior during the twentieth century. Although it would have been desirable to select one stream in each of the subregions, the choices were limited to streams that met three criteria:

1. A long period of record (most of the twentieth century) maintained in the Hydro-Climatic Data Network (HCDN) Streamflow Data Set 1874–1988 (Slack et al. 1992, 1993).
2. A record of high quality. Only five of the eight subregions had gaging stations that met these quality criteria, and one station in each was selected.
3. Absence of major human impacts (for example, diversions and dams) upstream from the gaging stations.

The five rivers that met all of the criteria are the Boise (Idaho), Yellowstone (Montana and Wyoming), Humboldt (Nevada), Blacksmith Fork (Utah), and San Juan (Colorado, New Mexico, Utah) (Fig. 1.1).

For each stream, successive 30-year moving averages of annual streamflows, measured in mean annual cubic feet per second (cfs), were calculated, plotted as time series,

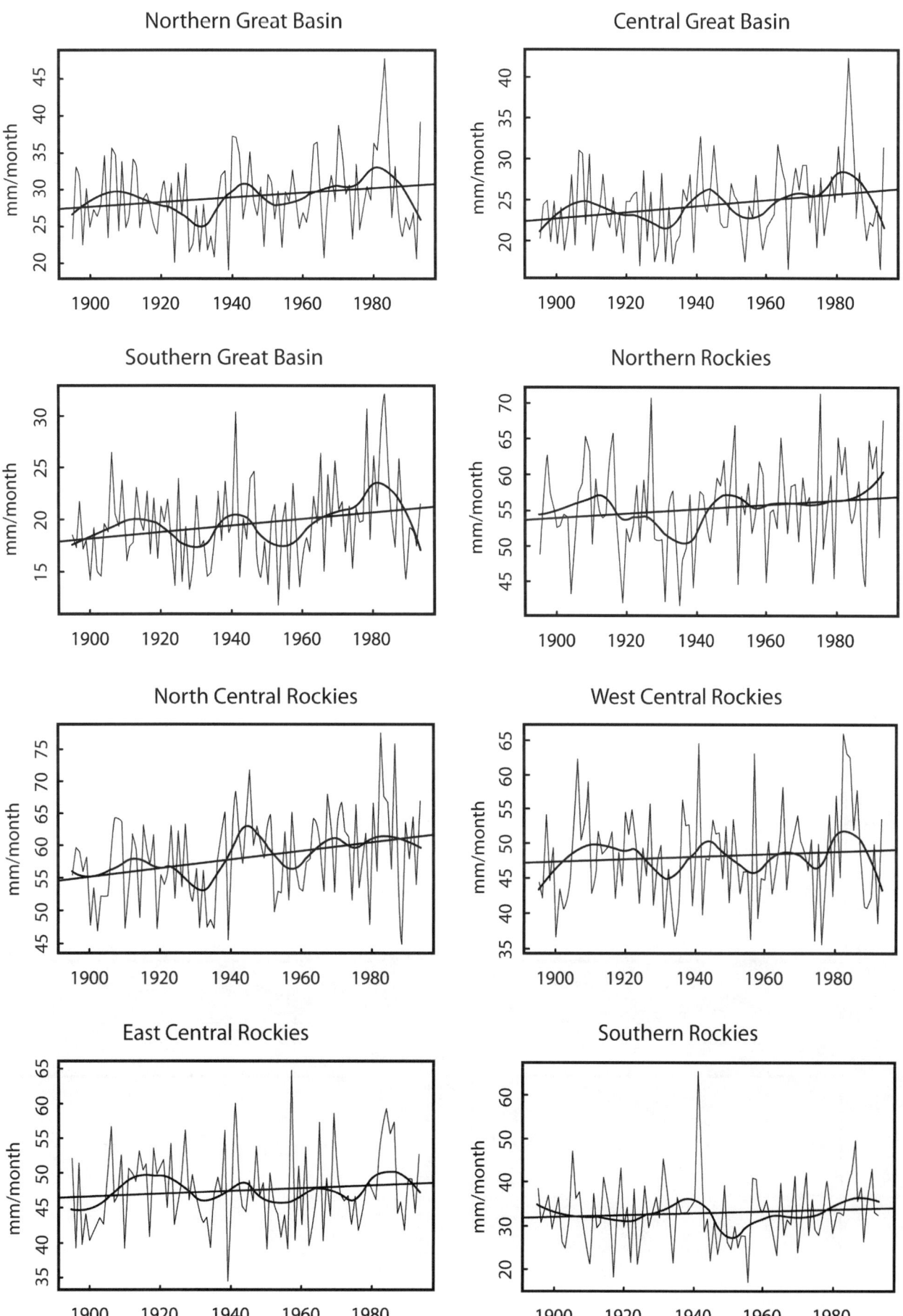

Figure 1.3. Time series mean of mean annual precipitation in eight subregions (Fig. 1.1). Sinuous lines are Lowess smooths of the annual values; straight lines are slopes produced by least-squares regressions. The latter are significant at $p < 0.05$ for NGB, CGB, SBG, NCR; at $p < 0.10$ for NR. P-values for the remaining three are > 0.10. (Redrawn from Fig. 3.23 in Baldwin 2003.)

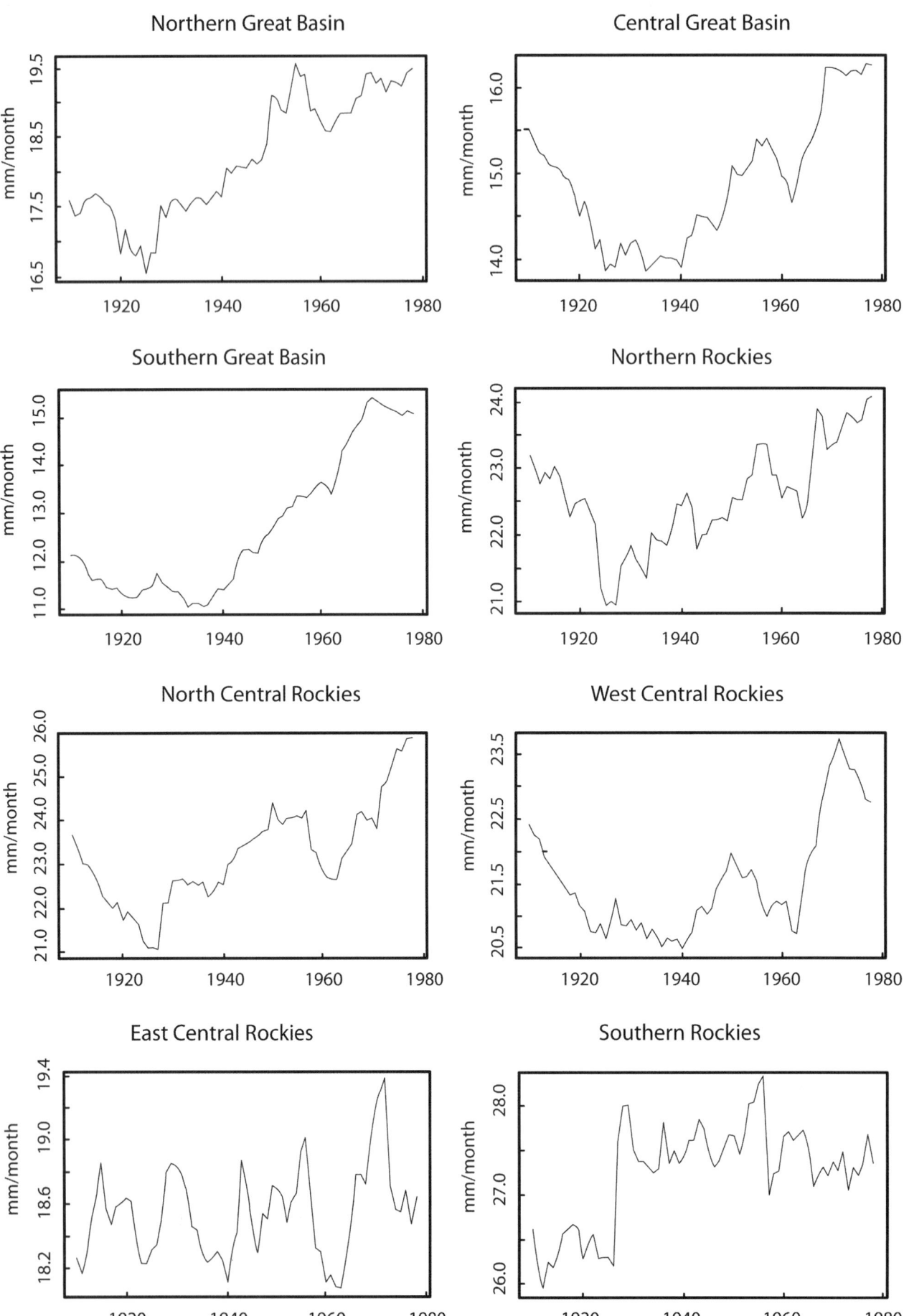

Figure 1.4. Time series of standard deviations around 30-year moving averages of mean annual precipitation for eight subregions (Fig. 1.1). Subregions showing rising trends through the 1900s are those shown in Figure 1.3 as having had significant twentieth-century precipitation increases. (Redrawn from Fig. 3.24 in Baldwin 2003.)

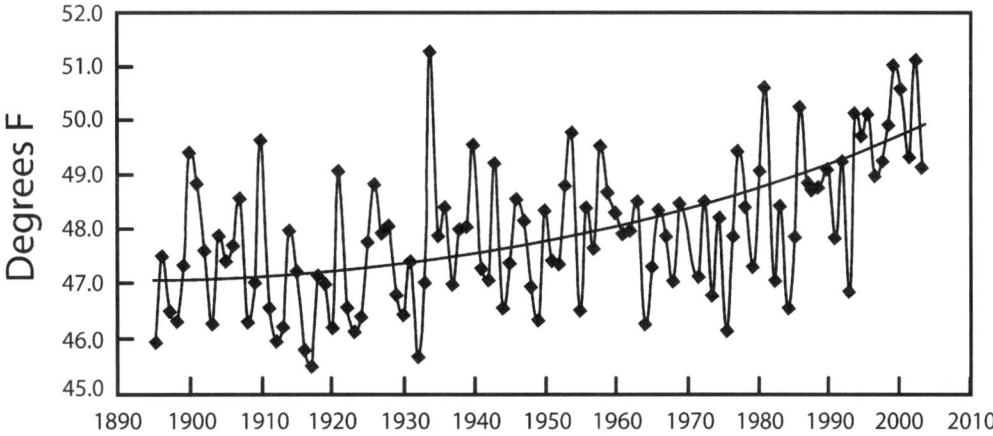

Figure 1.5. Mean annual temperatures for 1895–2004 averaged over all Utah weather stations. (Graph provided by Brian McInerny, National Weather Service, Salt Lake City, Utah.)

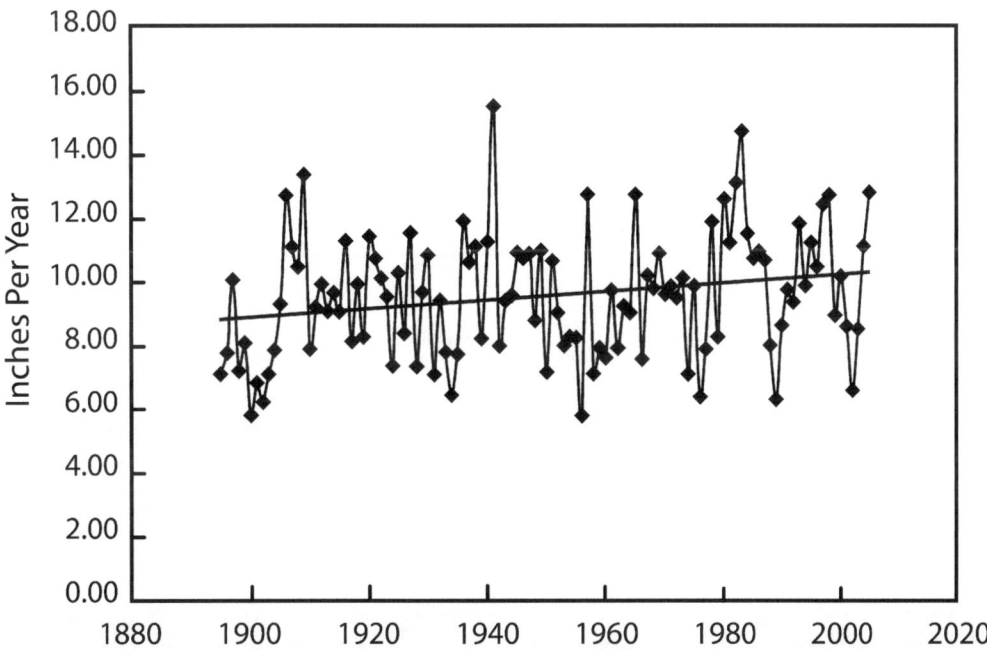

Figure 1.6. Mean annual precipitation for 1895–2004 averaged over all Utah weather stations. (Graph provided by Brian McInerny, National Weather Service, Salt Lake City, Utah.)

and regressed over time. The regressions were tested with the seasonal Kendall test.

In three of the five streams—the Boise River, Humboldt River, and Blacksmith Fork—the flow increased significantly ($p < 0.05$) during the 1900s. The trend in the Yellowstone was similar, but the regression was short of significance at 0.05. The rate of increase in the three with significant trends ranged from 12 percent/100 years in the Boise to 33 percent/100 years in the Blacksmith Fork. The latter has the smallest watershed situated entirely in

the Wasatch Mountains of Utah with no irrigation withdrawals above the gaging station. The other two have larger basins with some irrigation withdrawals above the gaging points.

The increased flow in the Boise and Humboldt, and possibly the Yellowstone, occurred in three subregions (Northern Great Basin, Central Great Basin, and North Central Rockies) with increased precipitation during the 1900s (Fig. 1.3). The increase in the Blacksmith Fork, arising in the West Central Rockies, occurred in a subregion in which no significant precipitation increase was found. Moreover, the major increases occurred largely in the second half of the twentieth century. The absence of any discernable trend in the San Juan also occurred in a sub-region (East Central Rockies) in which no statistically significant twentieth-century precipitation increase was detected.

Interannual variability of flows in the five streams was also analyzed by calculating standard deviations around the 30-year moving averages of flow, posing these as time series, and testing the trends over time. These trends coincide with increases in flow, indicating that interannual variability increased as annual streamflow increased. Thus the frequency of extremely high flows increased, while the frequency of low flows declined. (The increasing variance in streamflow also coincides with increasing variance in annual precipitation in these subregions, as discussed above.)

These patterns agree with those detected in other data sets. Analyzing hydrologic trends in the entire United States, Lettenmaier et al. (1994) found strong increases in streamflow in the RMGB region, particularly in winter and spring, during the latter half of the twentieth century. They also noted that precipitation had increased in the region over the same period.

A second analysis focused on the Colorado River and its tributaries, four of which originate in the East Central Rockies, where no twentieth-century precipitation increase was detected. As noted above, flow of the fifth tributary, the San Juan, did not increase.

The Colorado's flow trend for the twentieth century was compiled by Susan Selby, of the Las Vegas Valley Water District, using U.S. Bureau of Reclamation (BOR) measurements made at Lee's Ferry, Arizona. A regression fit to the entire series portrays a downward trend from ~18 million acre feet at the beginning of the century to 15 million in 1996, a decline of nearly 17 percent. The trend was corrected for human withdrawals and was therefore presumably induced by factors other than human use.

Drawing a firm conclusion about the trend in the full data set may be misleading, however, because the set is a combination of estimated and observed values. BOR flow values prior to the mid-1920s were reconstructed from tree-ring proxies. Later values came from actual gage measurements at Lee's Ferry. Although the slope for the entire series, which combines the two data sets, is negative, regression on the observed values only (1920s to date) results in no significant slope.

To examine the origins of the river's behavior at Lee's Ferry and further explore the circumstances surrounding the absence of trend in the San Juan, twentieth-century flow of the Colorado's five major tributaries were examined: the Green River, arising in the southwest quadrant of Wyoming; the Yampa of northwestern Colorado; the Colorado and Gunnison, arising in central Colorado; and, once again, the San Juan, in southern Colorado. Data were provided by Tom Ryan (1996) of the Upper Colorado Basin Office of the Bureau of Reclamation in Salt Lake City. The flow data were "reconstructed" or "virgin" flows corrected for human withdrawals. They were calculated, or measured, for the years 1905–1990 at the following stations, respectively, of the above rivers: Green River, Utah; Maybell, Colorado; Cisco, Colorado; near Grand Junction, Colorado; and Bluff, Utah.

The streamflows for the five tributaries have been gaged on the first of each month, and thus total annual flows cannot be calculated. However, mean monthly flows were calculated for each water year. When the 1926–1990 flows were regressed over time, as with the Colorado at Lee's Ferry, none of the slopes was significantly different from zero. The conservative interpretation is that the Colorado and its headwaters in the East Central Rockies subregion have experienced no net change in flow since 1926, despite great interannual variation. This is the subregion in which both the RMGB assessment and Kittel et al. (2002) found no precipitation increase during the twentieth century.

Discussion of Rocky Mountain–Great Basin Climate Trends

The purpose of the climate-change assessment ordered by Congress in 1991 was to project the potential effects of climate change on the nation during the twenty-first century, a change expected to be driven by a twofold increase in CO_2. Researchers in the Rocky Mountain/Great Basin region, like those in the other 18 regions, selected a group of socioeconomic sectors most likely to be affected by the change.

To make such projections, it was necessary to formulate reasonable climate-change scenarios for the twenty-first century that could then be interfaced with the sectors. One source of scenarios was the simulations of two general circulation models, the British Hadley (HadCM2) and the Canadian (CGCMl), programmed with a twofold increase in CO_2, and decided upon at the beginning of the national assessment because of their advanced stage of development.

The Hadley and Canadian models are global models, structured to project climate change for the earth's entire climate system. The regional assessments, however, needed scenarios at regional levels, and Mearns (2003) adapted the two models to make climate-change projections for the nine-state RMGB region. While useful for the purpose, it was not clear how precisely these adaptations would simulate the regional changes, and they were taken as hypotheses of unknown precision. With these reservations, Mearns conducted the simulations, which projected sizeable increases in temperatures and precipitation by 2100 for the region as a whole.

Since atmospheric CO_2 increased by a third during the twentieth century, we postulated that if the regional climate had changed during the century in the same directions (although not necessarily in magnitudes) as the model projections, it would be tentative and qualitative support for the latter. Although not of the same magnitude, analyses of twentieth-century RMGB weather records showed a statistically significant increase in annual averages of daily minimum temperatures in five of eight subregions, and in annual mean temperatures in three subregions.

Other investigators' analyses have shown similar twentieth-century trends: Karl et al. (1996) for the entire region; Kittel et al. (2002) for the northern and central Rockies, but not for the southern; and McInerney (pers. comm., 2006) for all Utah weather stations (Fig. 1.5). In the latter case, mean annual Utah temperatures rose, on average for the state, 3°F.

Subsequent to the RMGB assessment, modeling efforts worldwide were continuing to project significant temperature increases globally. Authors of the fifth report of the Intergovernmental Panel on Climate Change (Albritton et al. 2007) were continuing to project significant twenty-first-century warming:

> The globally averaged surface temperature is projected to increase by 1.4 to 5.8°C [2.5 to 10.4°F]… over the period 1990 to 2100. These results are for the full range of 35 SRES scenarios, based on a number of climate models.

Three other patterns emerging from the RMGB twentieth-century weather records give further concordance with the model projections. The models project that night-time minimum temperatures will rise more strongly than daytime maxima, and thus day-night temperature differentials will narrow as temperatures rise. This is precisely what the RMGB records show, and what Kittel et al. (2002) found for the Rockies.

The models also project that interannual variability of temperatures will decline as temperatures rise. Baldwin's (2003) analyses of the RMGB weather records showed this also to have been the case during the twentieth century.

Third, the models project that temperature increase will become progressively more pronounced at progressively higher latitudes. While they do not unequivocally support this projection, our analyses found statistically significant temperature rise in the northern portion of the RMGB region, but not in the southeastern portion. Also, Kittel et al. (2002) found significant increase in the northern and central portions of the Rockies, but none in the southern.

The models also project global increase in precipitation, but it is a more complex phenomenon than temperature change, and climatologists are more cautious in making regional projections. Mearns (2003), who modified the Hadley and Canadian models, projected increased precipitation for the RMGB region as a whole; however, recent models projecting northward shift of the Hadley Cell now project reduced precipitation in the southerly portion of the western United States and increased precipitation in the northerly portion (Seager 2007). Where the transition zone will occur cannot be prescribed exactly.

Here again, the twentieth-century weather record suggests that these changes may be occurring. In our RMGB analysis, annual precipitation increased in the westerly and northerly subregions during the twentieth century, but not in the three southeasterly subregions. Similarly, the Karl et al. (1996) analyses of NCDC precipitation data showed annual precipitation increases ranging from 5 to 10 percent per 100 years in the northern and western Rockies and Great Basin subregions, but a decline in the north-central Rockies, and no clear signal in the east-central and southern Rockies. The Kittel et al. (2002) analysis of HCN data on summer precipitation found 100-year increases of 29.5 percent and 32.8 percent in their northern and central Rockies subregions, but no significant increase in the southern Rockies.

In the Baldwin et al. (2003) streamflow analyses, annual flows in the three northerly streams with significant increases were 12, 16, and 33 percent per 100 years. The Yellowstone, while short of significance, increased by a comparable amount, but the trend in the San Juan, the most southerly stream, did not. Nor did annual flow increase in the five main tributaries of the Colorado River, or in the river itself, between 1926 and 1996.

Finally, the models project increased interannual variability in precipitation as precipitation increases. Our RMGB analysis showed this to have been the case during the 1900s in those subregions where precipitation rose and in annual streamflows, again, in those subregions where precipitation increased.

Thus, evidence from a variety of sources shows twentieth-century changes in temperature and precipitation in the RMGB region qualitatively in the same direc- tions projected for the twenty-first century by increasingly sophisticated models developed over more than a decade. This evidence cannot, of course, be taken as proof of the exact model projections. Nonetheless, it strengthens the probability that at least the directions of projected change are correct.

References

Albritton, D. L., et al. 2007. *Summary for Policymakers: A Report of Working Group I of the Intergovernmental Panel on Climate Change.* I.P.C.C., Shanghai, PRC.

Backland, P., S. Bassow, R. Bierbaum, N. Lapham, J. Melillo, and F. Sharples. 1999. *Climate Change: State of Knowledge.* Off. Sci. and Technol. Pol., Washington, D.C.

Baldwin, C. K. 1998. Does El Niño affect snowpack in Utah? *Proc. Ann. West. Snow Conf. 66.* Snowbird, UT.

———. 2003. Historical climate analysis. Pp. 58–72 in F. H. Wagner (ed.), *Preparing for a Changing Climate: The Potential Conse- quences of Climate Variability and Change, Rocky Mountain/ Great Basin Regional Climate-change Assessment.* Rept. for the U.S. Global Change Res. Prog. Utah State Univ. Press, Logan.

Baldwin, C. K., F. H. Wagner, and U. Lall. 2003. Water resources. Pp. 79–112 in F. H. Wagner (ed.), *Preparing for a Changing Climate: The Potential Consequences of Climate Variability and Change, Rocky Mountain/Great Basin Regional Climate-change Assessment.* Rept. for the U.S. Global Change Res. Prog. Utah State Univ. Press, Logan.

Cayan, D. R., and R. H. Webb. 1992. El Niño/Southern Oscillation and stream flow in the western United States. In H. F. Dias and F. Markgraf (eds.), *Historical and Paleoclimatic Aspects of the Southern Oscillation.* Cambridge Univ. Press, Cambridge, UK.

Gilbert, R. O. 1987. *Statistical Methods for Environmental Pollution Monitoring.* Van Nostrand Reinhold Co., New York.

Karl, T. R., R. W. Knight, D. R. Easterling, and R. G. Quale. 1996. Indices of climate change of the United States. *Bull. Amer. Meteor. Soc.* 77:279–292.

Kittel, T. G. F., P. E. Thornton, J. A. Royle, and T. N. Chase. 2002. Climates of the Rocky Mountains: Historical and future pat- terns. Pp. 59–82 in J. S. Baron (ed.), *Rocky Mountain Futures: An Ecological Perspective.* Island Press, Washington, D.C.

Lettenmaier, D. P., E. F. Wood, and J. R. Wallis. 1994. Hydrocli- matological trends in the continental United States. *J. Clim.* 7:586–607.

Mantua, N. J., S. R. Hara, Y. Zhang, J. M. Wallace, and R. C. Francis. 1997. A Pacific interdecadal climate oscillation with impacts on salmon production. *Bull. Amer. Meteor. Soc.* 78:1069–1079.

Mearns, L. O. 2003. GCM scenarios for the RMGB region. Pp. 72– 73 in F. H. Wagner (ed.), *Preparing for a Changing Climate: The Potential Consequences of Climate Variability and Change, Rocky Mountain/Great Basin Regional Climate-change Assessment.* Rept. for the U.S. Global Change Res. Prog. Utah State Univ. Press, Logan.

Miller, K. A. 1997. *Climate Variability, Climate Change, and Western Water.* Rept. West. Water Policy Rev. Advisory Comm. N.T.I.S., Springfield, VA.

Mote, P. 1999. *Impacts of Climate Variability and Change in the Pacific Northwest.* Rept. Pac. Northwest Reg. Assess. Group., Univ. of Washington, Seattle.

Ryan, T. E. 1996. *Development and Application of a Physically-Based Distributed Parameter Rainfall Runoff Model in the Gunnison River Basin.* U.S. Dept. Int., Bur. Reclam., Denver, CO.

Seager, R. 2007. *An Imminent Transition to a More Arid Climate in Southwestern North America.* http://www.ldeo.columbia.edu/ res/div/ocp/drought/science.shtml.

Slack, J. R., A. M. Lumb, and J. M. Landwehr. 1992. *Hydro-climatic Data Network (HCDN); A U.S. Geological Survey Streamflow Data Set for the United States for the Study of Climate Variations, 1874–1988.* U.S. Geol. Surv. Open File Rept. 92-129. Reston, VA.

———. 1993. *Hyro-climatic data network (HCDN) Streamflow Data Set, 1974–1988* [CD-ROM]. U.S. Geol. Surv. Water-Resources Invest. Rept. 93-4076. Reston, VA.

Stohlgren, T. J. 2003. Climatologists' workshop on scenarios: Abstracted and annotated by Thomas J. Stohlgren. Pp. 38–58 in F. H. Wagner (ed.), *Preparing for a Changing Climate: The Potential Consequences of Climate Variability and Change, Rocky Mountain/Great Basin Regional Climate-change Assessment.* Rept. for the U.S. Global Change Res. Prog. Utah State Univ. Press, Logan.

Thompson, D. W. J., and J. M. Wallace. 1998. Arctic oscillation sig- nature in the winter-time geopotential height and temperature fields. *Geophys. Res. Lett.* 25:1297–1300.

Timmerman, A., J. Oberhuber, A. Bacher, M. Erch, M. Latif, and E. Roeckner. 1999. Increased El Niño frequency in a climate model forced by greenhouse warming. *Nature* 398:694–697.

Wagner, F. H. (ed.). 2003. *Preparing for a Changing Climate: The Potential Consequences of Climate Variability and Change, Rocky Mountain/Great Basin Regional Climate-change Assessment.* Rept. for the U.S. Global Change Res. Prog. Utah State Univ. Press, Logan.

2

Surface Warming Inferred from Borehole Temperatures

Results from Utah and Comparisons with the Northern Hemisphere

David S. Chapman, Robert N. Harris, and Marshall Bartlett

Abstract

There is now compelling evidence from direct measurement of temperature and measures of the response of physical and biological systems to changes in temperature that planet Earth is warming to an extent and at a rate that exceeds natural variations seen in the recent past. Borehole temperature profiles are an important part of that evidence and complement other methods of reconstructing surface temperature histories at the decadal to century time scales. Borehole temperatures in Utah reveal temperature changes of 0.5 to 0.7°C through the 1990s; the extratropical Northern Hemisphere has warmed by at least 1.1°C in the last two centuries.

Introduction

There is growing recognition of unnatural increases in both atmospheric concentrations of greenhouse gases and mean global surface temperature (Prather et al. 2001, Folland et al. 2001, Jones and Moberg 2003, Jones and Mann 2004, Trenberth et al. 2007). Changes in the concentration of two important greenhouse gases, carbon dioxide and methane, over the last 500 years are shown in Figure 2.1. These records are derived from a combination of modern instrumental data and measurements of gas bubbles trapped in ice sheets in Greenland and Antarctica. Since the start of the industrial revolution, at about AD 1750, atmospheric concentration of carbon dioxide has increased from 280 ppm to 382 ppm and is predicted to reach 560 ppm in the twenty-first century, a doubling of its historic value. Methane concentration has more than doubled, from 700 to 1600 ppb. Both the levels of greenhouse gases and the current rate of increasing concentrations are greater than experienced on earth for the past

650,000 years, and probably much longer (Siegenthaler et al. 2005).

The surface temperature of the planet over the same time span is less well known. The best-studied signal of temperature change, and thus global warming, is the time series of surface air temperature (SAT) measured at weather stations (e.g., Jones and Moberg 2003). Globally averaged SATs indicate relatively constant temperatures between 1860 and 1910, warming to 1945, a weak cooling trend to 1970, and a strong warming trend to the present (Fig. 2.1, top). Overall, global SAT observations show a warming of about 0.8°C over the past 100 years.

One part of the debate over global warming hinges on whether or not this unprecedented warming is due primarily to increasing concentrations of greenhouse gases. With the exception of a few long records, mostly in urbanized Europe, the instrumental record of SATs extends back to only about 1860. A critical gap in our knowledge is the value of the baseline or reference temperature ca. 1750, when greenhouse gases were presumably at their background level. Was twentieth-century warming simply a natural variation, returning to normal after a cool period at the turn of the century? Or is it a continuation of a warming trend started when greenhouse gas concentration began to increase? If the warming trend is due primarily to increasing concentrations of greenhouse gases, what is the climatic sensitivity?

Because temperature variations at the earth's surface diffuse downward, where they are manifested as perturbations to the planet's background thermal regime, a novel source of information for reconstructing surface temperature is contained in the subsurface temperature field (Lachenbruch and Marshall 1986, Huang et al. 2000, Harris and Chapman 2001, Beltrami 2002).

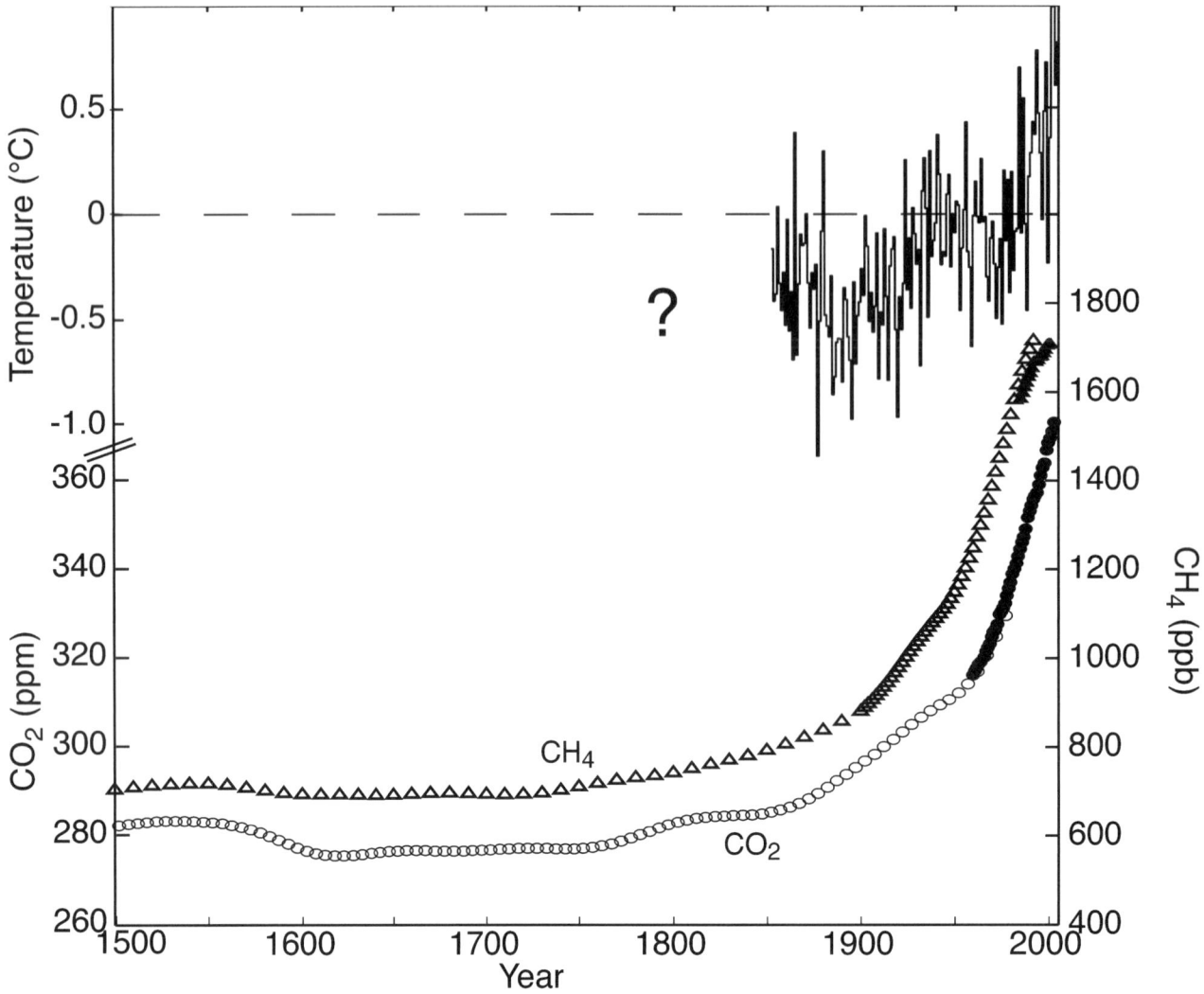

Figure 2.1. Global warming and greenhouse gases. Atmospheric concentrations of carbon dioxide and methane based on ice-core measurements (open symbols) grade seamlessly into real-time instrumental measurements (closed symbols). The upper instrumental surface air temperature (SAT) record (Jones and Moberg 2003) is plotted relative to the 1961–1990 mean. A geographically extensive SAT record does not exist prior to 1860.

Temperature-depth records measured in boreholes can be used to estimate baseline temperature and the magnitude and timing of changes on time scales of decades to centuries. We describe the use of temperature-depth profiles for climate reconstruction starting with individual profiles in Utah and then expand to the Northern Hemisphere data set. Finally we place our results in context with other temperature reconstructions based on proxy techniques.

Geothermics of Climate Change

The subdiscipline of reconstructing past surface temperature variations from borehole temperatures is called "the geothermics of climate change." The process starts with measuring temperature versus depth in boreholes,

and ends with an interpretation of surface temperature changes through time.

Figure 2.2 shows a site and a typical field setup for measuring borehole temperatures. In the middle ground is a 150 m deep borehole drilled into a granitic pluton. A 2.5 inch plastic casing extends above the rock surface. The borehole casing is capped on the bottom and filled with water to stabilize the measuring environment and facilitate a fast response time of the temperature transducer. The water at any depth is the same temperature as the surrounding rock; they both return to thermal equilibrium a few days following the drilling disturbance.

Subsurface temperatures are measured quite precisely. A thermistor probe is lowered over a pulley with a depth counter, pausing at an appropriate logging interval (typically 1 m) to record temperature. Temperatures are measured using a digital resistance meter recorded on a laptop computer. Temperature is measured to a precision of 2.5 mK and can be reproduced, in the borehole, point by point to a root mean squared (RMS) difference of better than 11 mK. The RMS difference arises from slight positioning errors, instrument noise, and the thermal stability of the borehole. One can also return to a site after several years and monitor the rate of change of temperature as a function of depth (e.g., Chapman and Harris 1993).

Interpreting temperature-depth profiles in terms of climate change involves understanding how temperature changes on the earth's surface diffuse, with time, to depth in the subsurface. Transient temperature in a homogeneous medium is governed by

$$\frac{\delta T}{\delta t} = \alpha \frac{\delta^2 T}{\delta z^2} \qquad (1)$$

where T is temperature, t is time, z is depth, and α is thermal diffusivity. Equation 1 shows that transient temperature variations are proportional to curvature in the subsurface temperature profile where the proportionality is thermal diffusivity. It implies that current spatial variations in subsurface temperature are a manifestation of past temporal variations of temperature at the earth's surface.

Figure 2.3a shows three hypothetical temperature histories at the earth's surface: a constant temperature scenario in which the surface temperature remains fixed at T_0, a warming scenario in which the surface warms gradually by an amount ΔT_0 after a distinct onset time τ and a symmetric cooling scenario. Each scenario results in a distinctly different subsurface temperature-depth profile (Fig. 2.3b). The linear profile results from the constant surface temperature scenario and depicts the background temperature field in the absence of a changing surface temperature. A steady increase in temperature with depth (constant geothermal gradient) is a consequence of heat flowing out of the earth's interior and is essentially a steady-state process for the time scale of climate studies. The background temperature field is estimated by extrapolating the profile from the undisturbed zone, below the climatic perturbations, to the surface. To highlight temperature perturbations, we plot reduced temperatures, defined as the difference between the observed temperature data and the background temperature field (Fig. 2.3c). The expanded scale produces a better display of the anomaly. An unchanging surface temperature condition through time results in a constant (zero) reduced temperature profile.

Changes in surface temperature through time, however, produce characteristic subsurface temperature profiles. The warming scenario shown in Figure 2.3a produces positive curvature in the temperature profile, and a new surface temperature intercept at $T_0 + \Delta T_0$ is established as the disturbance diffuses downward to a depth related to the duration of the warming. A cooling scenario produces the opposite effect in the subsurface temperature. In the simple scenarios illustrated here, warming and/or cooling yield two parameters: the magnitude of the warming or cooling, ΔT_0, and an onset time, τ, that is a function of thermal diffusivity and can be calculated from the depth of penetration of the thermal disturbance. Fortunately the thermal diffusivity of many rocks varies by less than 20 percent and is close to 1.0×10^{-6} m²/sec for many rock types. The depth of penetration scales with the square root of time such that a surface temperature change that started 100 years ago has now penetrated to a depth of about 80 m, and an event beginning 1,000 years ago has penetrated to about 250 m. The last thousand years of climatic history of the earth is thus contained in the achievable depth range of many drill holes, the upper 250 m of the earth.

Actual climatic variations in surface temperature are more complicated than the simple ramps shown in Figure 2.3, but the principle remains the same. The borehole temperature profile responds to continuous changes at the surface irrespective of seasons. The physics of heat diffusion serves as a low-pass filter for surface variations, attenuating high-frequency temporal variations. The daily variation is attenuated in less than 1 m; annual variations are attenuated in about 15 m. What remains is a robust temperature record of slow, decadal to century changes: exactly what is needed in studies of climate change.

Climatic Warming in Utah

Utah provides an ideal opportunity for geothermal reconstructions of climate because many boreholes suitable for this type of analysis exist. Additionally, due to early rural settlement in Utah, hundred-year temperature records exist in several remote sites that do not suffer from thermal contamination by urban heat islands. As we demonstrate, combining subsurface temperature-depth profiles with meteorological data adds confidence to the technique and produces a robust estimate of baseline temperatures tied to the meteorological record.

Borehole temperature logs used for climatic studies in Utah consist of high-quality measurements at six sites in western Utah and nine sites in the Colorado Plateau

Figure 2.2. Field equipment and setup for borehole temperature measurement (foreground). Temperatures are measured by lowering a probe in a water-filled plastic pipe that extends to the bottom of the drill hole. Background shows a weather station established at the borehole site to study how closely ground temperature tracks air temperature and how meteorologic variables (solar radiation, rain, wind, snow) affect ground temperature.

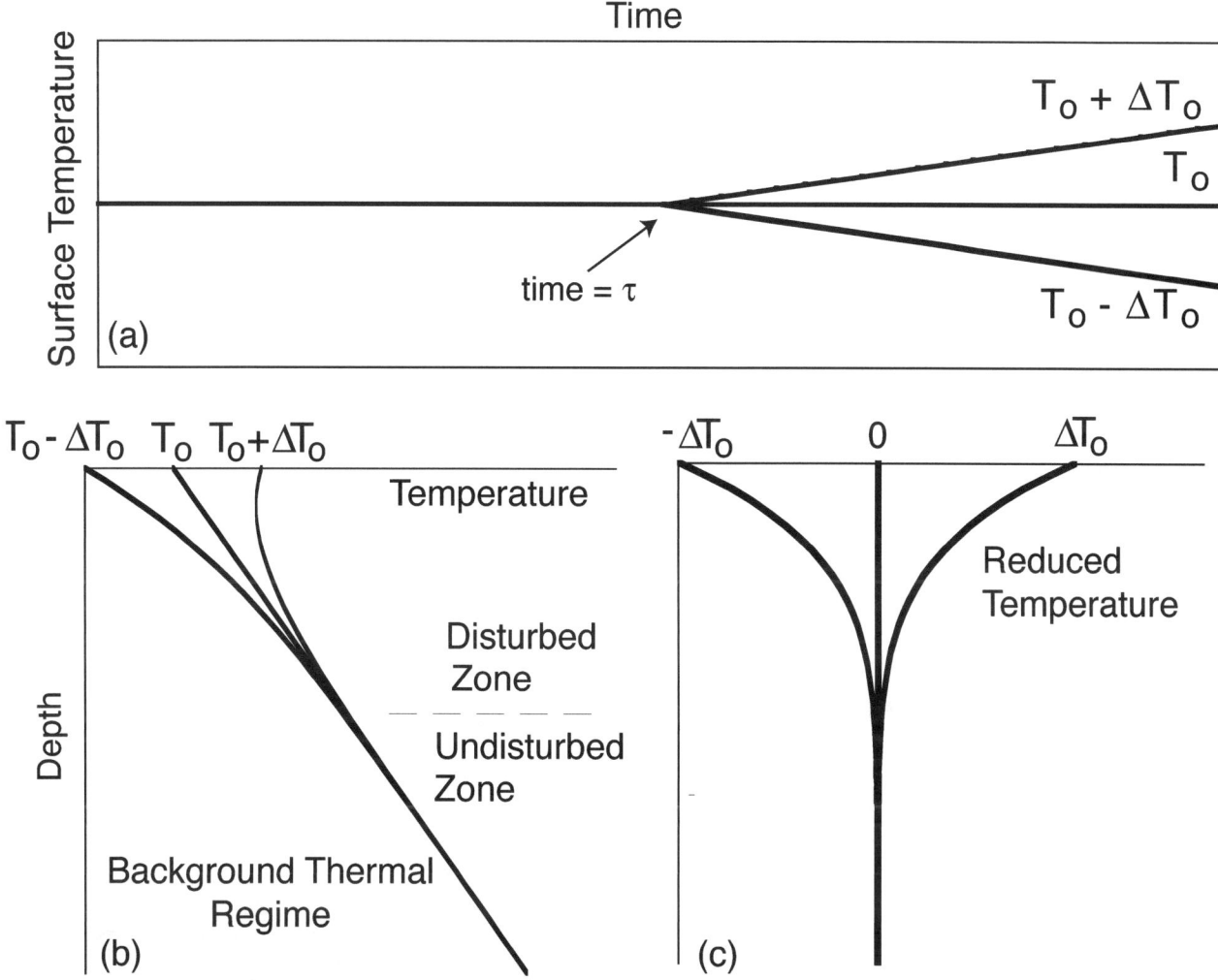

Figure 2.3. Geothermics of climate change: (a) Schematic diagram showing three hypothetical temperature-time scenarios for the ground surface and (b) the subsurface temperature-depth profiles that would result. (c) Reduced temperature is simply a transformed temperature with the steady-state background thermal gradient removed.

portion of southeastern Utah (Fig. 2.4). These data come from larger data sets and have been screened for studies of climatic change (Chisholm and Chapman 1992, Harris and Chapman 1995). The borehole sites in western Utah penetrate low-permeability granite, which limits possible spurious temperature perturbations caused by moving groundwater; those in southeastern Utah penetrate sedimentary sequences.

Raw temperature-depth profiles for western and southeastern Utah are shown in Figures 2.5a and 2.6a, respectively. The temperature profiles are offset to avoid overlap. The first-order effect exhibited in these profiles is a general increase in temperature with depth, consistent with an upward flux of heat toward the earth's surface. The profiles in western Utah appear to be linear in the deepest

part of each borehole. Temperature profiles in southeastern Utah are generally linear within formations; changes in trend are attributed to measured changes in thermal conductivity.

Figures 2.5b and 2.6b show reduced temperatures with an expanded scale that illustrates small departures from the background thermal field. High-frequency variations indicate the noise level of the data. At four of the six sites in western Utah (Fig. 2.5b) reduced temperatures near the surface are consistently positive, with amplitudes of 0.1 to 0.5°C, and extend to depths between 80 and 100 m. In southeastern Utah (Fig. 2.6b), reduced temperatures near the surface are consistently positive, with amplitudes between 0.2 and 0.5°C and depth extents between 100 and 200 m. These reduced temperatures are largely consistent

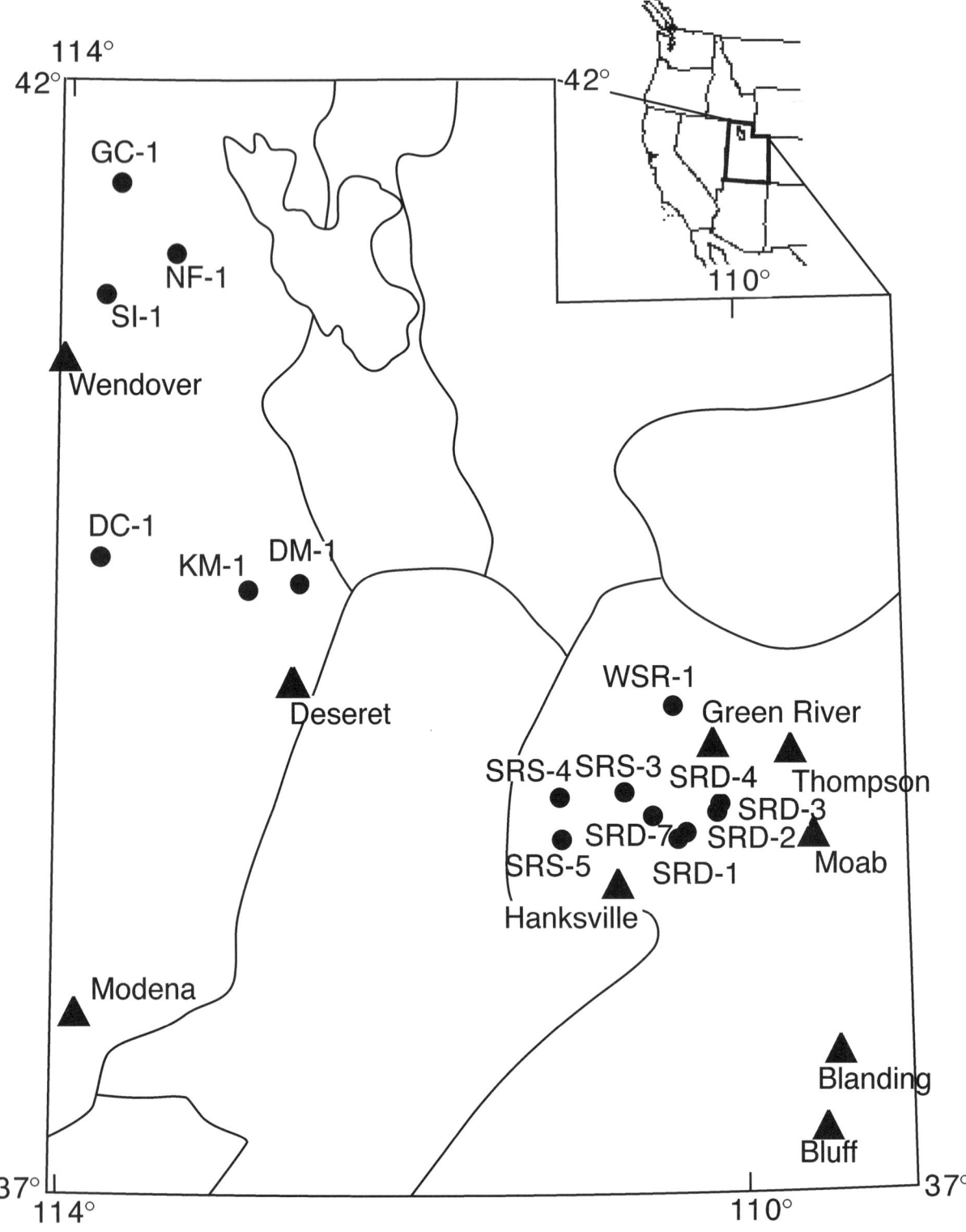

Figure 2.4. Location map of Utah showing borehole sites (circles) and meteorological stations (triangles) used in this study. Curved lines show climatic divisions in Utah that are thought to represent areas of relative climatic homogeneity.

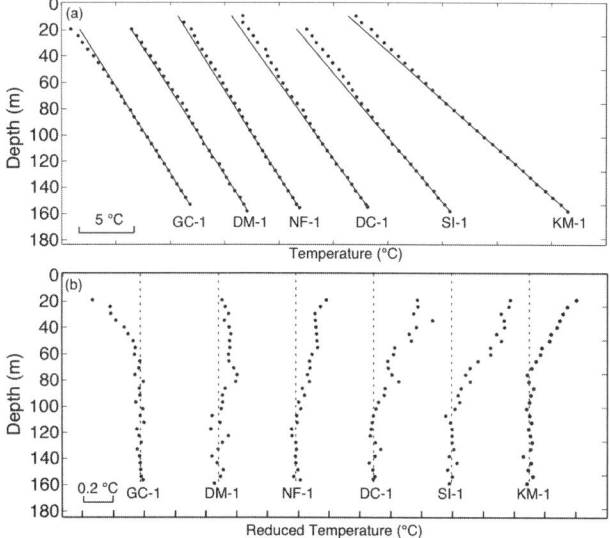

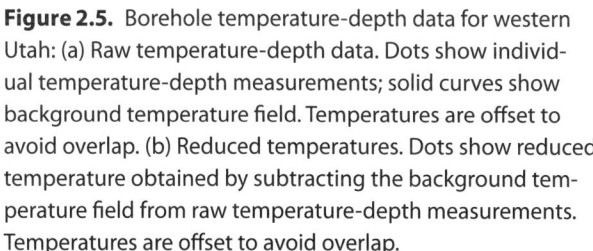

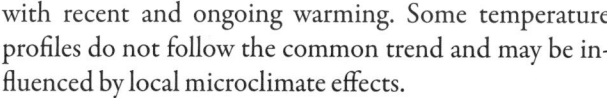

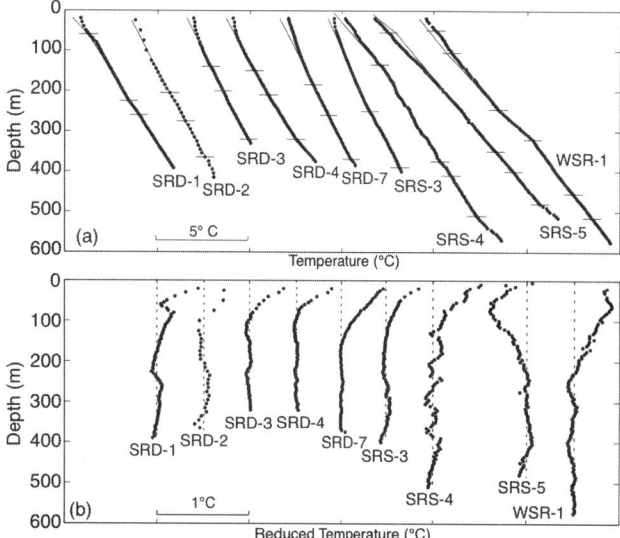

Figure 2.5. Borehole temperature-depth data for western Utah: (a) Raw temperature-depth data. Dots show individual temperature-depth measurements; solid curves show background temperature field. Temperatures are offset to avoid overlap. (b) Reduced temperatures. Dots show reduced temperature obtained by subtracting the background temperature field from raw temperature-depth measurements. Temperatures are offset to avoid overlap.

Figure 2.6. Borehole temperature-depth data for southeastern Utah: (a) Raw temperature-depth data. Dots show individual temperature-depth measurements; solid curves show background temperature field. Temperatures are offset to avoid overlap. (b) Reduced temperatures. Dots show reduced temperature obtained by subtracting the background thermal field from raw temperature-depth measurements. Temperatures are offset to avoid overlap.

with recent and ongoing warming. Some temperature profiles do not follow the common trend and may be influenced by local microclimate effects.

Utah meteorological data used are drawn from the U.S. Historical Climatology Network (HCN) (Easterling et al. 1996). These records are constructed from daily average temperature, defined as half the sum of the maximum and minimum, and are aggregated into annual mean temperatures for each station. Annual temperature differences can then be computed relative to the average for the entire time series for that station or, as is commonly done, relative to some arbitrary 30-year mean.

SAT records for nine locations—six on the Colorado Plateau (Hanksville, Green River, Blanding, Thompson, Moab, and Bluff) and three in western Utah (Modena, Wendover, and Deseret)—are shown in Figure 2.7. Linear fits to these time series serve to highlight overall warming or cooling trends. Western Utah sites yield warming from 0.1 to 0.5°C/100 yr, with an average of 0.3°C/100 yr (s.d. 0.2°C/100 yr), while Colorado Plateau sites yield greater warming, from 0.7 to 2.8°C/100 yr, with an average of 1.4°C/100 yr (s.d. 0.8°C/100 yr). These trends are highly variable, illustrating the difficulty in extracting

the long-term climatic signal, which is small compared to daily and seasonal temperature variations. For example, in Logan, Utah, the daily temperature swings by 15°C, and seasonally it swings by 30°C or 40°C. The climate-change inference we are making is approximately a tenth of a degree per decade in this widely varying signal.

The trends from the reduced temperature profiles and meteorological records are generally consistent, but more information about past temperature change can be extracted by linking these records (Fig. 2.8). Simply comparing trends is not entirely appropriate (Harris and Chapman 1998b) because of the different temporal and frequency contents in the two data sets. One appropriate method for comparing these data is to filter SAT records in the same way that earth filters SGT (surface ground temperature) signals. This process allows us to assess the consistency between the SAT record and borehole temperature profile, and determine a baseline temperature prior to the meteorological observations (preobservational mean [POM]).

Before one can use the SAT as a forcing function to produce a synthetic borehole temperature profile, the SAT record must be initialized with a background or

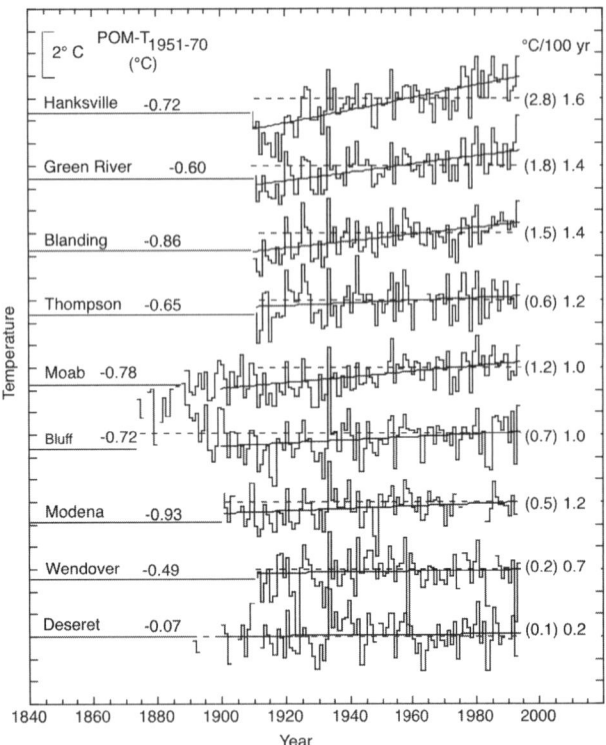

Figure 2.7. Surface air temperature (SAT) reconstructions for Utah. Each location has a SAT time series and a preobservational mean (POM) temperature. The top six time series are from southeastern Utah, and the bottom three from western Utah. Linear century trends fit to the SAT data are also shown. Numerical century trends (°C/100 yr) are computed by fitting the SAT data alone (parentheses) and by fitting the SAT data with the constraint of satisfying the POM at the start of the SAT time series. Values of the preobservational mean relative to the 1951–1970 mean SAT are given beside the location name.

starting temperature. As an example of this technique, we show the link between data from borehole SI-1 and the meteorological station at Wendover (Fig. 2.4). Three possible initial temperatures are labeled POM I, POM II, and POM III (Fig. 2.8). Each choice of POM yields a different subsurface reduced temperature profile. For POM I, a baseline value not very different from the 1951–1970 mean yields a synthetic reduced temperature profile with a minimum at 20 m depth in response to cooling around 1970, relative to the baseline. The synthetic temperature-depth profile produced using the POM I assumption is a poor fit to the actual reduced temperature profile.

In contrast, POM III is significantly lower than the mean observed temperature, and the synthetic reduced temperature-depth profile exhibits a warming trend at depths above 150 m. Again, the synthetic temperature-

depth profile is a poor fit to the data. On the other hand, POM II, intermediate between POM I and POM III, provides an excellent fit to the observed reduced-temperature profile (Fig. 2.8, inset). The best-fitting POM II is 0.5°C less than the 1951–1970 mean for Wendover and produces a small RMS (root mean square) misfit of 5 mK, which confirms that borehole and surface air temperatures are tracking each other over the common period of overlap; the sharp trough in the RMS misfit diagram indicates that the POM is a robust estimate of the baseline temperature. We estimate the warming for the SI-1/Wendover pair to be about 0.7°C, the sum of 0.5°C from the seventeenth-century baseline to the 1951–1970 mean and an additional 0.2°C to the present, or 0.7°C (Fig. 2.7).

Warming computed through this hybrid method provides for greater stability in the estimated warming because of the longer time period involved and the filtering effect of heat conduction on a noisy climate signal. Warming at the same sites on the Colorado Plateau determined from the hybrid method varies only from 0.60 to 0.78°C/100 yr (s.d. 0.09°C/100 yr), and data for the three western Utah sites indicate warming of 0.50°C/100 yr (s.d. 0.43°C/100 yr) (Fig. 2.7). With the success of the Utah example, we turn our attention to the Northern Hemisphere.

The Northern Hemisphere

Huang and Pollack (1998) scrutinized the global database of borehole temperature-depth data and found almost a thousand temperature logs suitable for climate-change studies. Figure 2.9 shows the location of, and composite reduced temperature profiles for, 687 sites in the Northern Hemisphere. Dots show positions of boreholes; the 5° × 5° grid elements containing borehole sites are shaded. The spatial distribution of boreholes shows that these data are most sensitive to extratropical land regions and dominantly fall between 30° and 60° north. Each gray line on the reduced temperature plot shows an individual temperature log, and as with the meteorological records, there is considerable scatter. Some areas show cooling, but most sites show warming. The bold reduced temperature profile is the weighted average of all sites (Harris and Chapman 2001).

A simple ramp increase of surface temperature that best explains the average reduced temperature profile is a temperature change of 0.8°C starting about 160 years ago. This simple ramp change produces a synthetic borehole temperature profile with an RMS misfit to the data of only

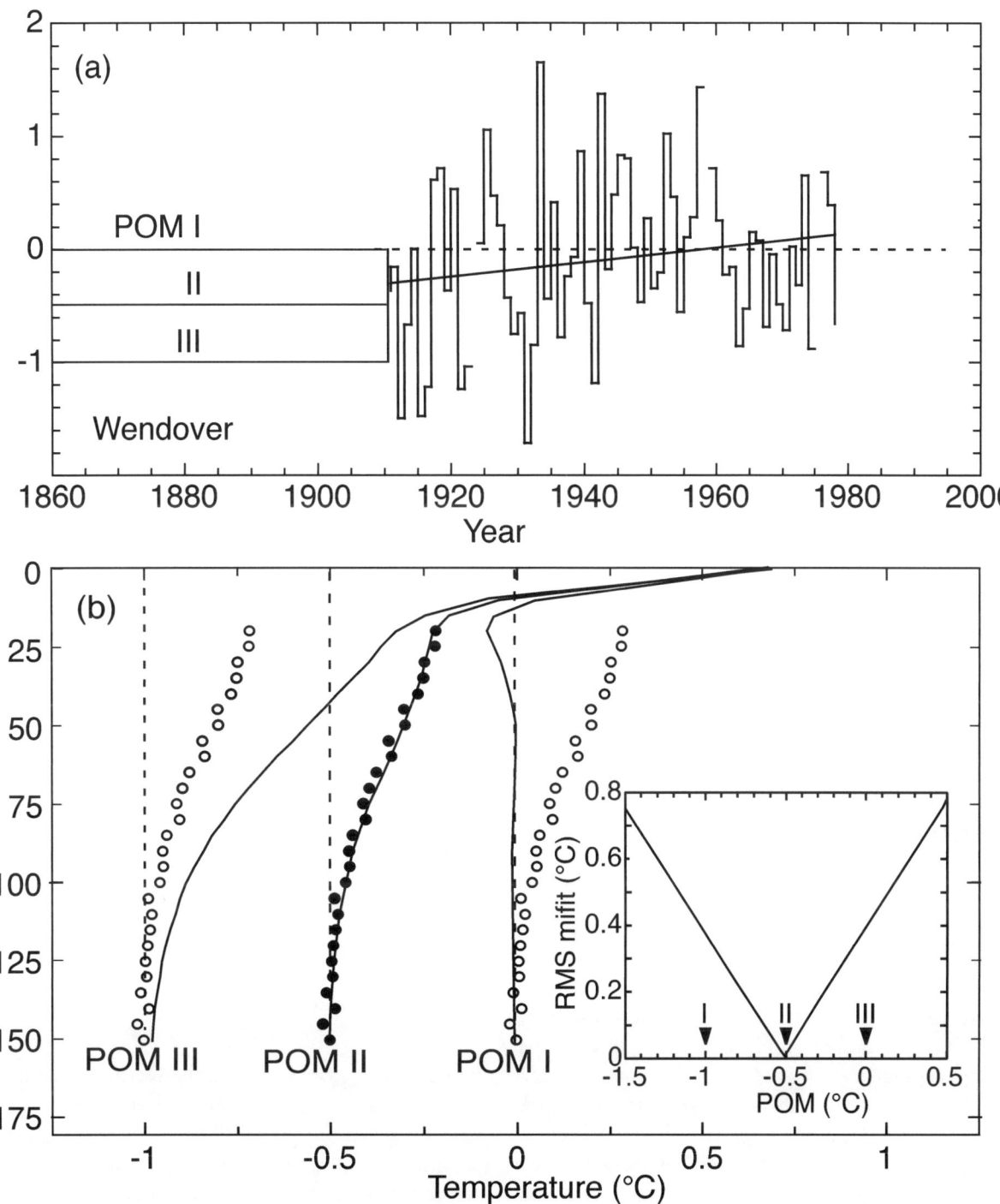

Figure 2.8. The hybrid geothermal method for climate change reconstruction: (a) Mean annual departures of surface air temperature (SAT) for Wendover over the period 1911–1978; dashed line is 1950–1971 mean. Linear fit to SAT data is also shown. Horizontal lines marked POM I, II, and III show three different choices of preobservational mean. The SAT data coupled with a particular choice of POM are used as a forcing function to compute synthetic temperature profiles. (b) Reduced temperatures for borehole SI-1 (circles). Three synthetic transient temperature-depth profiles constructed using SAT data from Wendover coupled with a corresponding choice of POM are shown. Reduced temperatures are plotted relative to POM. Inset shows RMS misfit as a function of the POM and illustrates the best fit for POM II.

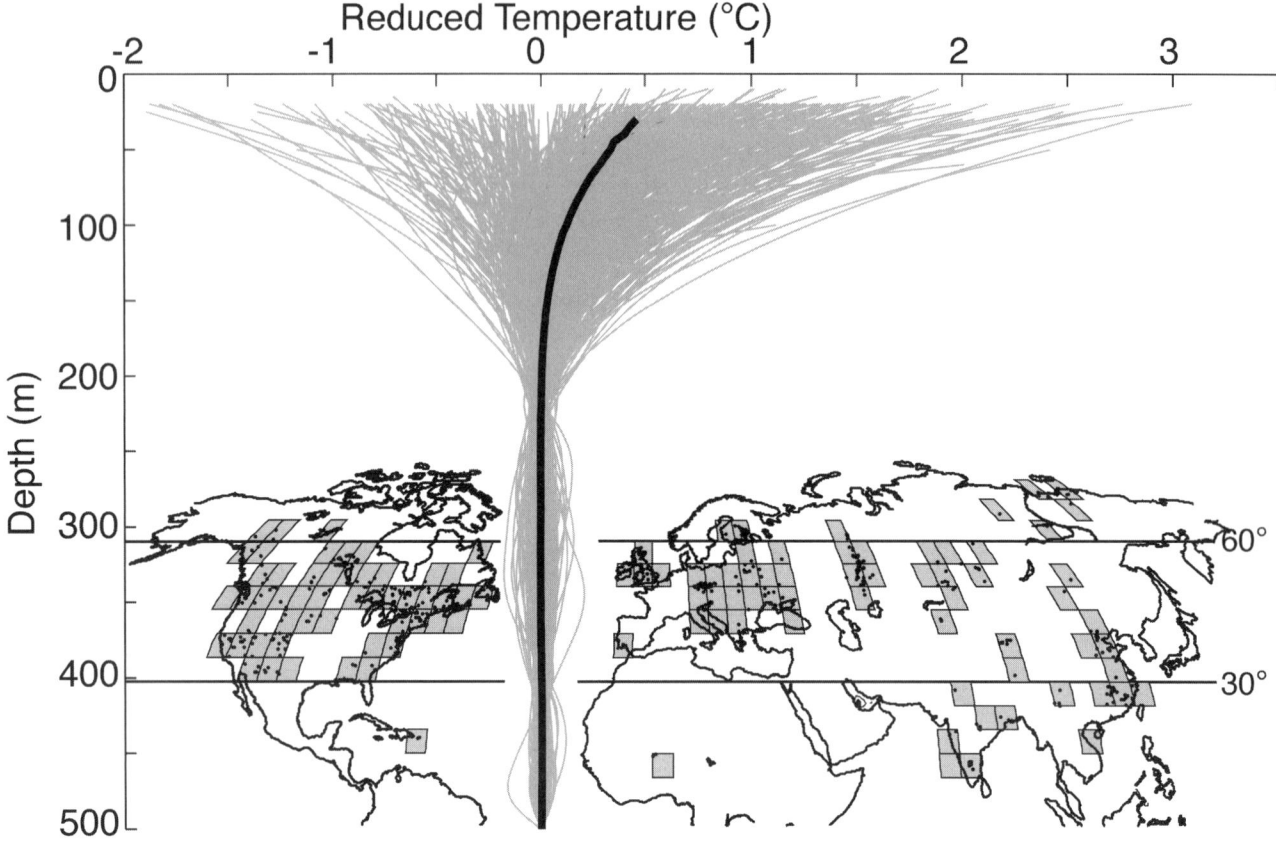

Figure 2.9. Northern Hemisphere reduced temperature profiles. Individual gray lines show 649 profiles whose locations are shown in the inset. Bold line shows the weighted average reduced temperature profile. The mean SAT record is based on data from the shaded 5° × 5° grid cells.

12 mK. For comparison purposes, gridded SAT data collocated with borehole data (shaded grid elements) are combined to form a weighted average. The best-fitting linear trend between 1856 and 2001 is 0.8°C, providing an excellent comparison to the simple ramp fit to the average reduced temperature profile. It is hard to change this result by very much.

We combined these two records of climate change through our hybrid method to assess the consistency between the SAT record and borehole temperature profile, and to determine a baseline temperature prior to the meteorological observations. The best-fitting, POM II, is 0.7°C less than the 1961–1990 mean for the Northern Hemisphere. Thus we estimate the current amount of warming in the Northern Hemisphere to be the sum of 0.7°C, from the seventeenth-century baseline to the 1961–1990 mean, and 0.4°C, from the 1961–1990 mean to present—or 1.1°C. These results indicate that the SAT warming observed in the meteorological record is not simply a re-

covery from a cool period prior to 1900 but reflects a real departure from preindustrial baseline conditions.

Constructing a Temperature History for the Past 500 Years

The examples from Utah and the Northern Hemisphere demonstrate that borehole temperature profiles contain a valuable signal for estimating the magnitude and timing of surface warming since preindustrial time. We now place our results in the context of other surface temperature reconstructions and comment on the relative strengths and limitations of each method.

The most intuitive method, and the one most commonly used to establish climate-change trends for this century, relies on SAT measurements at weather stations around the world (e.g., Folland et al. 2001, Karl et al. 1986, Karl and Williams 1987, Jones et al. 1999, Jones and Moberg 2003) (Fig. 2.10a). There is broad consensus that the globe has warmed by about 0.8°C over the last century, with most of the warming occurring in two periods: from about 1890 to 1940, and from 1975 to the present. Global mean SAT decreased slightly from 1940 to 1975.

There are several advantages to the instrumental SAT record of climate change: (a) it is a direct measurement of temperature; (b) it has annual time resolution; and (c) it has regionally good spatial coverage. Limitations include the relatively short time period for which there are widespread instrumental records (e.g., Hansen and Lebedeff 1987). Since about 1960, land regions in both hemispheres have been well covered. Prior to about 1930, most meteorological stations were in the Northern Hemisphere, and coverage before 1860 is limited to Europe and the British colonies. Other disadvantages of the SAT record are that the long-term climatic increase in temperature is small compared to daily and seasonal temperature variations, and nonclimatic factors may be biasing the record. These potential biases include changes in daily measurement times, data gaps, and station moves (Karl and Quayle 1988). Many researchers (e.g., Karl and Williams 1987, Peterson and Easterling 1994, Jones et al. 1999) have worked to eliminate these systematic biases.

Thus while the SAT record remains the anchor for modern climate-change estimates, it is important to have other, independent estimates of recent warming—in particular, temperature information going back at least to the initial modern increase in greenhouse gases (Fig. 2.1).

The principal method for reconstructing surface temperatures prior to instrumental measurements is based on proxies for temperature. A proxy is some measurable quantity that, although not a direct measure of temperature, can be statistically transformed to temperature through an empirical calibration during the period for which the proxy and instrumental data overlap. Common proxies include dendrochronological measurements of tree-ring widths and late-wood density (e.g., Fritts 1976, 1991; Schweingruber 1988; Briffa et al. 2002), corals (e.g., Evans et al. 2002, Hendy et al. 2002) ice cores (e.g., O'Brien et al. 1995, Meeker and Mayewski 2002) and lake sediments (e.g., Hughen et al. 2000). Because each proxy has different climatic sensitivities and geographic coverage, they may be combined into multiproxy assemblages (e.g., Bradly and Jones 1993, Overpeck et al. 1997, Mann et al. 1999, Crowley and Lowery 2000, Folland et al. 2001).

Fig. 2.10b shows five often-quoted Northern Hemisphere temperature reconstructions. These curves are similar in that they all show unprecedented late-twentieth-century warming. However, prior to about 1900 the reconstructed temperature curves diverge somewhat and differ from each other routinely by 0.2 to 0.3°C, and by as much as 0.6°C in certain periods. The multiproxy reconstruction that is most commonly quoted is from Mann et al. (1999) and is referred to as "the hockey stick." In contrast to the hockey stick curve, which shows little change between 1500 and 1900, the Esper et al. (2002) curve shows greater variability. This curve is notable for its preservation of low-frequency trends through age-band decomposition, which is often lost through detrending practices (e.g., Cook et al. 1995) and is typically 0.3°C cooler from 1500 to 1850, and as much as 0.5°C cooler in the early part of the seventeenth century. A recent summary of all surface temperature reconstructions for the last 1,000 years (NRC 2007) acknowledges the historic contributions of Mann et al. (1999) but shows a preference for the more recent reconstructions that contain longer period fluctuations and imply greater warming from the time of the industrial revolution.

The advantages to proxy reconstructions are the long time span over which temperatures can be estimated with annual or seasonal resolution. The biggest disadvantage of proxy reconstructions is that statistical techniques are relied on to transform the proxy indices to temperature. Imprecise knowledge about the relationship between the proxy and the climatic variable being sought can lead to a number of limitations. Proxies are likely responding to an ensemble of climatic and nonclimatic variables that are complexly intertwined, and the relative influence of each climatic and nonclimatic variable may change over time.

A necessary assumption to extrapolate temperatures prior to the period for which the transform function is developed and validated is that the transform function remains the same. However, these transform functions are developed from instrumental records from the twentieth century, when anthropogenic forcing appears to have played a dominant role in climate. The relationship between the proxy data and the climatic variable may vary due to the different relative influences of dominant climatic forcing. For example, recent data suggest that the relationship between tree rings and temperature is degrading (Briffa et al. 1998, Jones et al. 2003). Also, while these proxies provide annual resolution, reconstructions over multicentennial time scales are controversial and rare (Cook et al. 2004, Briffa and Osborn 2002, Mann and Hughes 2002).

Finally, different proxies are differently sensitive to seasonal temperatures and climatic forcings, and provide limited geographic coverage. For example tree-ring data

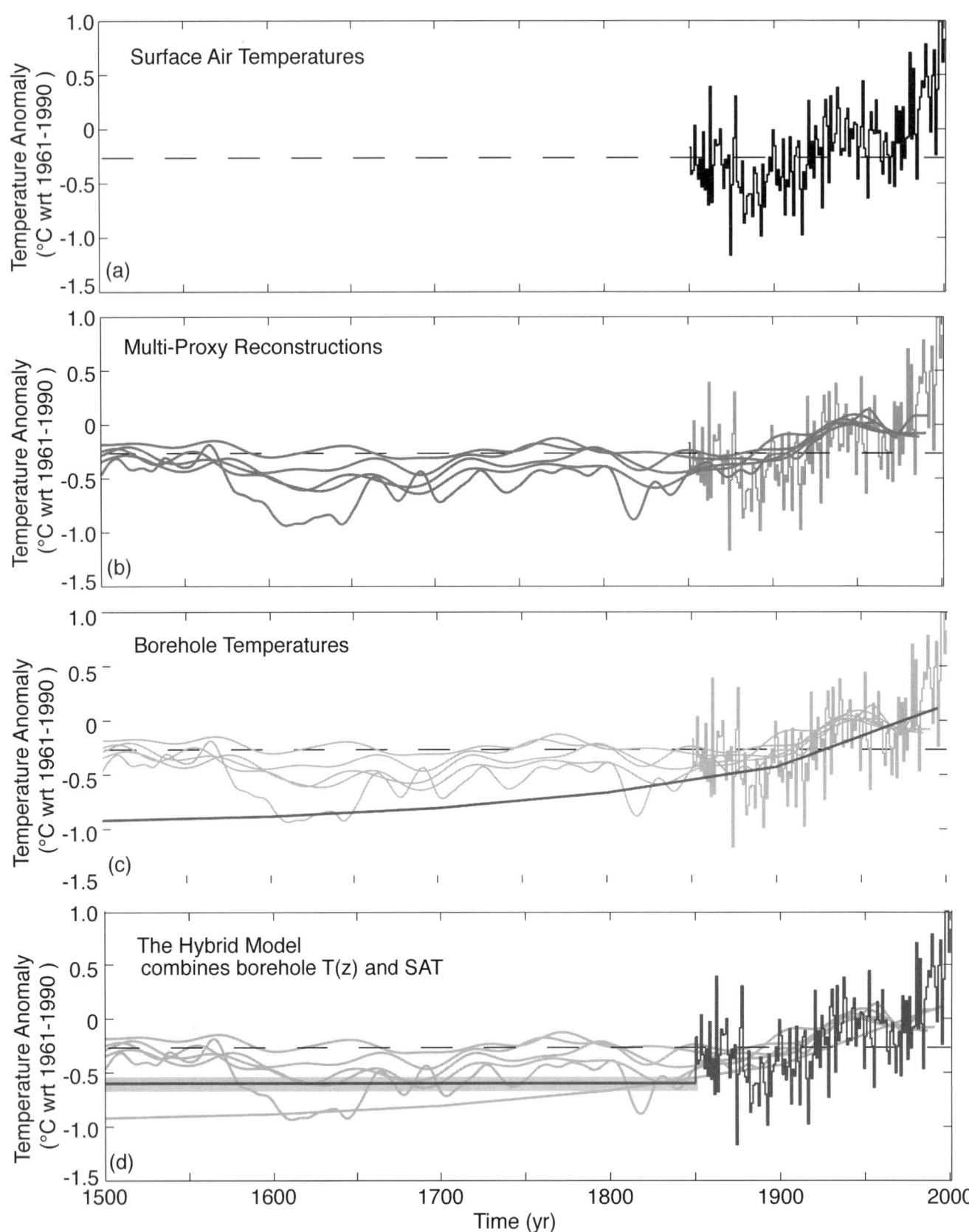

Figure 2.10. Reconstructing surface temperature change for the past 500 years: (a) Instrumental record of surface air temperature (SAT). (b) Multiproxy reconstructions. (c) Borehole temperature reconstruction. (d) Hybrid reconstruction combining borehole temperature-depth profiles and the SAT record. All temperatures are plotted relative to the 1961–1990 mean temperature.

provide information only from subpolar terrestrial regions and are weighted to the warm season (e.g., Briffa and Osborn 2002, Jones et al. 2003), whereas coral data represent tropical to subtropical marine environments (e.g., Evans et al. 2002, Hendy et al. 2002). While multi-proxy reconstructions are designed to take advantage of the relative strengths of each record, it is difficult to know how the different variables interact in the solution. Despite these potential limitations, proxy methods give insight into the nature of climatic variability over the millennium time scale.

As we have shown, borehole temperature-depth profiles also contain advantages and limitations. Figure 2.10c shows the Northern Hemisphere surface temperature reconstruction of Huang et al. (2000). Their solution, which is parameterized in terms of connected centennial trends, shows unprecedented warming in the twentieth century. For comparison with the meteorological record of warming, Huang et al. (2000) superimpose their curve, based on borehole measurements, on the SAT by matching the midpoint of their twentieth-century trend to the trends computed from the SAT record.

This comparison neglects some of the important differences in the two records and is qualitative at best. The smoothing parameterization used by Huang et al. (2000) to construct surface temperature histories has a particular problem for the last part of the twentieth century. Many of the boreholes were logged for temperature in the 1960s and 1970s, and because many of the logs started at 20 or 30 m depth (to avoid surface effects), the borehole temperature profiles contain little useful information for the period about 1960 to the present, when significant warming has occurred. It is fortuitous that the twentieth-century trend established by Huang et al. (2000) based on the first half of the century extrapolates not unreasonably to the last half of the century (Fig. 2.10c). Nevertheless, it is an imprecise procedure to match the borehole temperature construction of Huang et al. (2000) with the SAT record or the multiproxy construction.

The most important observation from the borehole reconstruction is the significantly greater temperature change inferred since ca. 1500. Whereas the multiproxy curves suggest 0.5°C of warming since 1500, the Huang et al. (2000) construction suggests more than 1.0°C of warming. This discrepancy has important implications for the sensitivity of the earth's climatic system.

The borehole method has several advantages. It is sensitive to surface temperatures over several hundred years. The earth continuously records surface temperature, day and night throughout the year, and therefore the borehole method is sensitive to the annual rather than seasonal temperature change (Harris and Chapman 2005). Because it is firmly rooted in the physics of heat diffusion and provides a direct measure of temperature, this technique does not need a statistical transform to temperature such as that required by proxy methods.

The principal limitation of borehole climatic reconstructions is that temporal resolution decreases the further back in time the reconstruction is made. The method can reveal that last year was warm or cold by the temperature a few meters below the surface, but it cannot detect that June 14 of the previous year was a warm day or cold day because that high-frequency information diffuses away. The "spread" rule of thumb for borehole temperature studies is that if one looks back τ years in time by interrogating temperatures at an appropriate depth, the paleo temperature one constructs from the borehole reconstruction represents an average over ± 40 percent of τ (e.g., Clow 1992, Harris and Chapman 1998a) centered on τ. For example, if one looks back 200 years to ca. 1800, the temperature calculated for that time has a spread of ± 80 years, and the temperature plotted for that time is actually an average from about 1720 to 1880. Consequently, the borehole reconstruction looks smooth over time.

Fig. 2.10d shows our Northern Hemisphere surface temperature reconstruction based on a hybrid method joining borehole science and the meteorological SAT record (Harris and Chapman 1998b, Harris and Chapman 2001). The baseline temperature is 0.7°C below the 1961–1990 mean SAT temperature. With the addition 0.4°C after 1990, we estimate a total surface warming of 1.1°C. This result is consistent with the warming since ca. 1800 found by Huang et al. (2000) and with the baseline temperature found by Esper et al. (2002). The small RMS misfit between a reduced temperature profile for the Northern Hemisphere and the synthetic reduced temperature computed for the hybrid reconstruction (Fig. 2.10d) indicates that ground and air temperatures track each other through time, thus validating the use of borehole temperatures in reconstructions of surface temperature change.

The advantages of the hybrid method mimic the advantages of the borehole temperature method: direct and continuous measure of temperature, extension of surface temperature prior to instrumental observations, reasonable spatial coverage on continents, and sensitivity to an annual rather than seasonal temperature. Unlike the straight borehole method, however, the hybrid method ties the baseline temperature directly to the SAT record.

The disadvantage of the hybrid method is the loss of temporal resolution in time.

Fig. 2.10d illustrates both the successes of surface temperature construction for the most recent five centuries and the challenges. Multiproxy methods are available to construct temperatures back many centuries, but different investigators have arrived at significantly different results. The differences may be a result of sampling differences and analysis techniques—most critically, the ability to discern longer-period variations. Borehole temperature studies suffer from poor time resolution, but they robustly predict greater warming since the start of the industrial revolution than most multiproxy studies. It is interesting to note some convergence between the multiproxy methods that claim to resolve long-period variations (Esper et al. 2002, Moberg et al. 2004) and the borehole temperature results.

Despite the differences in multiproxy and borehole temperature surface temperature reconstructions, the most recent large-scale analysis (NRC 2006) finds considerable evidence from ice cores, tree rings, borehole temperatures, glacial-length records, and historical documents for a relatively cold period, termed the "Little Ice Age," from roughly 1550 to 1850 (centered on AD 1700) and a time of relatively warm conditions centered around AD 1000 and termed "the Medieval Warm Period." Neither of these two longer-term climatic episodes denies the interpretation that the last 25 years have been warmer than any comparable period in the preceding millennium, and that the current rate of climate change, consistent with climate models, is highly unusual (NRC 2006).

Conclusions

There is mounting evidence from direct measurement of temperature (Jones and Moberg 2003) and measures of the response of physical and biological systems to changes in temperature (this volume) that planet Earth is warming at a significant and unusual rate. The most likely cause of this warming is the increase in concentration of atmospheric greenhouse gases since the beginning of the industrial revolution ca. 1750. Based on the widespread work of others and our own work on the geothermics of climate change, we offer the following observations and conclusions:

1. Direct measurement of surface air temperature (SAT) shows that the globe has warmed about 0.8°C since 1860. The warming has been uneven in time, with most occurring in two periods: between 1900 and 1940, and since 1980. The warming is also uneven in space, with high-latitude regions warming more than the tropics. A global SAT record does not exist prior to 1860.

2. Studies in western Utah and in the Colorado Plateau region using borehole temperatures and SAT information reveal consistent patterns of warming. The borehole temperature information, in particular, extends the temperature reconstructions prior to 1910, which is a common starting time for the local SAT records. Warming above baseline (ca. 1700–1800) temperatures on the Colorado Plateau has been 0.7 (plus recent) °C; warming in the Basin and Range region of northwest Utah has been slightly less, at 0.5 (plus recent) °C.

3. Borehole temperatures can also be used to reconstruct temperature changes on the surface over the last several centuries. These reconstructions are limited to very low-frequency changes but are a direct measure of temperature and are sensitive to annual rather than seasonal temperature changes. Inversion of borehole temperatures for extratropical Northern Hemisphere sites yields 1.0°C of warming since about AD 1500.

4. A hybrid method of temperature reconstruction combining borehole temperature-depth information and SAT data has produced a baseline temperature from which the last century of warming can be referenced. For the extratropical Northern Hemisphere, our hybrid method indicates 1.1°C of warming since ca. 1750.

5. Indirect measures, or proxies, for temperature can be used to extend the surface temperature record back to about AD 1000. Several multiproxy reconstructions of temperature show about 0.5°C of warming since ca. 1750, although reconstructions by different groups differ by as much as 0.4°C. More recent reconstructions using methods that recover multidecade trends (i.e., lower frequency) in surface temperature infer greater temperature change since the industrial revolution and are more consistent with borehole estimates of surface warming.

References

Beltrami, H. 2002. Climate from borehole data: Energy fluxes and temperatures since 1500. *Geophys. Res. Lett.* 29:2111. doi:10.1029/2002GL015702.

Bradley, R. S., and P. D. Jones. 1993. "Little Ice Age" summer temperature variations: Their nature and relevance to recent global warming trends. *Holocene* 3:367–376.

Briffa, K. R., and T. J. Osborn. 2002. Blowing hot and cold. *Science* 295:2227–2228.

Briffa, K. R., T. J. Osborn, F. H. Schweingruber, P. D. Jones, S. G. Shiyatov, and E. A. Vaganov. 2002. Tree-ring width and density data around the Northern Hemisphere: Part 1, Local and regional climate signals. *Holocene* 12:737–757.

Briffa, K. R., F. H. Schweingruber, P. D. Jones, T. J. Osborn, S. G. Shiyatov, and E. A. Vaganov. 1998. Reduced sensitivity of recent tree-growth to temperature at high northern latitudes. *Nature* 391:678–682.

Chapman, D. S., and R. N. Harris. 1993. Repeat temperature measurement in borehole GC-1, northwestern Utah: Towards isolating a climate-change signal in borehole temperature profiles. *Geophys. Res. Lett.* 20:1891–1894.

Chisholm, T. J., and D. S. Chapman. 1992. Climate change inferred from analysis of borehole temperatures: An example from western Utah. *J. Geophys. Res.* 97:14, 155-14, 176.

Clow, G. D. 1992. The extent of temporal smearing in surface-temperature histories derived from borehole temperature measurements. *Global Planet. Change* 6:269–281.

Cook, E. R., K. R. Briffa, D. M. Meko, D. S. Graybill, and G. Funkhouser. 1995. The "seqment length curse" in long tree-ring chronology development for paleoclimatic studies. *Holocene* 5:229–237.

Cook, E. R., J. Esper, and R. D. D'Arrigo. 2004. Extra-tropical Northern Hemisphere land temperature variability over the past 1000 years. *Quat. Sci. Rev.* 23:2063–2074.

Crowley, T. J., and T. S. Lowery. 2000. How warm was the medieval warm period? *Ambio.* 29:51–54.

Easterling, D. R., T. R. Karl, E. H. Mason, P. Y. Hughes, and D. P. Bowman. 1996. *United States Historical Climatology Network (U.S. HCN) Monthly Temperature and Precipitation Data.* ORNL/CDIAC-87, NDP-019/R3, 280 pp. Carbon Dioxide Inf. Anal. Cent. Oak Ridge Nat. Lab., Oak Ridge, TN.

Esper, J., E. R. Cook, and F. H. Schweingruber. 2002. Low-frequency signals in long tree-ring chronologies for reconstructing past temperature variability. *Science* 295:2250–2253.

Evans, M. A., A. Kaplan, and M. A. Cane. 2002. Pacific sea surface temperature field reconstruction from coral $\delta^{18}O$ data using reduced space objective analysis. *Paleoceanography* 17:1007, doi:10.1029/2000PA000590.

Folland, C. K., T. R. Karl, J. R. Christy, R. A. Clarke, G. V. Gruza, J. Jouzel, M. E. Mann, J. Oerlemans, M. J. Salinger, and S. W. Wang. 2001. Observed climate variability and change. Pp. 99–181 in J. T. Houghton et al. (eds.), *Climate Change 2001: The Scientific Basis.* Cambridge Univ. Press, New York.

Fritts, H. C. 1976. *Tree Rings and Climate.* Academic Press, London.

———. 1991. *Reconstructing Large-Scale Climatic Patterns from Tree-Ring Data.* Univ. of Arizona Press, Tucson.

Hansen, J. E., and S. Lebedeff. 1987. Global trends of measured surface air temperature. *J. Geophys. Res.* 92:13, 345-13, 372.

Harris, R. N., and D. S. Chapman. 1995. Climate change on the Colorado Plateau of eastern Utah inferred from borehole temperatures. *J. Geophys. Res.* 100:6367–6381.

———. 1998a. Geothermics and climate change: Part 1, Analyses of borehole temperatures with emphasis on resolving power. *J. Geophys. Res.* 103:7363–7370.

———. 1998b. Geothermics and climate change: Part 2, Joint analysis of borehole temperatures and meteorological data. *J. Geophys. Res.* 103:7371–7383.

———. 2001. Mid-latitude (30°–60°N) climatic warming inferred by combining borehole temperatures with surface air temperatures. *Geophys. Res. Lett.* 28:747–750.

———. 2005. Borehole temperatures and tree rings: Seasonality and estimates of extratropical Northern Hemisphere warming. *J. Geophys. Res.* 110:F04003, doi:10.1029/2005JF000303.

Hendy, E. J., M. K. Gagan, C. A. Alibert, M. T. McCulloch, J. M. Lough, and P. J. Isdale. 2002. Abrupt decrease in tropical Pacific sea surface salinity at end of Little Ice Age. *Science* 295:1511–1514.

Huang, S., and H. N. Pollack. 1998. *Global Borehole Temperature Database for Climate Reconstruction.* IGBP PAGES/Word Data Center-A for Paleoclimatology Data Contribution Series No. 1998-044. NOAA/NGDC Paleoclimatology Program, Boulder, CO.

Huang, S., H. N. Pollack, and P. Y. Shen. 2000. Temperature trends over the past five centuries reconstructed from borehole temperatures. *Nature* 403:756–758.

Hughen, K. A., J. T. Overpeck, and R. Anderson. 2000. Recent warming in a 500-year paleoclimate record from Upper Soper Lake, Baffin Island, Canada. *Holocene* 10:9–19.

Jones, P. D., K. R. Briffa, and T. J. Osborn. 2003. Changes in the Northern Hemisphere annual cycle—implications for paleoclimatology? *J. Geophys. Res.* 108:4588, doi:10.1029/2003JD003695.

Jones, P. D., and M. E. Mann. 2004. Climate over past millennia. *Rev. Geophys.* 42:2003RG2002/2004, doi:10.1029/2003RG000143.

Jones, P. D., and A. Moberg. 2003. Hemispheric and large-scale surface air temperature variations: An extensive revision and an update to 2001. *J. Climate* 16:206–223.

Jones, P. D., M. New, D. E. Parker, S. Martin, and I. G. Rigor. 1999. Surface air temperature and its changes over the past 150 years. *Rev. Geophys.* 37:173–199.

Jones, P. D., T. J. Osborn, and K. R. Briffa. 2001. The evolution of climate over the last millennium. *Science* 292:662–667.

Karl, T. R., and R. G. Quayle. 1988. Climate change in fact and in theory: Are we collecting the facts? *Clim. Change* 13:6–17.

Karl, T. R., and C. N. Williams, Jr. 1987. An approach to adjusting climatological time series for discontinuous inhomogeneities. *J. Climate Appl. Meteor.* 26:1744–1763.

Karl, T. R., C. N. Williams, Jr., P. J. Young, and W. M. Wendlad. 1986. A model to estimate the time of observation bias associated with monthly mean maximum, minimum, and mean temperatures for the United States. *J. Clim.* 7:1114–1163.

Lachenbruch, A. H., and B. V. Marshall. 1986. Changing climate: Geothermal evidence from permafrost in the Alaskan Arctic. *Science* 234:689–696.

Mann, M. E., R. S. Bradley, and M. K. Hughes. 1999. Northern Hemisphere temperatures during the past millennium: Inferences, uncertainties and limitations. *Geophys. Res. Lett.* 26:759–762.

Mann, M. E., and M. K. Hughes. 2002. Tree-ring chronologies and climate variability. *Science* 296:848.

Meeker, L. D., and P. A. Mayewski. 2002. A 1400-year high resolution record of atmospheric circulation over the North Atlantic and Asia. *Holocene* 12:257–266.

Moberg, A., D. M. Sonechkin, K. Holmgren, N. M. Datsenko, and W. Karlen. 2004. Highly variable Northern Hemisphere temperatures reconstructed from low- and high-resolution proxy data. *Nature* 433, doi:10.1038/nature03265.

NRC (National Research Council of the National Academies). 2006. *Surface Temperature Reconstructions for the Last 2,000 Years*. National Academies Press, Washington, D.C.

O'Brien, S. R., P. A. Mayewski, L. D. Meeker, D. A. Meese, M. S. Twickler, and S. I. Whitlow. 1995. Complexity of Holocene climate as reconstructed from a Greenland ice core. *Science* 270:1962–1964.

Overpeck, J., K. Hughen, D. Hardy, R. Bradley, M. Case et al. 1997. Arctic environmental change of the last four centuries. *Science* 278:1251–1256.

Peterson, T. C., and D. R. Easterling. 1994. Creation of homogeneous composite climatological reference series. *Int. J. Climatol.* 14:671–680.

Prather, M., D. Ehhalt, F. Derwent, E. Dlugokencky, E. Holland, I. Isaksen, J. Katima, V. Kirchoff, P. Matson, P. Midgley, and M. Wang. 2001. Atmospheric chemistry and greenhouse gases. Pp. 239–287 in J. T. Houghton et al. (eds.), *Climate Change 2001: The Scientific Basis*. Cambridge Univ. Press, New York.

Rutherford, S., M. E. Mann, T. J. Osborn, R. S. Bradley, K. R. Briffa, M. K. Hughes, and P. D. Jones. 2005. Proxy-based Northern Hemisphere surface temperature reconstructions: Sensitivity to method, predictor network, target season, and target domain. *J. Clim.* 18:2308–2327.

Schweingruber, F. H. 1988. *Tree Ring Basics and Applications of Dendrochronology*. Kluwer, Dordrecht.

Siegenthaler, U., T. Stocker, E. Monnin, D. Luthi, J. Schwander, B. Stauffer, D. Raynaud, J. M. Barnola, H. Fischer, V. Masson-Delmotte, and J. Jouzel. 2005. Stable carbon cycle-climate relationship during the late Pleistocene. *Science* 310:1313–1317.

Trenberth, K. E., Jones, P. D., and others. 2007. Observations: Surface and atmospheric climate change. Pp. 235–336 in S. Solomon, D. Qin, M. Manning, Z. Chen, M. Marquis, K. B. Averyt, M. Tignor, and H.L. Miller (eds.), *Climate Change 2007: The Physical Science Basis*. Contribution of the Working Group 1 to the Fourth Assessment Report of the IPCC. Cambridge Univ. Press, Cambridge, U.K., and New York.

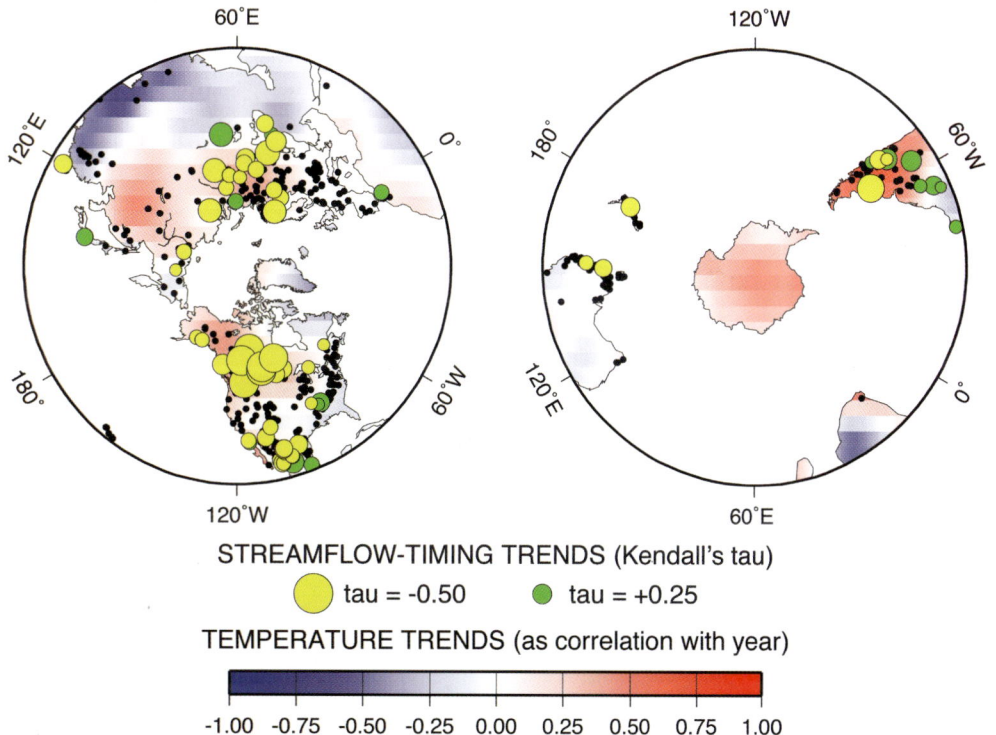

"CENTER-OF-MASS" STREAMFLOW-TIMING TRENDS, 1945-93
(Sites with 20+yrs of record, Kendall tau: p<0.01)
Yellow dot = EARLIER, green = LATER, black = no trend

STREAMFLOW-TIMING TRENDS (Kendall's tau)

tau = -0.50   tau = +0.25

TEMPERATURE TRENDS (as correlation with year)

-1.00 -0.75 -0.50 -0.25 0.00 0.25 0.50 0.75 1.00

Figure C.1. Global trends in the timing of streamflow, as measured by the CT, the flow-weighted average day of flows, in extratropical, snowmelt-dominated rivers. Red-filled circles denote retrogression of the annual hydrograph. (Source of data: Dettinger and Diaz 2000.)

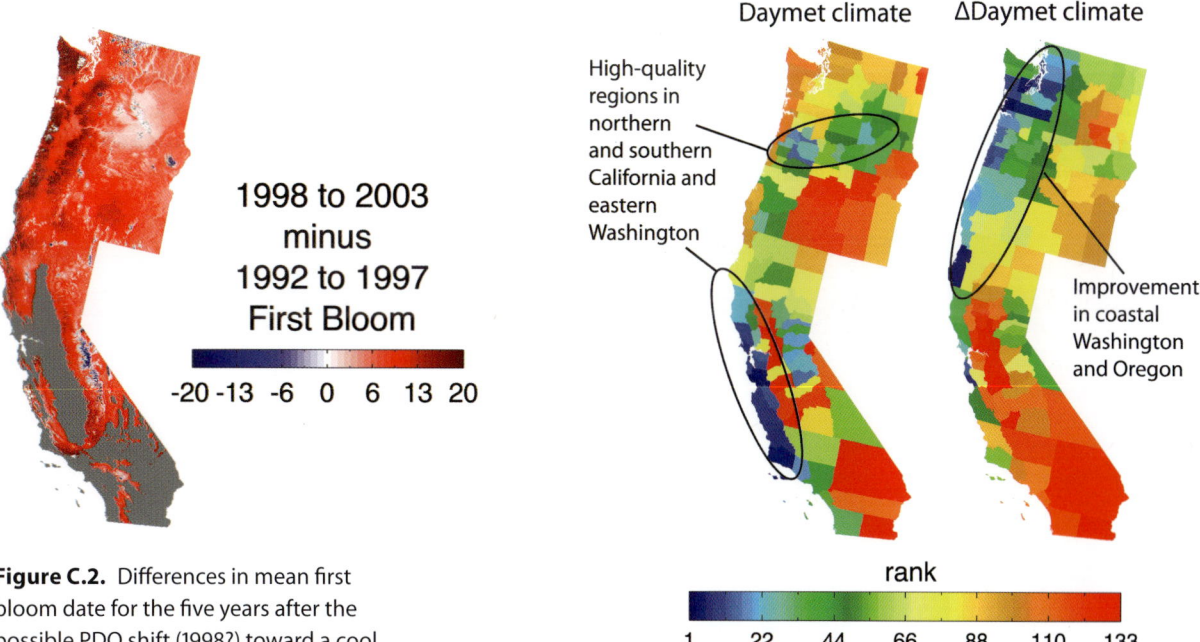

1998 to 2003
minus
1992 to 1997
First Bloom

-20 -13 -6 0 6 13 20

Figure C.2. Differences in mean first bloom date for the five years after the possible PDO shift (1998?) toward a cool phase versus the mean first bloom for the prior five years (1992–1997). Red colors indicate that first bloom, a spring phenological indicator, was earlier in the later five-year period.

Daymet climate   ΔDaymet climate

High-quality regions in northern and southern California and eastern Washington

Improvements in coastal Washington and Oregon

rank

1   22   44   66   88   110   133

Figure C.3. Ranking of climate index values for all counties in Washington, Oregon, and California for the Daymet (late twentieth century) and ΔDaymet (late twenty-first century) climates. Colors correspond to ranks: dark blue is the best quality; red is the worst. The index is based on a four-factor thermal assessment of average temperature, growing degree summations, extreme heat, and extreme cold. Only the upper quartile of climate index values within each county was used in calculations. See text for details.

Figure C.4. A. (*above*) This photo of the Angle Lakes area, Teton Wilderness Area, Bridger-Teton National Forest was taken by Jane Pargiter, EcoFlight Inc., in summer 2007. All of the red whitebark pine in this photograph were killed during one season (summer 2006) by the mountain pine beetle. The gray stems were killed during previous summers. This landscape is typical for a large percentage of high elevation whitebark pine forests in the Greater Yellowstone Ecosystem. **B.** (*top right*) Massive tree mortality due to mountain pine beetles is currently occurring in British Columbia, as shown in this 2002 photo taken near Lake Ootsa, near the town of Prince George.

Figure C.5. (*bottom right*) Taken from the overlook on the north side of Galena Summit, this photograph shows the Stanley Basin with the Sawtooth Mountains to the west and the headwaters of the Salmon River winding its way along the valley floor. The interesting thing about this photo is not so much the striking scenery, but the number of red, beetle-killed trees. All the red trees in this photograph were killed by the mountain pine beetle during the summer of 2002. See the text for a discussion of why mountain pine beetle mortality of this magnitude is unusual for the Stanley Basin.

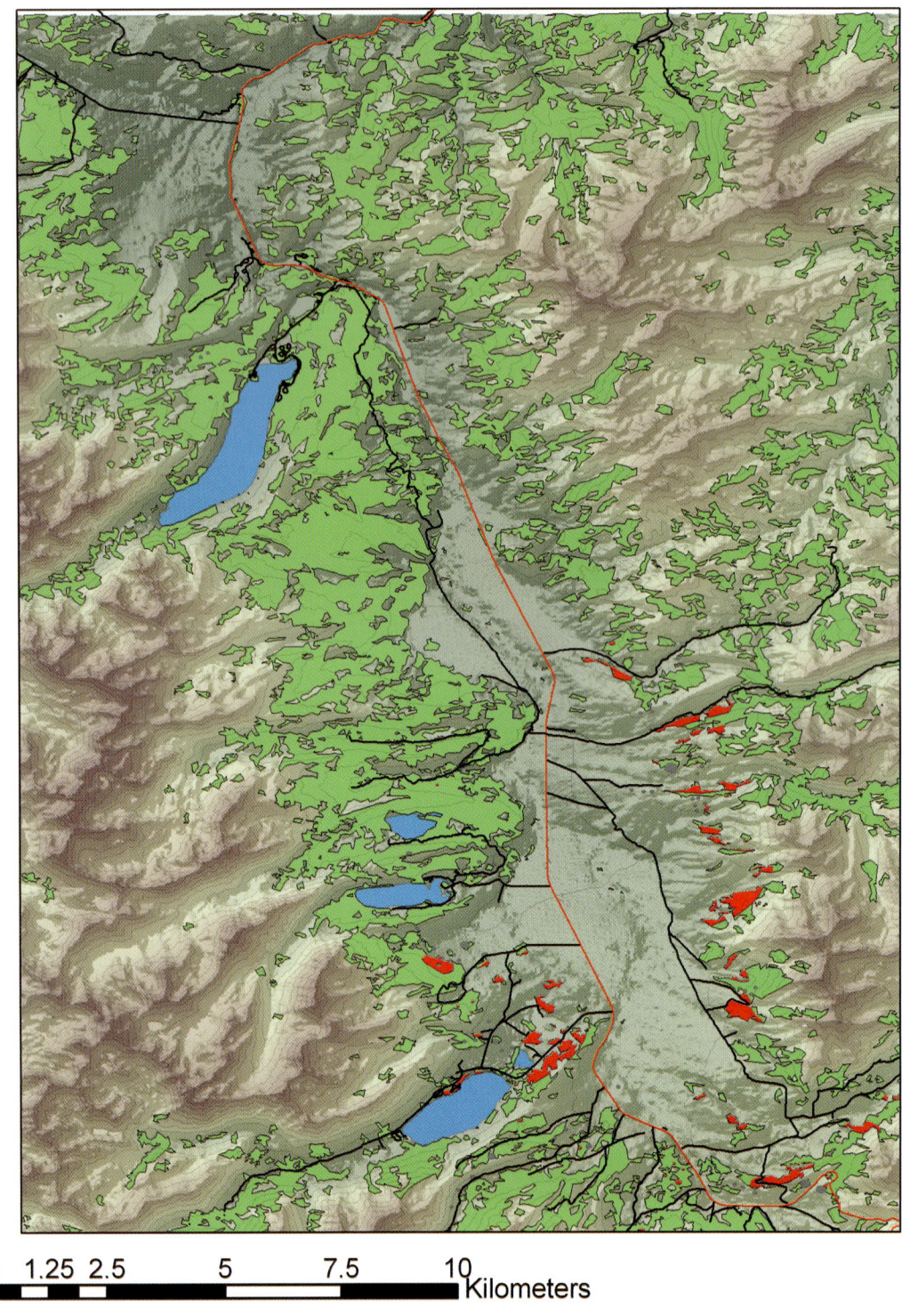

0   1.25 2.5   5   7.5   10

Kilometers

Figure C.6A–D. USDA Forest Service aerial detection survey (ADS) data for the Stanley Basin. Lodge-pole pine distribution is shown in light green. Red Fish Lake is in the upper left, with Stanley, Idaho, at the junction of State Highway 75 (red) with State Highway 21 (black) in the extreme upper left. Lodge-pole pine areas detected as killed the previous year (red during the current year) are shown in red; areas infested two years before (longer) are shown in gray, except for 2003, for which cumulative mortality is shown. **Figure C.6A.** (*above*) Survey year 1990 with a sub-outbreak population indicated in the warmer west-facing aspects and southern end of the Stanley Basin.

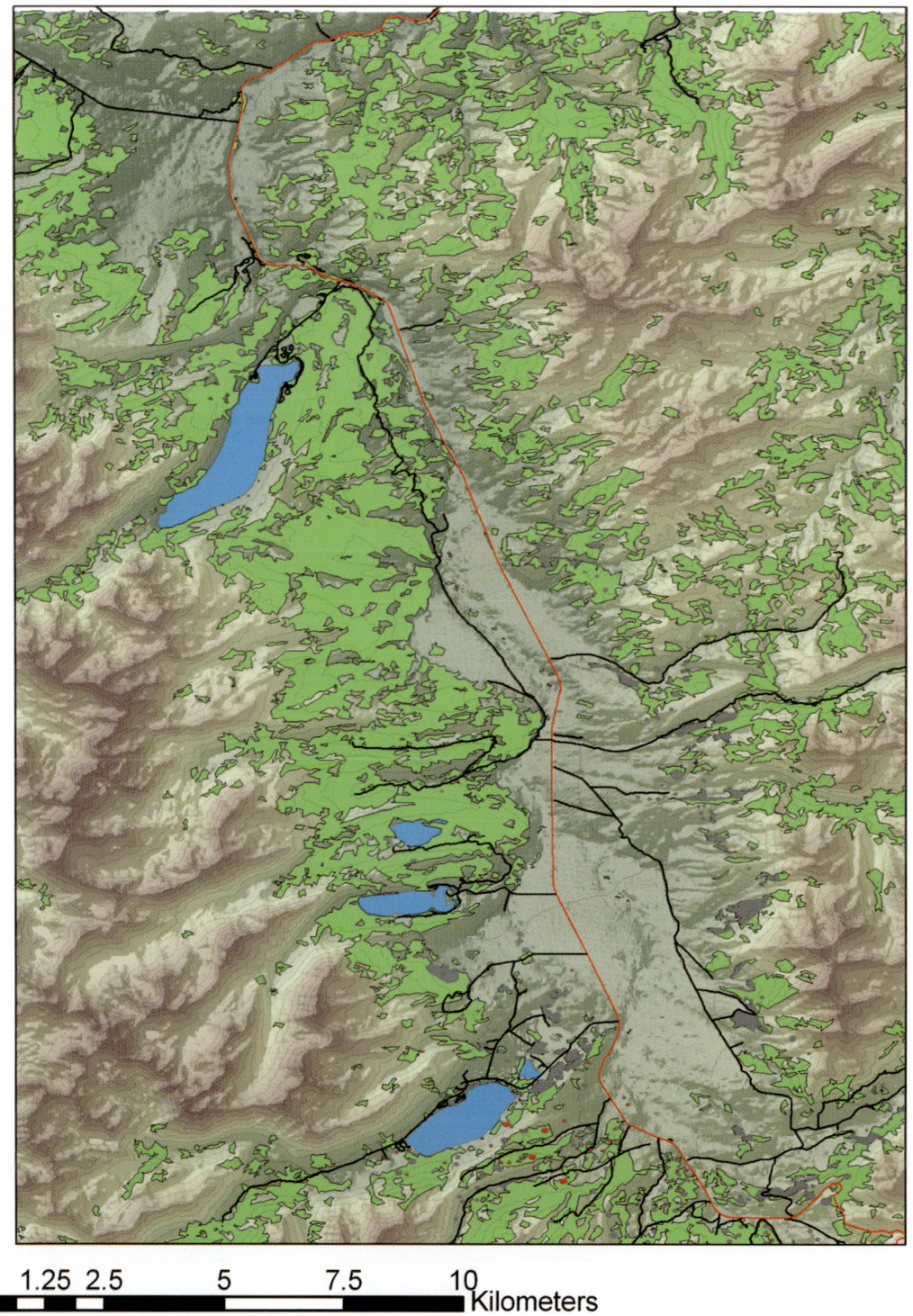

0   1.25 2.5   5   7.5   10
                                        Kilometers

Figure C.6B. Survey year 1994, following the summer that Mount Pinatubo erupted, indicating an almost complete collapse of the incipient outbreak.

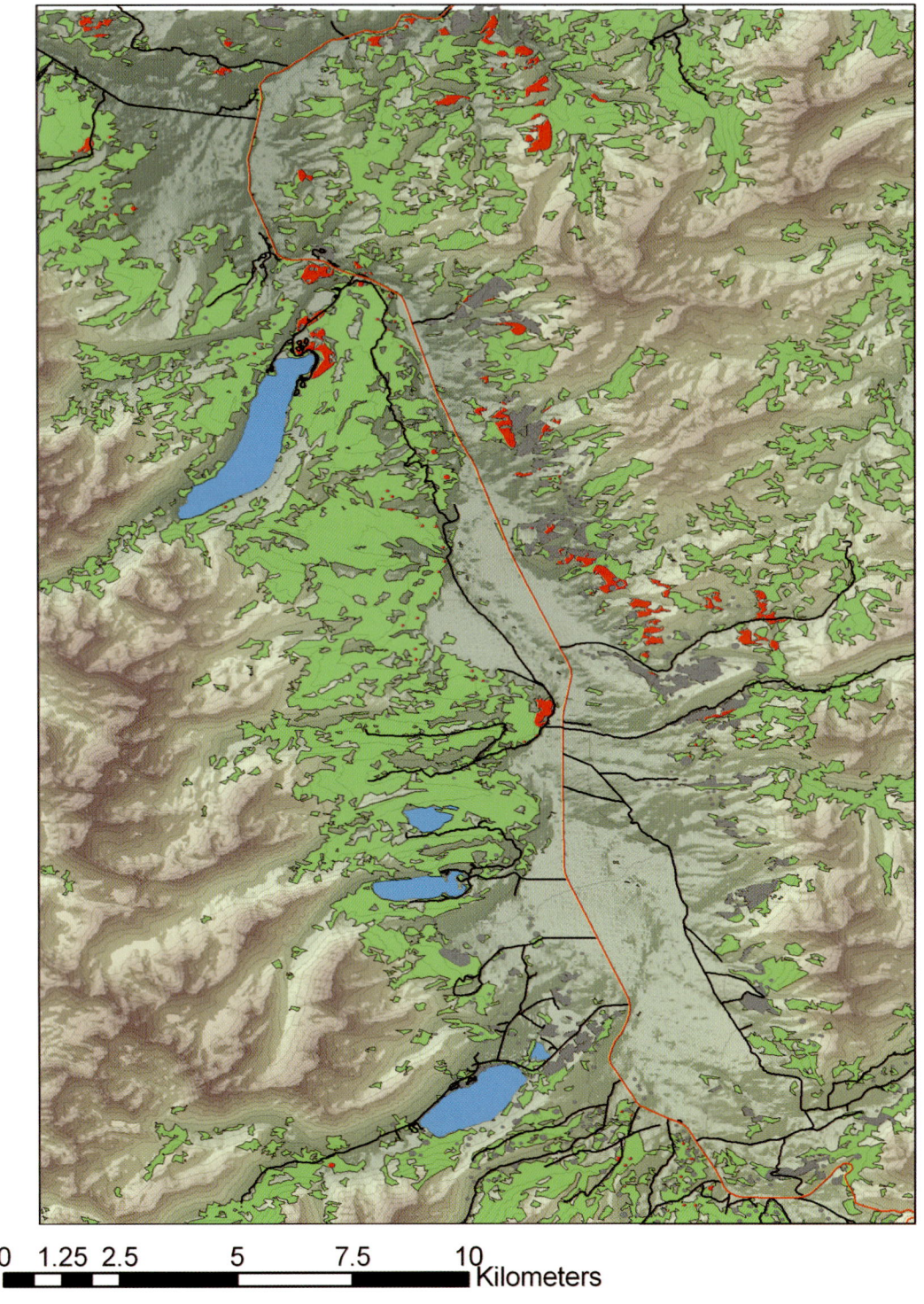

0 1.25 2.5   5   7.5   10
                                      Kilometers

Figure C.6C. Survey year 2000, by which time the population had recovered and was building on warm west-facing aspects, but also farther north in the valley.

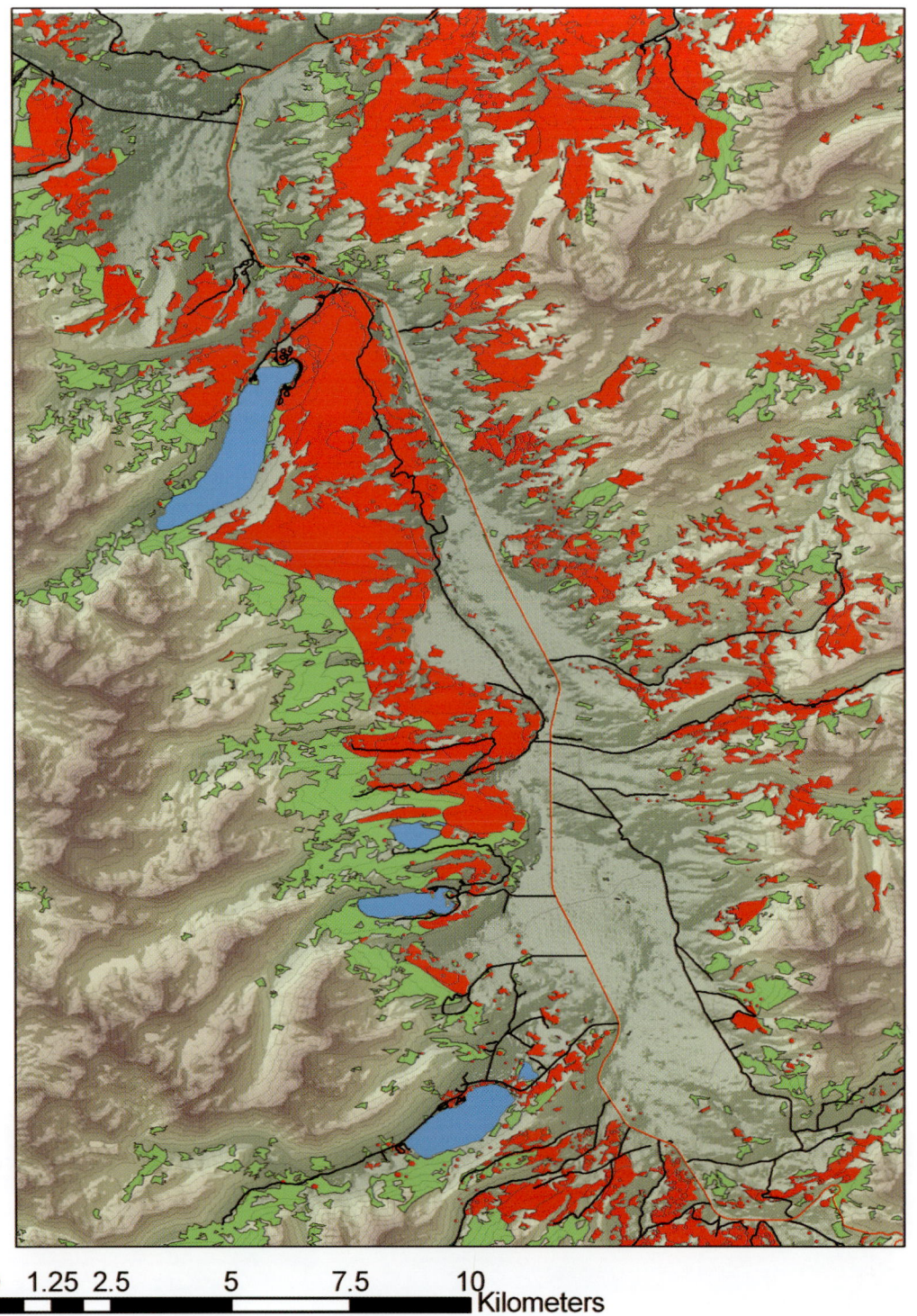

0   1.25 2.5   5   7.5   10
                                          Kilometers

Figure C.6D. Total area impacted by survey year 2003, the last year that data were available. The area shown in this image does not include trees killed during the summers of 2003 and 2004.

Figure C.7. This photograph was taken near the Railroad Ridge north site in the White Cloud Mountains of central Idaho. The red trees were killed in the summer of 2003, with significant mortality in nearby trees from the resulting adult mountain pine beetles that emerged during the summer of 2004.

Figure C.8. This photograph was taken above Sylvan Pass, near the northwest entrance of Yellowstone National Park, Wyoming. All of the red-topped trees were killed during the summer of 2003. (Photograph courtesy of Jeff Hicke.)

3

Ice-Core and Flood-Flow Evidence of Rapid Climate Change, Fitzpatrick Wilderness Area, Wind River Range, Wyoming

David L. Naftz, Liz Oswald, Paul F. Schuster, and Kirk Miller

Abstract

Site-specific transfer functions relating delta oxygen-18 ($\delta^{18}O$) values in snow to the average air temperature (T_A) during storms on Upper Fremont Glacier (UFG) were used in conjunction with $\delta^{18}O$ records from UFG ice cores to reconstruct long-term (270-year) trends in air temperature from alpine areas in the Wind River Range, Wyoming. On the basis of this reconstruction, an increase in T_A of approximately +2.1°C has occurred on UFG from the so-called Little Ice Age to the early 1990s, with approximately half of this temperature increase occurring from 1960 to 1990. A possible consequence of this accelerated warming since 1960 was a glacial outburst flood (*jokulhlaup*) that occurred on Grasshopper Glacier in September 2003. A natural ice dam at the head of Grasshopper Glacier failed, resulting in the instantaneous release of approximately 3.2 million m³ of water. The *jokulhlaup* peaked at a maximum streamflow of about 35.7 m³/sec at a downstream USGS streamflow-gaging station, resulting in the instantaneous peak streamflow for water year 2003. Continued warming and melting of glaciers in the Wind River Range will likely result in future, and perhaps more frequent, outburst floods.

Introduction

Ice-core records from glaciers within the continental United States do not contain reliable paleoenvironmental and paleoclimatological data. Unlike those collected from the polar regions, a variety of post-depositional processes can modify the original chemical and isotopic signals in ice cores collected from mid-latitude areas, including the western United States. For example, meltwater percolation can change the original chemical composition of

the ice by preferential elution processes during summer melting cycles (Wagenbach 1989). Even more significant, increasing air temperatures are threatening the future existence and integrity of mid-latitude glaciers throughout the world (Meier 1998). For example, Hall and Fagre (2003) have projected the disappearance of all of Glacier National Park's namesake glaciers by 2030.

As of 2005, the Upper Fremont Glacier (UFG) in northwestern Wyoming (Fig. 3.1) was the only glacier within the continental United States where ice cores have been documented to contain reliable paleoenvironmental and paleoclimatological records (Cecil and Vogt 1997, Naftz 1993, Naftz et al. 1994, Naftz et al. 1996, Naftz et al. 2002, Naftz et al. 2004, Schuster et al. 2000 and 2002). This particular glacier has a combination of unique characteristics conducive to preserving paleoenvironmental signals. Background glaciological data on the glacier were collected during 1990–1991 (Naftz and Smith 1993). Potential drill-site altitudes exceed 4,000 m to minimize meltwater modification of the snow and ice chemistry. Ice thickness in the upper half of the glacier ranges from 60 to 172 m, providing longer-term paleoenvironmental records. Densification processes proceed rapidly at the site, with densities exceeding 8.5×10^2 kg/m³ at depths 14 m below the surface, and net accumulation rates of 0.96 m of water equivalent per year. Ice velocities decrease in a downslope direction, ranging from 0.8 to 3.1 m/year. Average annual air temperature on the Upper Fremont Glacier during a five-year monitoring period was −7°C.

Ice cores exceeding 160 m in length were recovered from the glacier in 1991 and 1998 (Naftz et al. 1996, 2002). The 1991 ice core was estimated to contain 250 years of record as determined from carbon-14 age dating of an insect leg recovered from near the bottom of the core

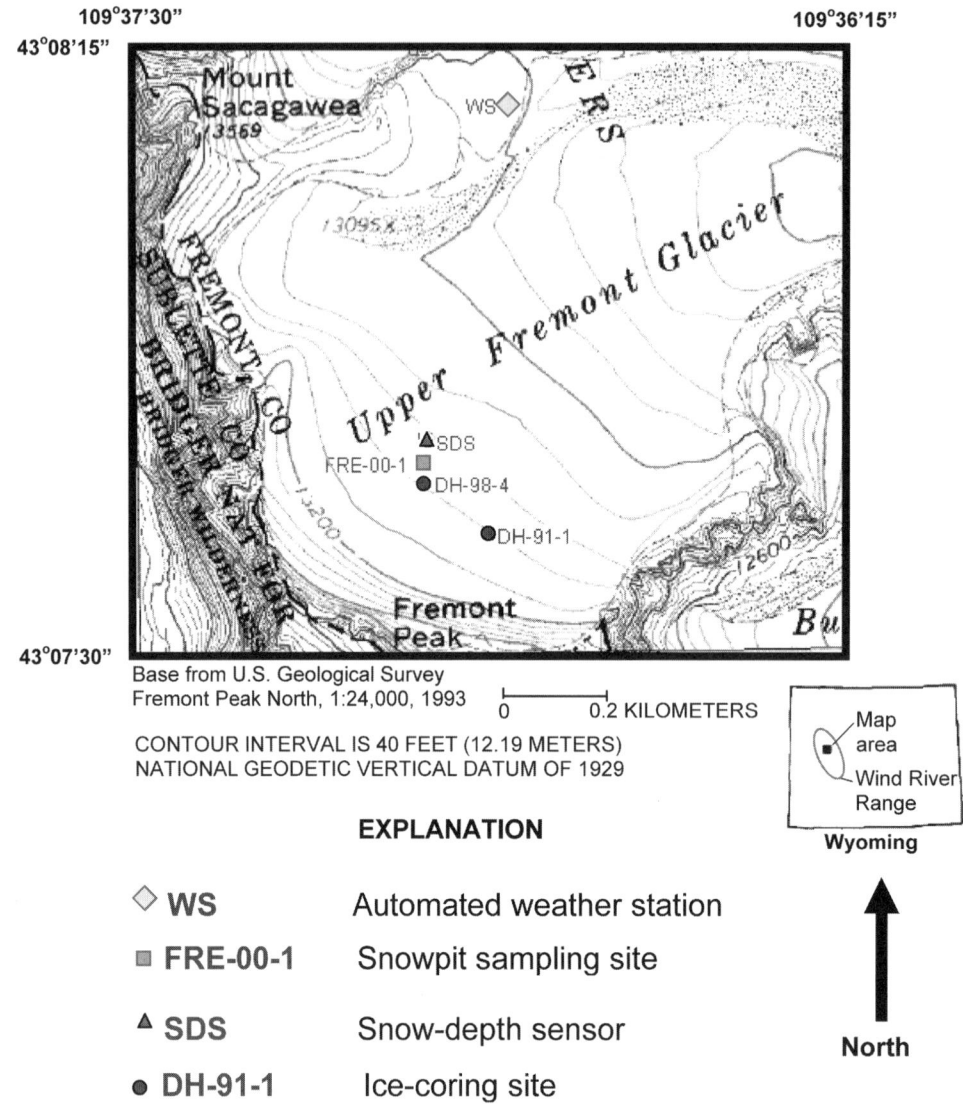

109°37'30"
43°08'15"

109°36'15"

43°07'30"

Base from U.S. Geological Survey
Fremont Peak North, 1:24,000, 1993

0   0.2 KILOMETERS

CONTOUR INTERVAL IS 40 FEET (12.19 METERS)
NATIONAL GEODETIC VERTICAL DATUM OF 1929

Map area
Wind River Range

Wyoming

EXPLANATION

◇ **WS**   Automated weather station

▣ **FRE-00-1**   Snowpit sampling site

▲ **SDS**   Snow-depth sensor

● **DH-91-1**   Ice-coring site

North

Figure 3.1. Location of meteorological station, snowpit sampling site, snow-depth sensor, and ice-coring sites, Upper Fremont Glacier, Wind River Range, Wyoming.

(Naftz et al. 1996). This length of record was subsequently confirmed by a higher-resolution chronology established with continuous electrical conductivity measurements (Schuster et al. 2000). On the basis of these data, ice at the bottom of UFG in proximity to the drilling sites (Fig. 3.1) was probably deposited as snow prior to AD 1710. Subsequent sections of this chapter include discussions of (1) the ice-core chronology and trends in isotopic composition; (2) development of a transfer function relating air temperature to the isotopic composition of snow and ice; (3) reconstruction of air temperature changes during the past 250 years in high-elevation areas of northwestern

Wyoming; and (4) outburst-flood response to recent changes in air temperature.

Background Information and Methodology

Ice-Core Data

Ice cores were collected in 1 and 2 m segments from the Upper Fremont Glacier (Fig. 3.1) in 1991 and 1998 using a thermal drill. On-site processing of the cores included visual inspection, logging, and density determinations by personnel wearing Tyvek suits and powder-free Latex gloves. The cores were sealed in polyethylene bags, placed in plastic core tubes, and stored in snow vaults until removal from the site to a freezer truck via a ten-minute helicopter flight.

Meteorological and Snow-Depth Data

Site-specific air temperature adjacent to the Upper Fremont Glacier was monitored with an automated weather station (WS) installed in a boulder field approximately 100 m to the north of the glacier at an altitude of 3,960 m (Fig. 3.1). Sensor height was approximately 2 m above land surface, and the sensor was shielded with an RM Young 12-plate grill radiation shield. The temperature sensor was factory calibrated and checked periodically for accuracy at the field site with a hand-held digital thermocouple probe.

The weather station was operated from July 1990 through August 1991, and from September 1997 through March 2001. Relative humidity, wind speed and direction, and solar radiation also were measured each minute and compiled into hourly averages. Changes in the glacier's snow depth and air temperature were monitored continuously with a snow-depth sensor installed at an altitude of 3,990 m in September 1997 (Fig. 3.1). This ultrasonic depth sensor, suspended on a horizontal bar above the snow surface, emitted a sonic signal. The travel time for the sonic pulse to the snow surface and return was measured, and the distance to the surface was calculated after correcting for air temperature effects on the speed of sound. The sensor recorded snow depth and air temperature every hour.

Snowpit Data

Snow samples were collected on the Upper Fremont Glacier from a vertical trench face in a snowpit excavated to the firn layer from the previous annual accumulation cycle. Snow samples of 7–10 cm were composited in precleaned 500 mL, wide-mouthed Nalgene bottles. Snow density was determined by collecting snow at 10 cm intervals from trench faces in 1,000 cm³ samplers and weighing the samples on a portable electronic balance.

Sample Processing and Analysis

The ice-core samples were melted according to strict protocols to minimize contamination. Ice cores were subsampled by using a bandsaw operated in walk-in freezers (air temperature between −10 and −24°C). The sections were split lengthwise, with one split being archived for future research. The surface ice from each subsample was scraped away with a stainless steel microtome. Each subsample was thoroughly rinsed with ultrapure (18.0 megohm) deionized water and placed in a prerinsed and covered plastic container.

Each sample was allowed to melt at room temperature for about one hour (or until about 15 ml of meltwater had accumulated). The sample was then rinsed in the accumulated meltwater, and the melt was discarded to eliminate any remaining isotopic signature from the rinse water. The remaining sample was allowed to melt in the covered plastic container at room temperature. After complete melting, the samples were filtered (0.45 micrometer [μm]), placed in a glass vial, sealed with a polyseal cap, and coated with ParaFilm.

The snow samples were processed in a similar manner, except they were not scraped or rinsed with ultrapure deionized water, and the initial meltwater was not discarded. Because of the study site's remote location, snow samples collected during the summer months could not be kept frozen during transport to the trailhead. Summer snow samples were allowed to melt in thick-walled plastic bottles with tape-sealed lids to eliminate isotopic exchange with outside air.

The $\delta^{18}O$ value of each sample was determined by using the method developed by Epstein and Mayeda (1953) at the U.S. Geological Survey Stable Isotope Laboratory in Menlo Park, California, and reported relative to standard mean ocean water (SMOW) in permil notation. Tritium concentration was determined by electrolytic enrichment/liquid scintillation counting (Thatcher et al. 1977) at the U.S. Geological Survey Tritium Laboratory, also in Menlo Park.

Results and Discussion

Ice-Core Record

A high-resolution ice-core chronology is needed to interpret paleoclimatic information contained in the cores collected from the Upper Fremont Glacier. Dust layers in the ice core were occasionally too faint to yield reliable annual stratigraphic markers; therefore, tritium, chloride-36, and carbon-14 age-dating methods (Naftz et al. 1996, Cecil and Vogt 1997) were combined with the timing of major volcanic eruption signals that were evident in the core (Schuster et al. 2000) to establish a high-resolution ice-core chronology. The polynomial fit for the refined age-depth profile (Schuster et al. 2000) was determined to be

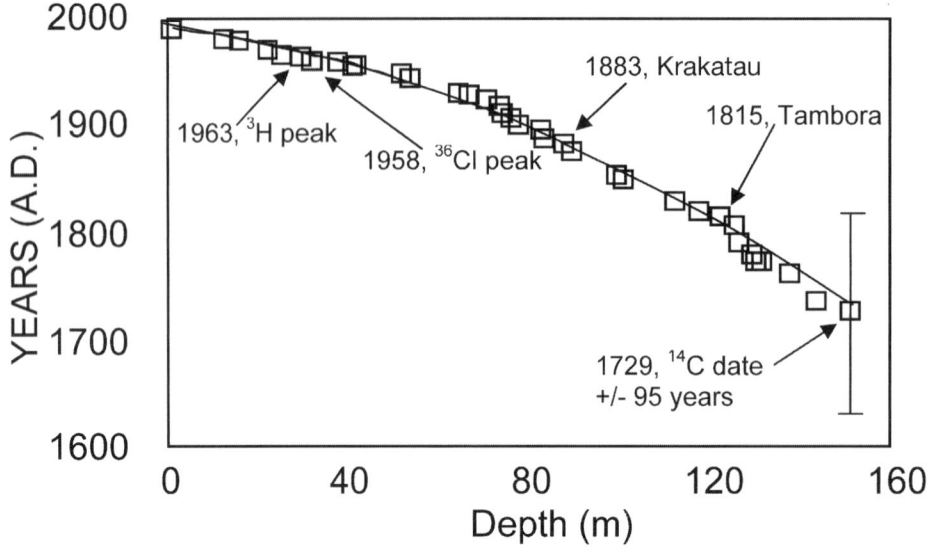

Figure 3.2. Plot of reported volcanic events and isotopic age dates used to generate a polynomial fit for an age-depth profile of the Upper Fremont Glacier ice core collected in 1991 (from Schuster et al. 2000), Wind River Range, Wyoming.

$$\text{Age (in years)} = 0.00739(D)^2 + 0.5558(D) \qquad (1)$$

where D is depth below glacier surface. This age-depth profile yielded excellent agreement with known time horizons in the ice core that include (1) the 1963 tritium and 1958 chloride-36 peaks from above-ground nuclear testing; (2) the 1883 Krakatau volcanic eruption; (3) the 1815 Tambora volcanic eruption; and (4) 1729 ($\pm$ 95 years) carbon-14 age date of the insect leg (Fig. 3.2).

The $\delta^{18}O$ profile was determined from 760 equally spaced samples along the length of the ice core collected in 1991 and compared to the ice-core chronology (Fig. 3.2). The $\delta^{18}O$ values in ice cores have been used extensively to reconstruct past climate trends. Covey and Haagenson (1984) and Charles et al. (1994) demonstrated that the $\delta^{18}O$ composition of precipitation may depend strongly on storm trajectories and moisture sources; however, most mid- and high-latitude isotopic data sets contain a positive correlation to average air temperature (T_A).

For example, a decreasing trend in the $\delta^{18}O$ values would be indicative of cooler T_A. Lorius et al. (1990) have demonstrated a positive correlation between trends in $\delta^{18}O$ values in ice cores from polar regions to variations in greenhouse gases since the last glacial maximum, 18,000 years before present. Borehole temperature profiles measured in central Greenland indicate that $\delta^{18}O$ values provide good proxy indicators of long-term T_A during the past 500 to 600 years (Cuffey et al. 1994). On the basis of

selected global circulation model (GCM) results, Jouzel et al. (1997) concluded that ice-core isotopic records can be used to interpret local temperature changes in polar regions.

Comparison of the $\delta^{18}O$ values relative to the mean value of the entire ice core (−18.90 permil, n = 760) provides insight into qualitative changes in T_A (Fig. 3.3). The core section from AD 1740 to 1720 (near the bottom of the core) contains a substantial enrichment in $\delta^{18}O$ values (mean = −16.22 permil) relative to the mean $\delta^{18}O$ of the entire core (−18.90 permil). This isotopic enrichment could represent post-depositional melting/freezing processes at the ice-bedrock interface, or more likely a rapid shift to warmer climatic conditions in the mid-1700s. Reconstructed temperature trends from tree-ring information in nonalpine settings from the western United States indicate a warming period from AD 1650 to 1740 (Fritts and Shao 1992).

Ice samples representing the time interval from AD 1740 to 1845 are isotopically depleted relative to the entire core. The mean $\delta^{18}O$ in this section of the core is −19.85 permil, which is 0.95 permil lighter than the mean core value (Fig. 3.3). In Europe, the time period from the mid-1700s to mid-1800s was characterized by cool temperatures (Jones and Bradley 1992) and is generally referred to as the Little Ice Age (LIA) (Thompson 1992). The large oscillations in $\delta^{18}O$ values during the LIA may reflect increased seasonality or better preservation of the annual signal as a result of cooler T_A during the LIA. The isotopically enriched horizons in the interval associated with the LIA probably represent summer snowfall preserved in the ice core as a result of cooler summer temperatures during that period.

In contrast, the isotopically depleted values in ice deposited as snow during the LIA are probably representative of much colder T_A during the winter. The abrupt decrease in the large-amplitude oscillations at AD 1845 (Fig. 3.3) indicates a sudden termination of the LIA at this site. This abrupt decrease in $\delta^{18}O$ oscillations is corroborated by a statistically significant decrease in the variance of the direct current electrical-conductivity measurement log for the same core section (Schuster et al. 2004).

From 1860 to 1950, $\delta^{18}O$ values in ice samples from the core are similar to the mean isotopic value for the entire core (Fig. 3.3). From 1950 to 1991 (top of core) there is an accelerated rate of isotopic enrichment (mean $\delta^{18}O$ = −18.16 permil) relative to the mean values for the entire core (−18.90 permil) and relative to snow deposited during the LIA (−19.85 permil). It is likely that this isotopic enrichment is caused by increasing T_A at this high-altitude site.

Although trends in the $\delta^{18}O$ values in the ice core can provide qualitative information on the direction of change in T_A, additional data are needed to convert these trends into more quantitative changes in T_A. Site-specific transfer functions are used to relate the actual air temperature during snowfall events with the $\delta^{18}O$ of the deposited snow. The following section discusses the development of a transfer function for the Upper Fremont Glacier site that is then used to convert $\delta^{18}O$ values in ice-core samples to representative average air temperatures or T_A.

Development of a Transfer Function

Only limited data link the $\delta^{18}O$ values in ice and snow samples to on-site air temperature at high-altitude, mid-, and low-latitude ice-coring sites (Davis et al. 1995, Yao et al. 1996, Yao et al. 1999, Naftz et al. 2002). It is relatively easy to remotely monitor air temperature during snowfall events; however, the remote setting of the Upper Fremont Glacier makes it impossible to physically collect discrete snow samples from individual storm events for $\delta^{18}O$ analysis. Instead, snowpits were excavated and used to sample the accumulated snowpack, and an ultrasonic depth sensor was used to determine the timing, magnitude, and redistribution of snowfall events on the glacier.

Snow depth was recorded continuously by the elec-

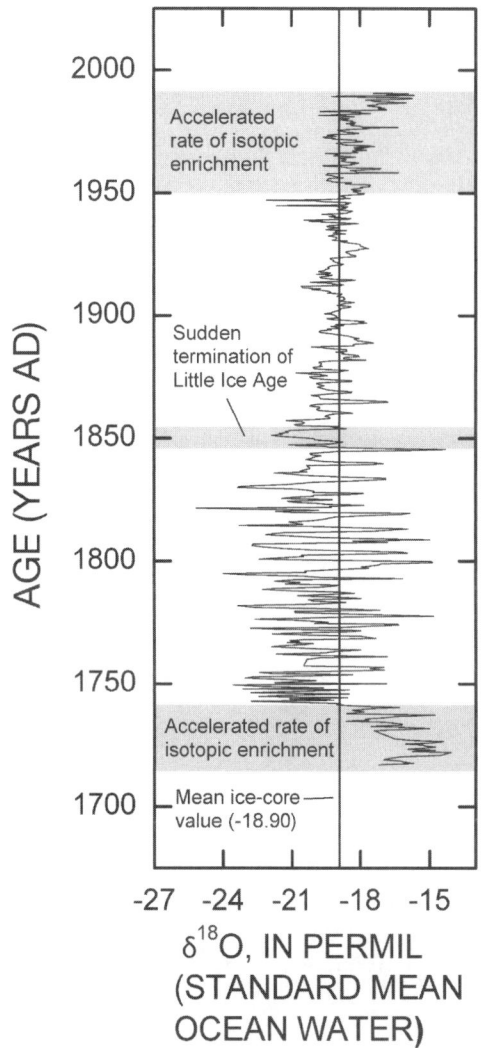

Figure 3.3. Delta oxygen-18 ($\delta^{18}O$) values in ice samples from the Upper Fremont Glacier ice core collected in 1991 relative to the year of deposition.

tronic depth sensor from late September 1999 through early May 2000 (Fig. 3.4). On May 5, 2000, a snowpit adjacent to the depth sensor was excavated and sampled for $\delta^{18}O$ values in 10 cm increments (Fig. 3.4). The measured depth in the snowpit was 137 cm and represented the accumulated snowpack from late September 1999 through early May 2000. The accumulated snow depth measured by the depth sensor was 138 cm (Fig. 3.4), indicating that the period of accumulation was the same as that sampled in the snowpit.

On the basis of continuous on-site snow-depth measurements, seven discrete storm events were preserved in the snow pack (Fig. 3.4). These events were identified as periods with continuous increases in snow depth. Because

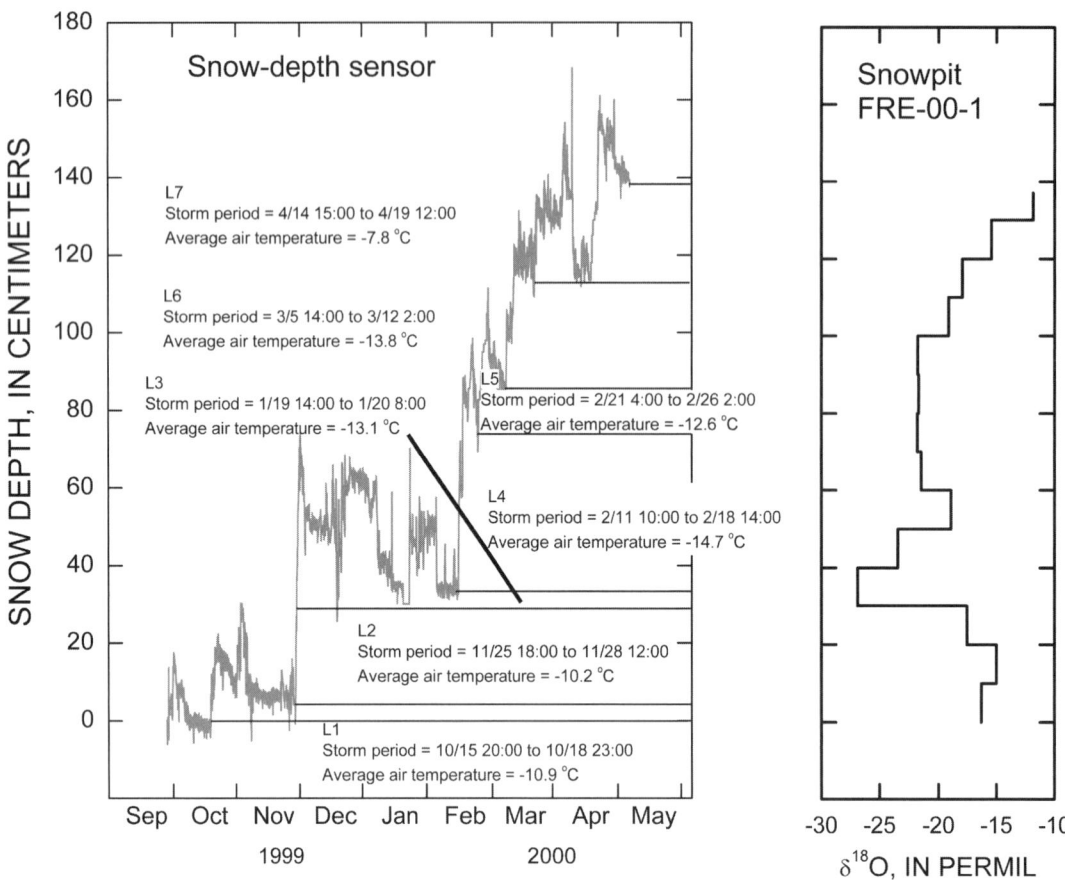

Figure 3.4. Cumulative snow depth and average air temperature during the 1999–2000 accumulation period in relation to changes in isotopic composition in snowpit FRE-00-1, Upper Fremont Glacier, Wyoming. Preserved snow layers designated as L1 through L7.

of wind removal, only the snow from seven accumulation periods remained intact. The mean air temperature during each of the seven accumulation periods was determined by averaging the hourly air temperature recorded at the depth sensor during each accumulation period. Changes in the $\delta^{18}O$ values in the snowpit samples agree with changes in the air temperature during each storm event (Fig. 3.4). For example, the most negative $\delta^{18}O$ value measured in the snowpit (30 to 40 cm depth interval) represented snow deposited during January and February, coinciding with some of the coldest air temperatures during the winter, ranging from −13.1 to −14.7°C. In contrast, the least negative $\delta^{18}O$ value measured in the snowpit (130 to 137 cm depth interval) was for snow deposited during mid-April, coinciding with the warmest air temperature (−7.8°C) recorded during the preserved snow accumulation period.

The $\delta^{18}O$ value of each preserved snow layer was determined by using the detailed depth accumulation data recorded by the on-site snow-depth sensor in combination with the density of each snowpit sample (Fig. 3.4). Each preserved snow layer (L1 to L7) was assigned a mean air temperature on the basis of measured air temperature on the glacier during the accumulation period. Because the snowpit was sampled in 10 cm composites, the $\delta^{18}O$ value for each of the seven snow layers was determined by using density-weighted averages accordingly (Fig. 3.5). After calculating the density-weighted $\delta^{18}O$ value for each storm event, the transfer function relating the $\delta^{18}O$ in snow or ice to air temperature during discrete storm events was developed:

$$\delta^{18}O = 1.350(T_A) - 3.48 \qquad (2)$$

The R^2 value for the transfer function was 0.71 and was statistically significant ($p = 0.0179$).

Reconstructed Air-Temperature Trends

The transfer function developed from on-site temperature and $\delta^{18}O$ values in snowfall was used to reconstruct

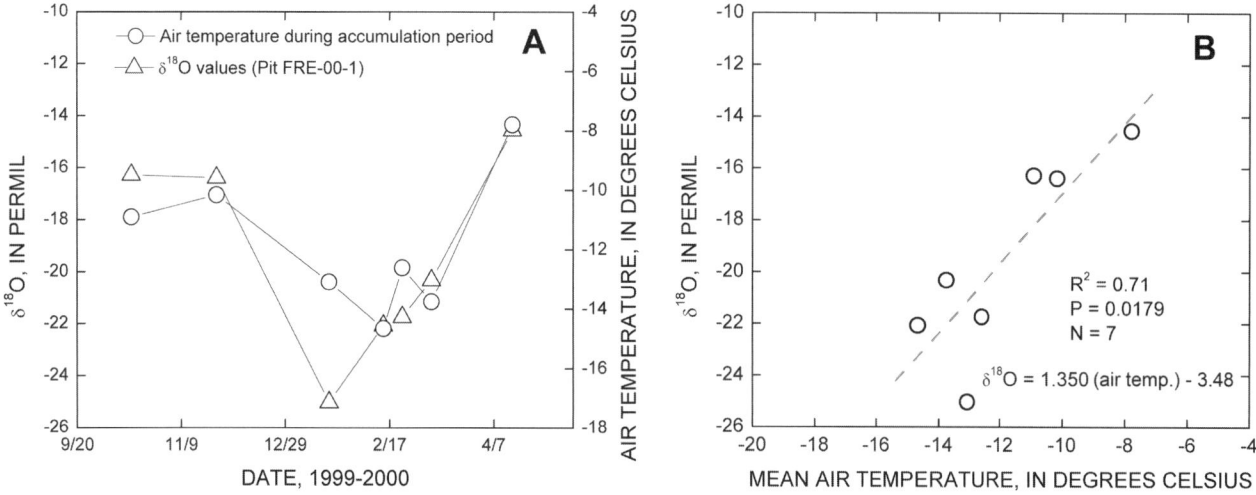

Figure 3.5. (A) Weighted mean $\delta^{18}O$ values from snowpit samples in relation to air temperatures measured during snowfall events by using on-site depth-sensor data, and (B) linear trends between mean air temperature and weighted mean $\delta^{18}O$ values from snowpit samples using on-site depth-sensor data.

Table 3.1. Average change in air temperature for selected time periods in the 1991 ice core relative to the calculated average air temperature during the Little Ice Age

Time period (AD)	Number of ice core samples	Mean delta oxygen-18 ($\Delta^{18}O$) (in permil)	Average air temperature (T_A) (in degrees Celsius)	Change in average air temperature relative to Little Ice Age (in degrees Celsius)
1740 to 1845 (Little Ice Age)	220	-19.85	−12.1	0.0
1860 to 1950	266	-19.00	−11.5	+0.6
1900 to 1991	364	-18.52	−11.1	+0.3
1950 to 1991	204	-18.16	−10.9	+1.2
1985 to 1991	39	-16.99	−10.0	+2.1
1720 to 1991	760	-18.90	−11.4	+0.7

and compare T_A in different sections of the ice core collected in 1991. Changes in T_A relative to the T_A during the Little Ice Age are shown in Table 3.1. The T_A from 1985 to 1991 increased by more than 2°C relative to the calculated T_A during the LIA time period. The T_A increase observed during 1985–1991 is much larger than the 0.6°C increase calculated for the 90-year period immediately after the close of the LIA (Table 3.1).

Additional insight into recent trends in T_A is provided by a second ice core collected in 1998 (DH98-4) about 200 m from the 1991 ice-coring location (Fig. 3.1). As of 2005, only the top 40 m of the 1998 core had been analyzed for $\delta^{18}O$ composition; however, this section of the core is useful for comparison with the 1991 core and for determining recent trends in $\delta^{18}O$ and T_A values. The tritium concentration spike (from above-ground nuclear

testing) in the DH98-4 ice core is about 1.5 m deeper than the spike in the DH91-1 ice core (Naftz et al. 2002). This offset is the result of additional snow deposited on the glacier between the 1991 and 1998 collection periods, and small-scale differences in snow accumulation and retention between the two sites. An ice-core chronology was assigned to the top 40 m of the DH98-4 ice core by using a slight modification of Equation 1.

A statistically significant ($p < 0.0001$) $\delta^{18}O$ enrichment was observed in the top sections of both ice cores (Fig. 3.6). Approximately 35 to 40 years are represented in both cores (the top 6.3 m of the DH 98-4 ice core was not recovered). The transfer function (Equation 2) was used to compare trends in air temperature between both cores (Fig. 3.6). A trend line was constructed with the calculated T_A values in each core for the period of 1960 through

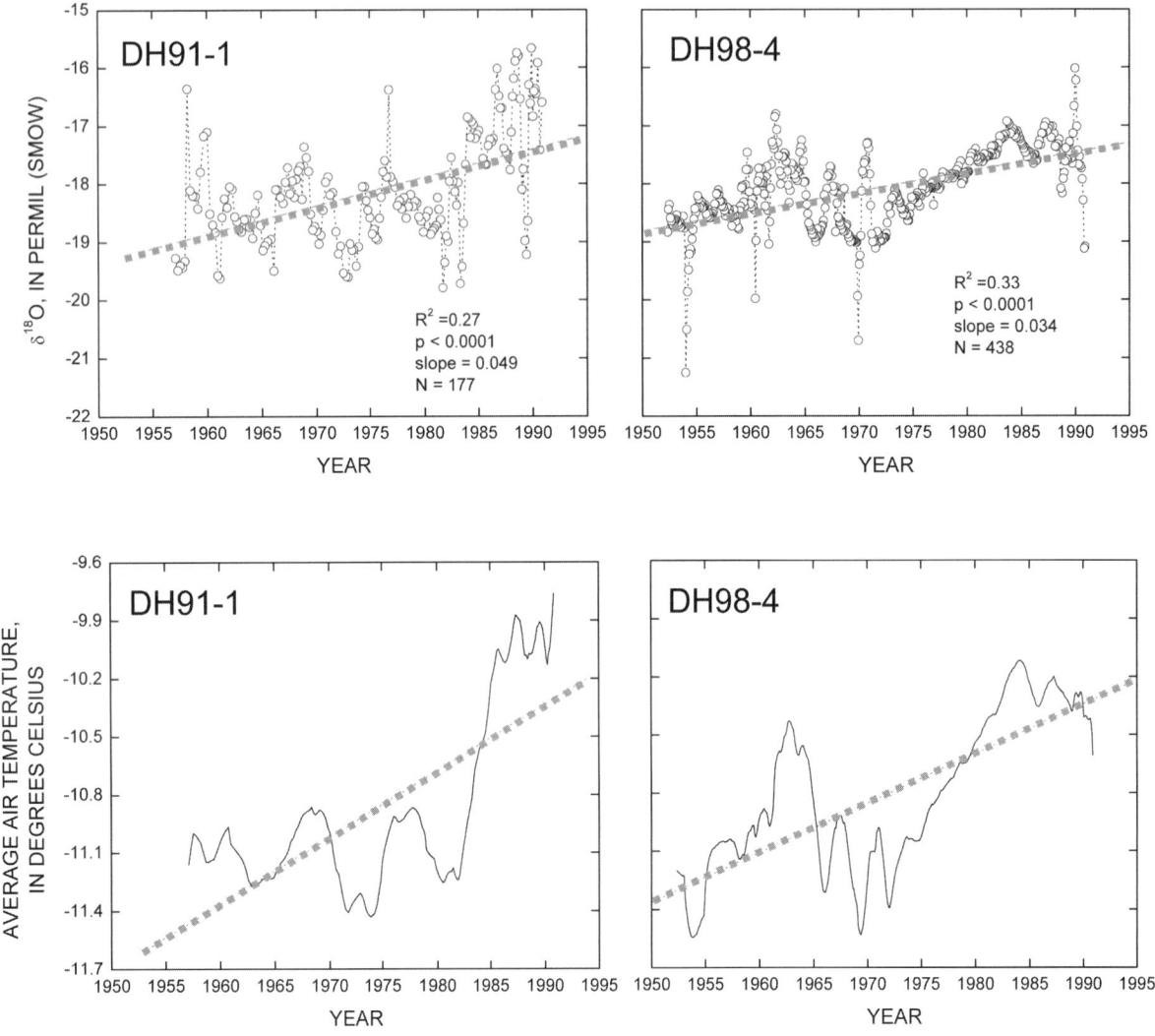

Figure 3.6. The $\delta^{18}O$ values and T_A (20-sample running mean) in the upper 40 m of two ice cores collected from Upper Fremont Glacier in 1991 and 1998. In the upper left graph, SMOW stands for "standard mean ocean water."

1990. On the basis of the trend for the T_A data, there is a +1.1°C increase in the 1991 core and a +0.8°C increase in the 1998 core (Fig. 3.6) during the 30-year time period between 1960 and 1990. This increase is much larger than the observed global mean temperature increase of approximately +0.7°C during the entire twentieth century.

The reconstructed increase in T_A from the Upper Fremont Glacier is comparable to other large temperature increases during the twentieth century, including +6°C for areas of Alaska from 1963 to 1998 (Hileman 1999); +2°C for sites in the European Alps since the early 1980s (Haeberli and Beniston 1998); +2.2°C at a high-altitude site in central Asia from 1962 to 1990 (Mikhalenko 1997); and +4°C in the western Arctic during the twentieth cen-

tury (Sin'kevich 1991). The ice-core data indicate that the Upper Fremont Glacier and adjacent alpine areas in the Wind River Range, Wyoming, may be warming at much faster rates than the global average, especially compared to T_A values present on the glacier during the Little Ice Age (Table 3.1).

In contrast to the reconstructed temperature increase at the UFG site, an instrumental temperature record from a high-altitude (3,048 m) forested site at Niwot Ridge in northern Colorado indicates a statistically significant decrease in mean annual temperature of 1.0°C from 1953 to 1994 (Williams et al. 1996). Knowledge of the long-term climatic characteristics of high-mountain regions are limited by sparse station networks and short records that seldom exceed 100 years (Williams et al. 1996). Local climate in alpine areas is often substantially different from regional and global conditions, and the coarse grid global circulation models cannot simulate climate change in these areas. The mechanism causing the different trends in air temperature observed at the Niwot Ridge and UFG sites is not known.

Glacial Outburst Flood

As a result of increased T_A since the close of the Little Ice Age, alpine glaciers in the Cascade Mountains in central Oregon have shrunk (O'Connor et al. 2001). The glacial melting and retreat have resulted in the exposure of a large volume of unconsolidated and unvegetated sediment and the formation of new lakes. Although many of these newly formed lakes are bound by bedrock rims and are stable, some lakes are dammed by unstable unconsolidated moraines that are susceptible to breaching, especially if the sediments are frozen in the center and then begin to melt (O'Connor et al. 2001).

One result of moraine breaching is the occurrence of a glacial outburst flood (Richardson 1968, Trabant et al. 2003), during which a large amount of water is suddenly released at the head of a steep mountain valley, with results that can be quite destructive. Peak discharges of the largest known historic glacial-lake outburst floods throughout the world have ranged from 10,200 to 112,400 cubic meters per second (m^3/sec) in 1986 at Russell Lake, Alaska (Trabant et al. 2003). The Russell Lake flood was about twice the peak recorded discharge ever measured for the Mississippi River (Trabant et al. 2003).

A possible consequence of the accelerated warming recorded in the ice-core record from UFG was a flood that occurred at Grasshopper Glacier, located about 14 km north of UFG (Fig. 3.7). A *jokulhlaup*, or glacial outburst flood (Richardson 1968), burst from an ice-dammed lake at the head of Grasshopper Glacier in early September 2003. The 12-hectare lake drained an estimated 3.2 million m^3 of water downslope, underneath the glacier, and down tributary valleys into the upper Wind River Valley. The outburst flood was recorded at a USGS streamflow gage (06221400; drainage area 245 km^2) approximately 33 km downstream from the ice-dammed lake.

We hypothesize that accelerated warming in the high-altitude areas of the Wind River Range caused melting and recession of the glaciers, which produced increased meltwater that eventually filled the ice-dammed lake perched on the bedrock floor above Grasshopper Glacier

and caused the glacial outburst flood. Satellite images since 1986 (John Amos, SkyTruth, pers. comm., 2004) also show the dramatic recession of glaciers and the increase of the lake area and volume in the small catchment at the head of the glacier.

It is probable that as the meltwater of the glacier, along with snowmelt and runoff, filled the lake to near-capacity, the hydrostatic pressure of the water volume overcame the ice overburden pressure, resulting in a flow of water underneath the ice into the subglacial conduits of the glacier's drainage system (Nye 1976, Walder and Costa 1996, Cenderelli 2000, Cenderelli and Wohl 2003). The outflow of water resulted in the enlargement of the conduits, the ice dam failure, and the catastrophic release of water from the lake down-valley into the Dinwoody drainage system. Toward the end of the lake's draining, the main conduit was apparently sealed by an ice collapse. An ice-collapse feature runs from the ice dam along the ice surface for about 0.5 km to the north before the glacier plunges down the valley.

Because of the glacier's remote location, no one witnessed the actual dam failure that caused the flood. Hank Williams, with the U.S. Forest Service, photographed the flood's aftermath at the ice-dammed lake in mid-October 2003. He reported an ice wall 18 m high on one shore of the drained lake, and icebergs that had been afloat were stranded on the former lake shore (Fig. 3.8). Williams estimated that 90 percent of the lake drained during the outburst flood.

The *jokulhlaup* peaked at a maximum streamflow of about 35.7 m^3/sec recorded on September 9, 2003, at USGS streamflow-gaging station 06221400 (Dinwoody Creek above lakes near Burris, Wyoming) (Fig. 3.9). This flood resulted in the instantaneous peak streamflow for water year 2003 (October 1, 2002, to September 30, 2003) (Fig. 3.9), and rose from a base flow of 7 m^3/sec on September 6 to its peak on September 9, and back to base flow on September 11. Annual peak streamflows at this stream gage generally occur in mid- to late June as a result of snowmelt runoff, and flows average 26.7 m^3/sec on the basis of 38 years of record for this site (Fig. 3.10).

Another consequence of the climatic warming that triggered the *jokulhlaup* was that sediment erosion and deposition shaped the river and its valley along the flood route. The outburst flood entrained subglacial sediment as it flowed for 2.8 km beneath Grasshopper Glacier. The sediment-laden flood then followed the steep, narrow valley of Grasshopper Creek to the Downs Fork and its confluence with Dinwoody Creek. Here, where the valley

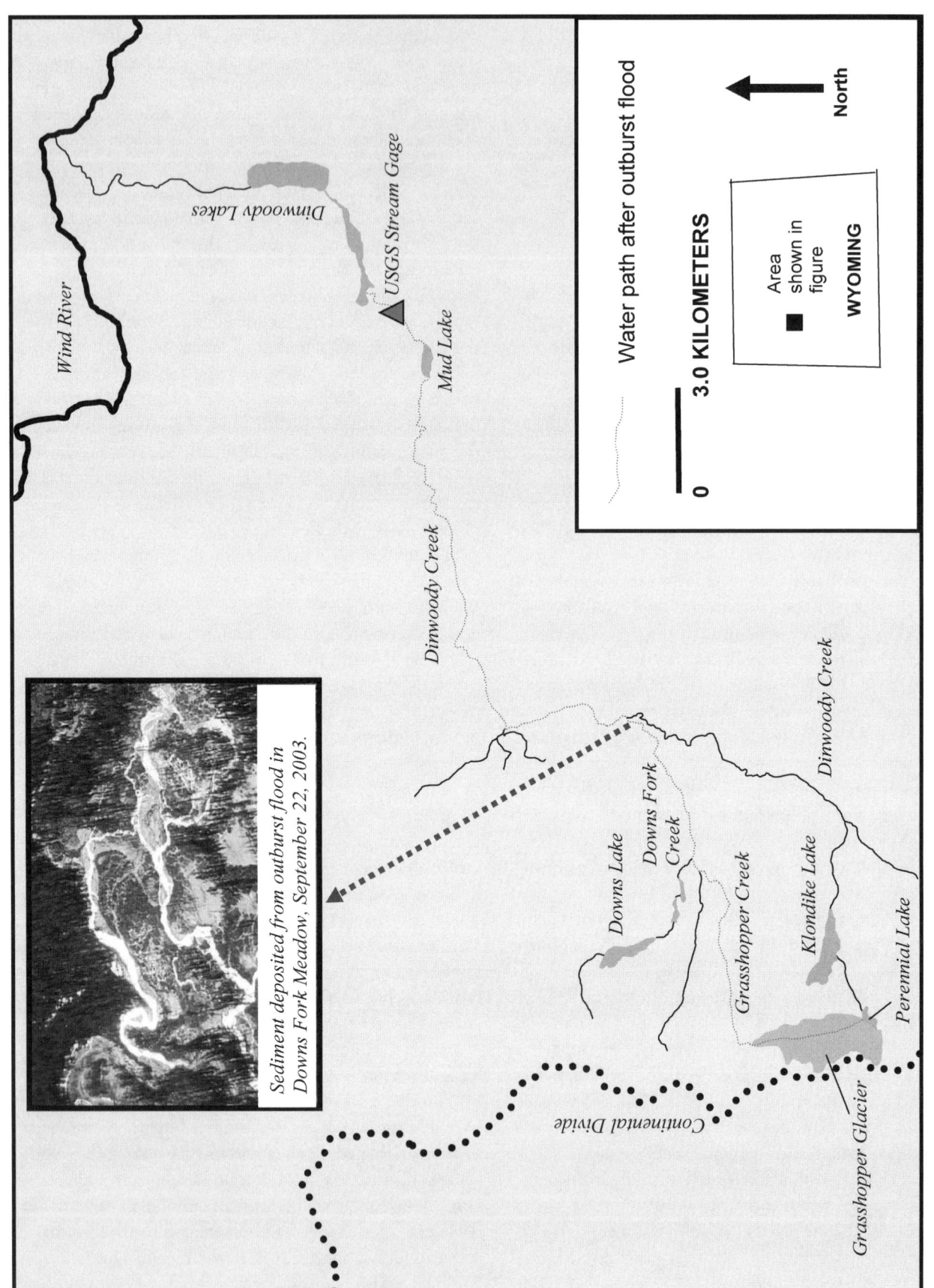

Sediment deposited from outburst flood in
Downs Fork Meadow, September 22, 2003.

Wind River

Dinwoody Lakes

USGS Stream Gage

Mud Lake

Dinwoody Creek

Downs Lake

Downs Fork
Creek

Grasshopper Creek

Dinwoody Creek

Klondike Lake

Perennial Lake

Continental Divide

Grasshopper Glacier

Water path after outburst flood

North

0   3.0 KILOMETERS

Area
shown in
figure

WYOMING

Figure 3.7. Map showing the path of the water after the outburst flood, the location of USGS streamflow-gaging station 06221400 (Dinwoody Creek above lakes near Burris, Wyoming), and photo of sediment deposition along Downs Fork Creek.

widens in a broad U-shaped glacial valley, the flood ponded in an ephemeral lake at the low-gradient Downs Fork Meadows, about 13 km downstream. The flood then continued to step down the valley for another 23 km to Mud Lake, the gage site, the Dinwoody Lakes, and on to the Wind River. Along most of this route, the flood traversed glaciated terrain of granites, migmatites, and gneiss. Paleozoic-age sedimentary rocks crop out near Mud Lake.

Reconnaissance surveys in the summer of 2004 noted five distinct sedimentation patterns in reaches downstream from the glacier. Due to the draining of the ice-dammed lake and subglacial entrainment of sediment, these reaches are characterized by:

- "outwash" deposition of expansion bars, logjams, streambed aggradation, and overbank deposition;
- channel avulsion and aggradation, and ponding in an ephemeral lake;
- in-channel depositional filling of riffles and pools; and
- silt and clay deposition in lakes and irrigation systems.

No known fatalities and only minor structural damage to the Downs Fork bridge and disruption of the Downs Fork outfitter camp occurred in the Fitzpatrick Wilderness as a result of this flood. There is an increased flood risk in the upper part of the flood reach because of aggradation of stream channels. Logjams formed during the flood created substantial geomorphic effects on the river valley. The *jokulhlaup* may be the largest maximum discharge event in this drainage. Irrigators of the upper Wind River Indian Reservation were impacted by the siltation of irrigation ditches and the diminished glacial reservoirs that supplement late-season irrigation supplies. It is likely that continued warming and melting of glaciers in the Wind River Range will result in future, and perhaps more frequent, outburst floods.

Conclusions

A transfer function, relating T_A and $\delta^{18}O$ values in recent precipitation on the Upper Fremont Glacier, was developed by using a snow-depth sensor to record continuous snow depth and a temperature sensor to record continuous air temperature. Snow depth and air temperature were recorded continuously on the glacier from late September 1999 through early May 2000. In early May 2000, a snowpit adjacent to the depth sensor was excavated and sampled for density and $\delta^{18}O$ content. The most negative $\delta^{18}O$ values in those samples corresponded to accumula-

Figure 3.8. Exposed lake bed and large iceberg left after the outburst flood at the head of Grasshopper Glacier. (Photograph taken on October 16, 2003, by Hank Williams, U.S. Forest Service.)

tion periods that occurred during the coldest air temperatures recorded by the on-site weather station. The transfer function, developed with on-site depth-sensor data, was used in combination with $\delta^{18}O$ values from two ice cores collected on UFG to reconstruct trends in T_A during the past 270 years. On the basis of this reconstruction, an increase in T_A of approximately $+2.1°C$ has occurred on UFG from the Little Ice Age to the early 1990s, with approximately half of this temperature increase occurring during 1960 to 1990.

A possible consequence of the accelerated warming recorded in the ice-core record from UFG was a glacial outburst flood (*jokulhlaup*) that occurred at Grasshopper Glacier in early September 2003. A natural ice dam at the head of Grasshopper Glacier failed, resulting in drainage of a 12-hectare lake, which equated to an instantaneous release of approximately 3.2 million m^3 of water. The *jokulhlaup* peaked at a maximum streamflow of about 35.7 m^3/sec, recorded at USGS streamflow-gaging station

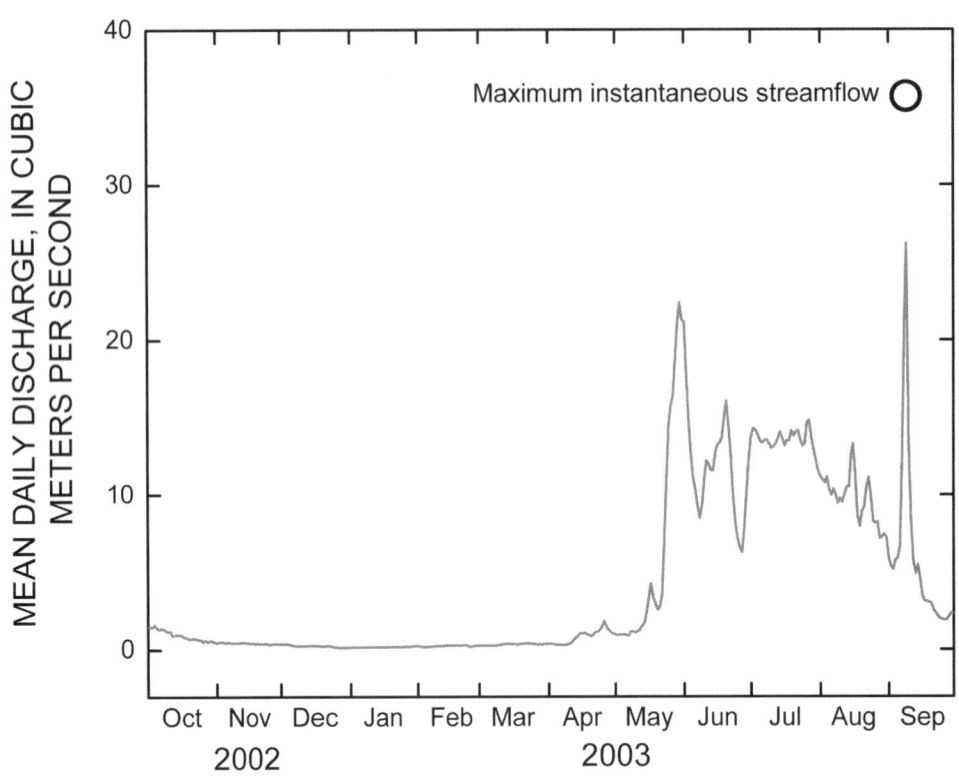

Figure 3.9. Mean daily discharge during water year 2003 at USGS streamflow-gaging station 06221400 (Dinwoody Creek above lakes near Burris, Wyoming) and estimated maximum instantaneous streamflow measured on September 9, 2003.

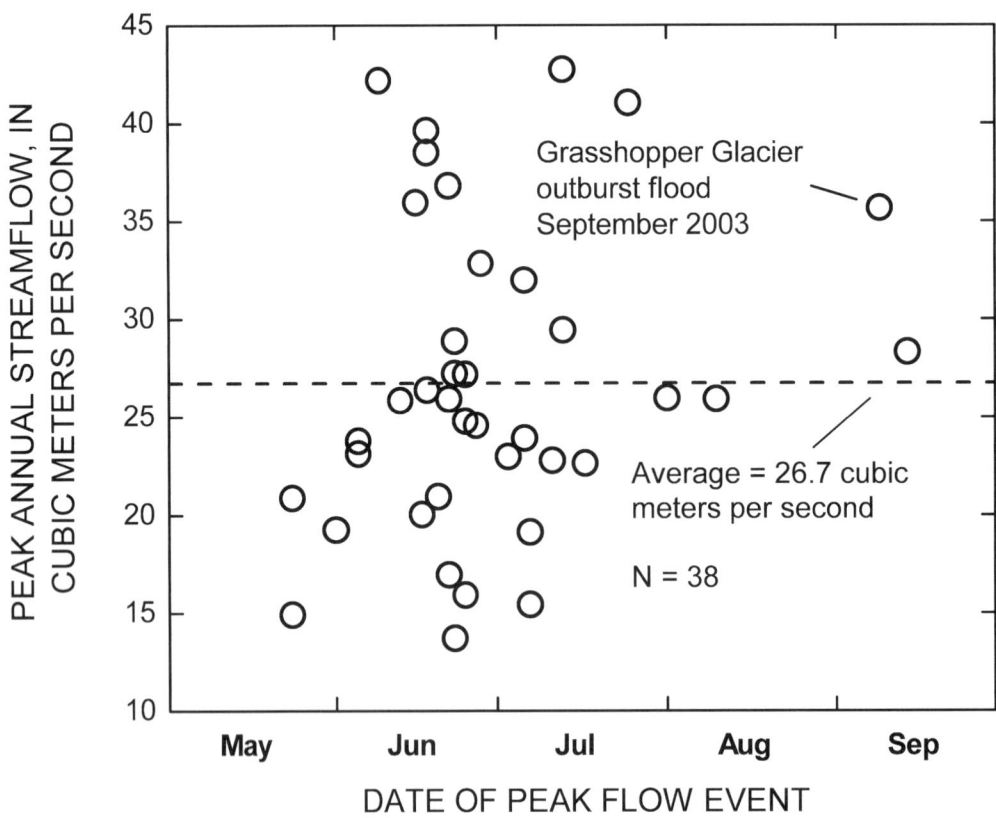

Figure 3.10. Amount and timing of peak annual flow measured at USGS streamflow-gaging station 06221400 (Dinwoody Creek above lakes near Burris, Wyoming) for the period of record 1958 through 2003 (water years).

06221400 (Dinwoody Creek above lakes near Burris, Wyoming), which was the instantaneous peak streamflow for water year 2003 (October 1 through September 30). Annual peak streamflows at this gage site generally occur during mid- to late June as a result of snowmelt runoff and average 26.7 m³/sec on the basis of 38 years of record for

this site. As a result of the *jokulhlaup*, there is an increased flood risk in the upper reaches of the drainage basin as a result of aggradation of the affected stream channels. Continued warming and melting of glaciers in the Wind River Range will likely result in future, and perhaps more frequent, outburst floods.

References

Cecil, L. D., and S. Vogt. 1997. Identification of bomb-produced chlorine-36 in mid-latitude glacial ice of North America. *Nucl. Instrum. Meth. B.* 123:287–289.

Cenderelli, D. A. 2000. Floods from natural and artificial dam failures. Pp. 73–103m in E. E. Wohl (ed.), *Inland Flood Hazards: Human Riparian, and Aquatic Communities.* Cambridge University Press, Cambridge.

Cenderelli, D. A., and E. E. Wohl. 2003. Flow hydraulics and geomorphic effects of glacial-lake outburst floods in the Mount Everest region, Nepal. *Earth. Surf. Processes* 28:385–407.

Charles, C. D., D. Rind, J. Jouzel, R. D. Koster, and R. G. Fairbanks. 1994. Glacial-interglacial changes in moisture sources for Greenland: Influences on the ice core record of climate. *Science* 263: 508–511.

Covey, C., and P. L. Haagenson. 1984. A model of oxygen isotope composition of precipitation; implications for paleoclimate data. *J. Geophys. Res.* 89, D3:4, 647-4,655.

Cuffey, K. M., R. B. Alley, P. M. Grootes, J. M. Bolzan, and S. Anandakrishnan. 1994. Calibration of the δ¹⁸O isotopic paleothermometer for central Greenland, using borehole temperatures. *J. Glaciol.* 40:341–349.

Davis, M. E., L. G. Thompson, E. Mosley-Thompson, P. N. Lin, V. N. Mikhalenko, and J. Dai. 1995. Recent ice-core climate records from the Cordillera Blanca, Peru. *Ann. Glaciol.* 25: 225–230.

Epstein, S., and T. Mayeda. 1953. Variation of the O-18 content of waters from natural sources. *Geochim. Cosmochim. Act.* 4:213–224.

Fritts, H. C., and X. M. Shao. 1992. Mapping climate using tree-rings from western North America. Pp. 269–295 in R. S. Bradley and P. D. Jones (eds.), *Climate Since AD 1500.* Routledge, New York, NY.

Haeberli, W., and M. Beniston. 1998. Climate change and its impacts on glaciers and permafrost in the Alps. *Ambio.* 27:258–265.

Hall, M. H. P., and D. B. Fagre. 2003. Modeled climate-induced glacier change in Glacier National Park, 1850–2100. *Bioscience* 53:131–140.

Hileman, B. 1999. Case grows for climate change. *Chem. Eng. News* 77:16–23.

Jones, P. D., and R. S. Bradley. 1992. Climatic variations over the last 500 years. Pp. 649–665 in R. S. Bradley and P. D. Jones (eds.), *Climate Since AD 1500.* Routledge, New York, NY.

Jouzel, J., R. B. Alley, K. M. Cuffey, W. Dansgaard, P. Grootes, G. Hoffmann, S. J. Johnsen, R. D. Koster, D. Peel, C. A. Shuman, M. Stievenard, M. Stuiver, and J. White. 1997. Validity of the temperature reconstruction from water isotopes in ice cores. *J. Geophys. Res.* 102:26, 471-26, 487.

Lorius, C., J. Jouzel, D. Raynaud, J. Hansen, and H. Treut. 1990. The ice-core record: Climate sensitivity and future greenhouse warming. *Nature* 347:139–145.

Meier, M. 1998. *Eos.* Langbein Lecture. *Trans Amer. Geo. Union* 79:S80.

Mikhalenko, V. N. 1997. Changes in Eurasian glaciation during the past century: Glacier mass balance and ice-core evidence. *Ann. Glaciol.* 24:283–287.

Naftz, D. L. 1993. Ice-core records of the chemical quality of atmospheric deposition and climate from mid-latitude glaciers, Wind River Range, Wyoming. Ph.D. thesis, Colorado School of Mines, Golden.

Naftz, D. L., R. W. Klusman, R. L. Michel, P. F. Schuster, M. M. Reddy, H. E. Taylor, T. M. Yanosky, and E. A. McConnaughey. 1996. Little Ice Age evidence from a south-central North American ice core, U.S.A. *Arctic Alpine Res.* 28:35–41.

Naftz, D. L., P. F. Schuster, and M. M. Reddy. 1994. Assessment of spatial variability of major-ion concentrations and delta oxygen-18 values in surface snow, Upper Fremont Glacier, Wyoming, U.S.A. *Nord. Hydrol.* 25:371–388.

Naftz, D. L., and M. E. Smith. 1993. Ice thickness, ablation, and other glaciological measurements on Upper Fremont Glacier, Wyoming. *Phys. Geogr.* 14:404–414. doi:10.1029/2001JD000621.

Naftz, D. L., D. D. Susong, L. D. Cecil, and P. A. Schuster. 2004. Variations between δ¹⁸O in recently deposited snow and onsite air temperature, Upper Fremont Glacier, Wyoming. Pp. 217–234 in L. D Cecil et al. (eds.), *Earth Paleoenvironments: Records Preserved in Mid- and Low-Latitude Glaciers,* vol. 9. Kluwer Academic Publishers, New York.

Naftz, D. L., D. D. Susong, P. F. Schuster, L. D. Cecil, M. D. Dettinger, R. L. Michel, and C. Kendall. 2002. Ice-core evidence of rapid air temperature increases since 1960 in alpine areas of the Wind River Range, Wyoming, United States. *J. Geophys. Res.* 107, D13.

Nye, J. F. 1976. Water flow in glaciers: Jokulhlaups, tunnels and veins. *J. Glaciol.* 17:181–207.

O'Connor, J. E., J. H. Hardison III, and J. E. Costa. 2001. *Debris Flows from Failures of Neoglacial-age Moraine Dams in the Three Sisters and Mount Jefferson Wilderness Areas, Oregon.* U.S. Geol. Surv. Prof. Paper 1606.

Richardson, D. 1968. *Glacier outburst floods in the Pacific Northwest.* U.S. Geological Survey Professional Paper 600-D.

Schuster, P. F., D. P. Krabbenhoft, D. L. Naftz, L. D. Cecil, M. D. Olson, J. F. Dewild, D. D. Susong, and J. R. Green. 2002. Atmospheric mercury deposition during the last 270 years: A glacial ice core of natural and anthropogenic sources. *Environ. Sci. Technol.* 36:2303–2310.

Schuster, P. F., D. L. Naftz, L. D. Cecil, and J. A. Green. 2004. Evidence of abrupt climate change and development of an historic mercury deposition record using chronological refinement of ice cores at Upper Fremont Glacier. Pp. 181–216 in L. D Cecil et al. (eds.), *Earth Paleoenvironments: Records Preserved in Mid- and Low-Latitude Glaciers,* vol. 9. Kluwer Academic Publishers, New York.

Schuster, P. F., D. E. White, D. L. Naftz, and L. D. Cecil. 2000. Chronological refinement of an ice core record at upper Fremont Glacier in south central North America. *J. Geophys. Res.* 105: 4657–4666.

Sin'kevich, S. A. 1991. Climate warming in the twentieth century as reflected in Svalbard ice cores: Glaciers-Ocean-Atmosphere: Interactions. *Int. Assoc. of Hydrol. Sci. Publication No. 208*:257–267.

Thatcher, L. L., V. J. Janzer, and K. W. Edwards. 1977. Methods for determination of radioactive substances in water and fluvial sediments. *U.S. Geol. Surv. Tech. of Wat.-Res. Invest.,* Book 5, Chap. A5.

Thompson, L. G. 1992. Ice core evidence from Peru and China. Pp. 517–548 in R. S. Bradley and P. D. Jones (eds.), *Climate Since AD 1500.* Routledge, New York.

Trabant, D. C., R. S. March, and D. S. Thomas. 2003. Hubbard Glacier, Alaska: Growing and advancing in spite of climate change and the 1986 and 2003 Russell Lake outburst floods. *U.S. Geol. Surv. Fact Sheet 001-03.* 4 pp.

Wagenbach, D. 1989. Environmental records in alpine glaciers. Pp. 69–83 in H. Oeschger and C. C. Langway, Jr. (eds.), *The Environmental Record in Glaciers and Ice Sheets.* John Wiley, New York.

Walder, J. S., and J. E. Costa. 1996. Outburst floods from glacier-dammed lakes: The effect of mode of lake drainage on flood magnitude. *Earth Surf. Proc. Land.* 21:701–723.

Williams, M. W., M. Losleben, N. Caine, and D. Greenland. 1996. Changes in climate and hydrochemical responses in a high-elevation catchment in the Rocky Mountains, USA. *Limnol. Oceanogr.* 41:939–946.

Yao, T., V. Masson, J. Jouzel, M. Stievenard, S. Weizhen, and J. Keqin. 1999. Relationships between $\delta^{18}O$ in precipitation and surface air temperature in the Urumqi River Basin, east Tianshan Mountains, China. *Geophys. Res. Lett.* 26:3473–3476.

Yao, T., L. G. Thompson, E. Mosley-Thompson, Y. Zhihong, Z. Xingping, and L. Ping-Nan. 1996. Climatological significance of the $\delta^{18}O$ in north Tibetan ice cores. *J. Geophys. Res.* 101:29, 531-29, 537.

PART II

Environmental Effects:
Twentieth-Century Observations and
Twenty-First Century Projections

4

Variability and Trends in Mountain Snowpacks in Western North America

Philip W. Mote

Abstract

Measurements of spring snowpack are examined here for climate-driven fluctuations and trends at various points in the snow season during the period 1940–2007. Much of the mountain West has experienced declines in spring snowpack, but declines in winter have been smaller and less widespread. The choice of starting year (between 1940 and 1970) makes little difference in the aggregate results. The largest decreases have occurred where winter temperatures are mild, especially in the Cascade Mountains and northern California. In most mountain ranges, relative declines grow from minimal at ridgetop to substantial at snowline. These results emphasize that the West's snow resources are already declining as Earth's climate warms.

Introduction

Temperatures have increased in western North America during the twentieth century (e.g., Trenberth et al. 2007), especially during the snow accumulation season (Fig. 4.1), and there is ample evidence that this widespread warming has produced changes in streamflow (Cayan et al. 2001, Regonda et al. 2005, Stewart et al. 2005), as is expected in a warming climate (Hamlet and Lettenmaier 1999). The warming of the West can now be attributed to human activity (Stott 2003).

Mote et al. (2005) analyzed snow-course data and output of a state-of-the-art hydrology model for the western United States over the period of record 1950–1997 for April 1 and showed substantial declines in snowpack at roughly 75 percent of locations. Relative losses depended on elevation in a manner consistent with warming-driven trends, and statistical regression on climate data also suggested an important role of temperature both in year-to-year fluctuations and in longer-term

trends at most locations. This chapter updates the observational section of Mote et al. 2005 through spring 2007, shows trends for four other dates in the snow season, and considers the effects of starting year in reporting period-of-record trends.

Data

Monitoring of snow conditions, largely for water supply forecasting, has been conducted since the early part of the twentieth century at "snow courses," which are locations that are carefully surveyed and marked at five to ten spots over a distance of tens to hundreds of meters. When new snow courses were established, surveyors had several goals in siting them, including minimizing the effects of blowing snow by selecting natural clearings on relatively level terrain, generally in bowls or benches. The snow surveyor, using a tool that resembles a slotted pipe, takes a core of snow, weighs it, and measures snow water equivalent (SWE) as well as snow depth. A few remote locations are monitored using aerial markers, for which snow depth is measured from aerial photos, and SWE is estimated using a typical density. Automated snow telemetry (SNOTEL) sites began to be used in the 1980s and in some cases replaced snow courses. SNOTEL sites measure the weight of overlying snow on a large, flat, fluid-filled "pillow."

Snow-course data through 2006 were obtained from the Natural Resources Conservation Service (NRCS) Water and Climate Center (www.wcc.nrcs.usda.gov/snow/snowhist.html) for most of the United States, from the Ministry of Environment for British Columbia (http://www.env.gov.bc.ca/rfc/archive/historic.html), and from the state Department of Water Resources through 2007 for California (cdec.water.ca.gov). I include only those snow courses for which data began by 1960, a total of 1,132

Trends in Nov-Mar temperature, 1950 to 2000

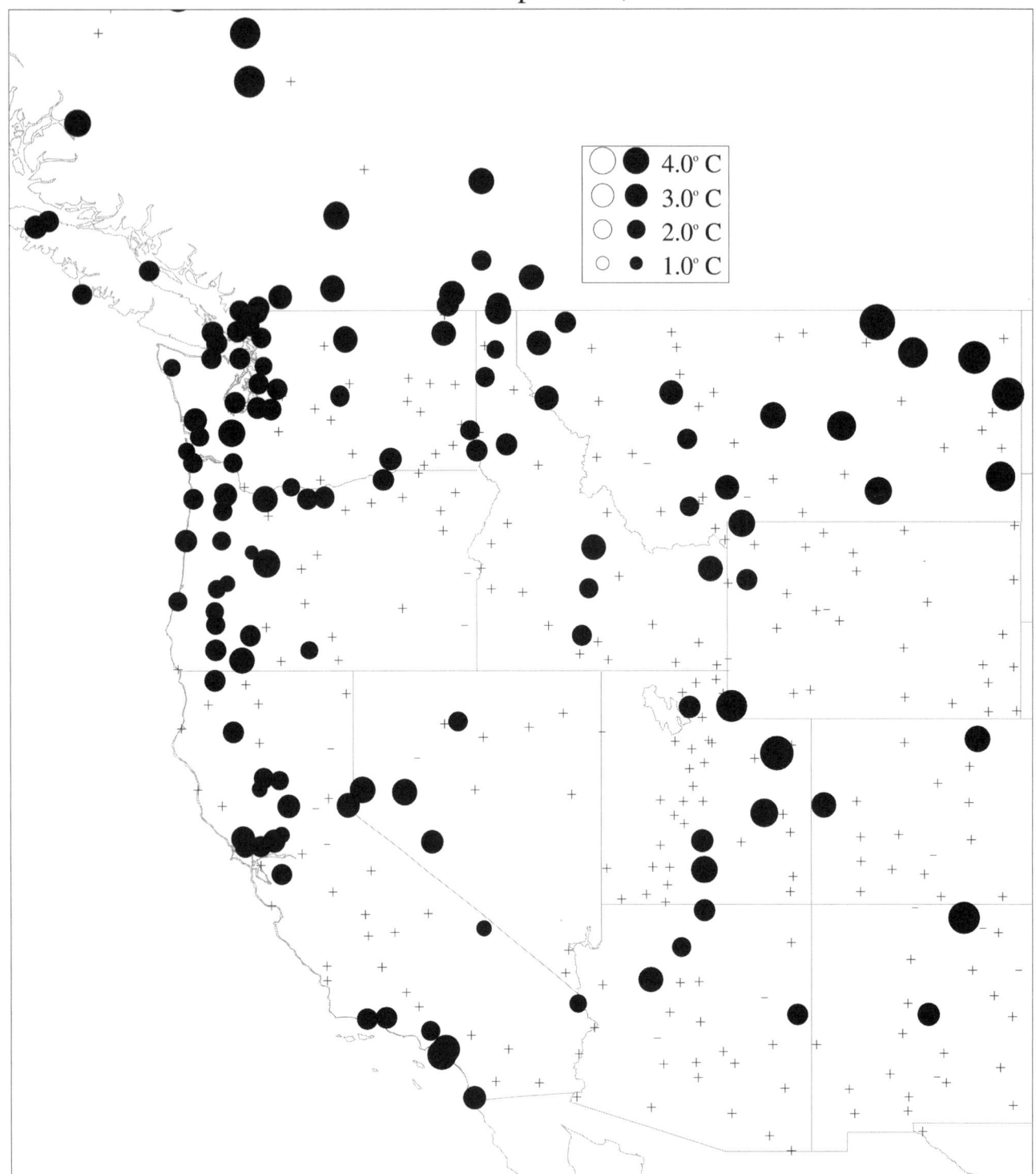

Figure 4.1. Linear trends in November–March mean temperature, 1950–2000. Statistically significant trends are shown by circles, and the size is indicated by the legend. Other trends are shown by plus or minus signs.

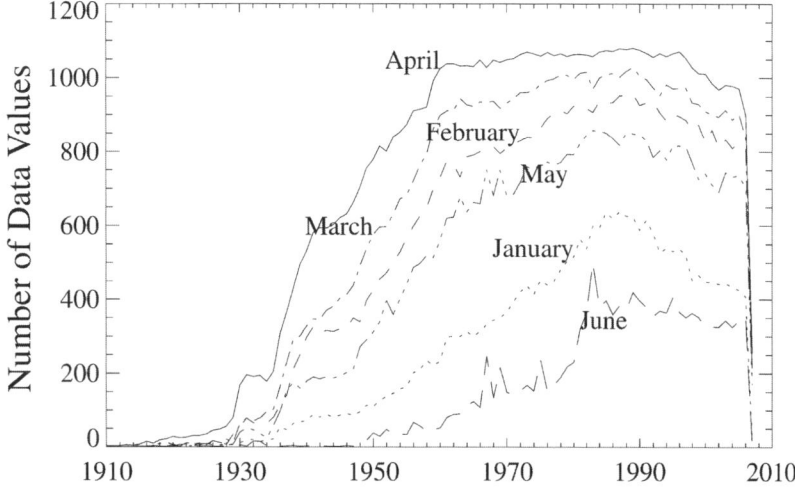

Figure 4.2. Number of data values per year for the first of each month indicated. Only snow courses with data collected on or before April 1, 1960, are included. Of the three data sources, only California DWR data included values for 2007 as of July 6, 2007.

sites. April 1 was the most common date of measurement, especially before 1950 (Fig. 4.2), because it is roughly the point at which many western mountain ranges reach their peak snow accumulation. March 1 was next most common sampling date, followed by February 1 and May 1.

Measurements are typically taken several days before the date indicated, so for example the "March 1" measurement could actually have been taken on February 25. A number of sites, especially in Arizona and New Mexico, also have measurements available mid-month (not shown in Fig. 4.2). The number of operational snow courses rose rapidly after the early 1930s. June measurements were scarcely made before 1950 and have a sampling bias, as we will see below. Of the available snow courses, those that have fewer than 75 percent of observations in the time span in question are not used, leaving a total of 926 snow courses for the 1956–2006 period of record.

The distribution of the snow courses is shown in Figure 4.3. The densest measurements are in California, with 267 long-term snow courses. Altitude of measurements roughly increases from north to south and west to east, with the exception of the southern Sierra Nevada, (indicated by + symbols), which includes some of the highest snow courses in the West.

At several locations a SNOTEL site has replaced or augmented a snow course with sufficient overlap that statistical estimation of the one value is possible in the absence of the other; in these cases NRCS provides two complete time series, with the estimated values flagged.

Both series are candidates for inclusion because the estimates of long-term trends may be slightly different. Figure 4.4 shows an example, Moore's Creek Summit in Idaho, objectively chosen as the location with the largest number of observations in the database owing to a long-term program of semimonthly measurements. The top panel (Fig. 4.4a) shows April 1 measurements directly from the snow course, the values reported for the SNOTEL as "estimated," and, after the installation of the SNOTEL in 1982, actually observed. NRCS estimates "missing" SNOTEL values (generally before the SNOTEL was installed) or "missing" snow course values using a linear regression between snow course and SNOTEL values during periods of overlap. Since in this case the snow course is complete to the present, there are no estimated snow course values.

The two time series have rather different linear trends owing primarily to the prevalence of underestimated high values in the early years. Figure 4.4b shows the scatterplot and linear fit of all coincident SNOTEL and snow-course observations; the correlation is 0.978, and the equation for the least-squares linear fit is SNOTEL $= -3.1 + 1.08 *$ snowcourse. The mean seasonal cycle is shown in Figure 4.4c; on average, the snowpack builds steadily from mid-October until mid-April and then melts rapidly. The end of the snow season is not well measured historically; the inset table in Figure 4.4c shows the number of measurements in each curve. For the first-of-month measurements, there is little difference between the snow-course and SNOTEL mean values (dashed and dotted curves),

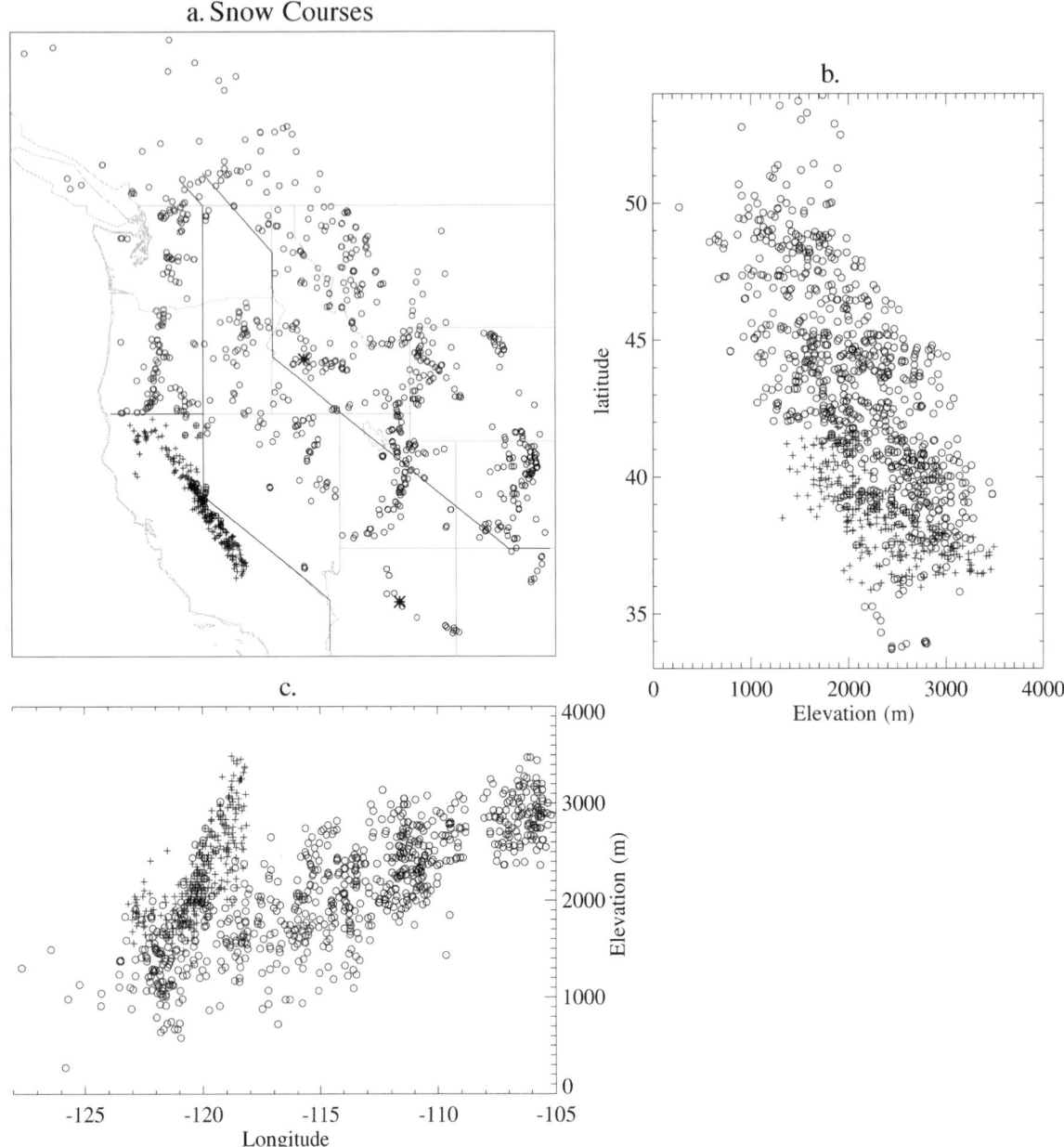

a. Snow Courses

b.

c.

Figure 4.3. Locations of snow courses with records dating to at least 1960: (a) mapped with lines dividing the West into four regions for analysis shown in Figure 4.9 and asterisks showing the locations of the snow courses featured in Figure 4.4; (b) latitude-altitude (latitudes are evenly spaced, unlike those in the Mercator projection of 4.3a); and (c) longitude-altitude. Plus symbols denote California snow courses.

though both tend to be lower for the 1982–2006 means than for the 1960–2006 mean seasonal cycle. This site always has snow until at least May 15. Toward the end of the season, sampling of snow courses can be sporadic: only three visits were made to the Moore's Creek Summit snow course on June 15 during 1982–2006, with only one of those occurring when no snow remained. In contrast, 68 percent of SNOTEL measurements indicate no snow

on June 15. The mid-month snow-course visits were rare at this site after the SNOTEL was installed.

For comparison, Figure 4.4c includes the most data-rich snow course in the Southwest, Mormon Mountain, Arizona (the dashed curve). In the average seasonal cycle, peak SWE occurs March 1 or 15 and is only about 20 percent of the peak SWE at Moore's Creek Summit. Even in the snowiest years, no snow is left on May 15.

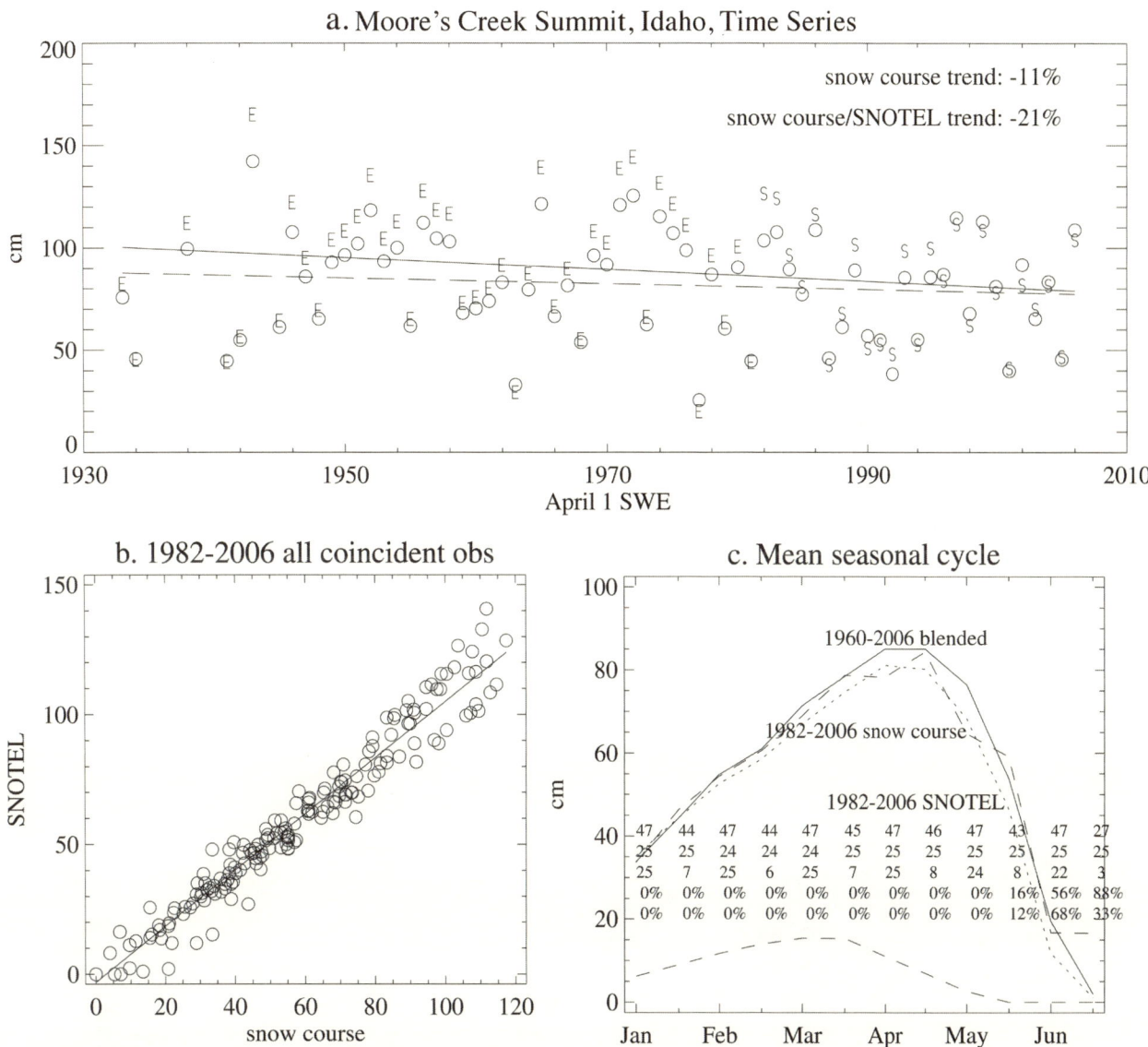

Figure 4.4. Data for Moore's Creek Summit, Idaho (see Fig. 4.3), elevation 1,860 m: (a) time series of April 1 SWE are shown for snow course observations (o), SNOTEL direct measurements (S), and SNOTEL estimated values (E). (b) scatterplot of all data (*n* = 183) for which both SNOTEL and snow course measurements were taken. (c) mean seasonal cycle at Moore's Creek Summit for 1960–2006 with blended snow course–SNOTEL data, and for snow course (dashed) and SNOTEL (dotted) data during the period of overlap. The table shows, for each observation date, the number of observations in each curve and (fourth and fifth rows) the percentage of observations of zero SWE for SNOTEL and snow course. The dashed curve below the table shows the 1960–2006 blended mean SWE for Mormon Mountain, Arizona (elev. 2,287 m; see Fig 4.3).

Trend Analysis

For each of the 926 snow courses with at least 75 percent of observations between 1956 and 2006, and at least one value before 1957 and after 2002, I calculated linear fits, expressed as percent change relative to the value of the line in 1956 (Fig. 4.5). Negative trends prevail at roughly 75 percent of snow courses. The largest relative losses (many in excess of 50 percent, some in excess of 75 per-

cent) occurred in western Washington, western Oregon, and northern California. The fraction of negative trends is almost the same as for the results of Mote et al. (2005) for 1950–1997 using both observations and simulations with a hydrologic model. Increases in SWE, some in excess of 30 percent, occurred in the southern Sierra Nevada of California, in New Mexico, and in some other locations in the Southwest.

Relative Trend in April 1 SWE, 1956-2006

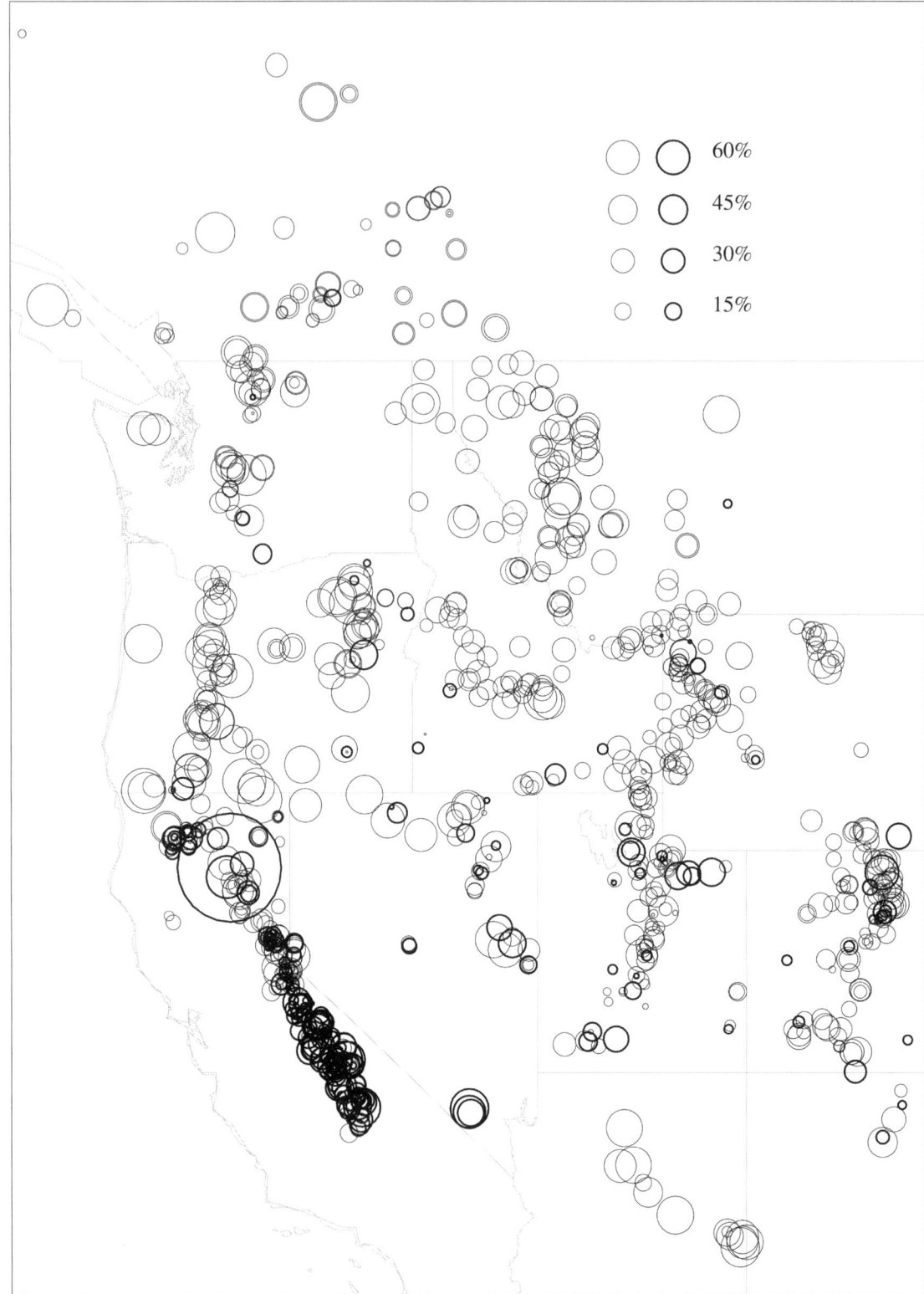

Figure 4.5. Map of relative trends in April 1 SWE over the 1956–2006 period. Gray circles indicate negative trends, and black circles positive trends. Of the 924 trends shown, 76 percent are negative.

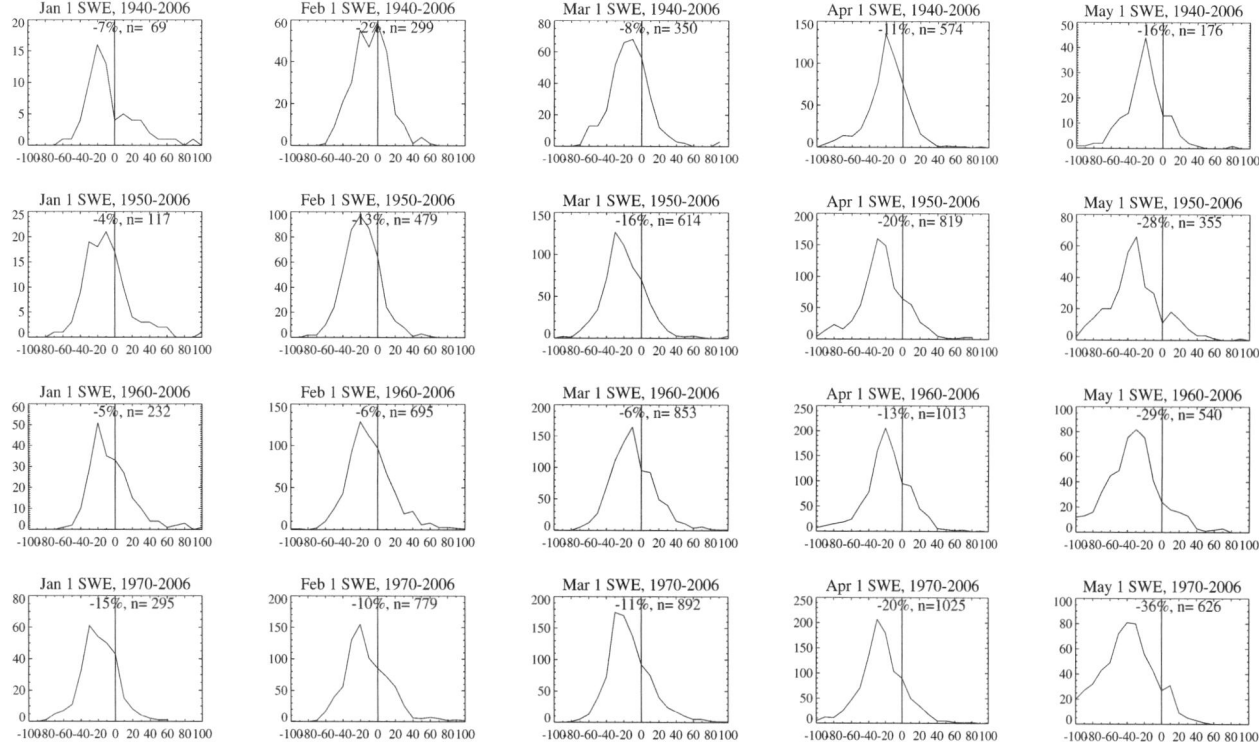

Figure 4.6. Histograms of relative trends for each month (columns) and for starting years by decade from 1940 to 1970, along with median value and number of snow courses shown.

These trends depend on both time of year and starting point, and Figure 4.6 shows histograms of the relative trends for all sufficiently complete snow courses for January 1 through May 1 and for starting points by ten-year intervals from 1940 to 1970. Each panel also shows the median trend, which is small for January 1 and February 1 but grows to a large maximum in May. The starting point makes a smaller difference, roughly a factor of two, in the median trend.

One of the most noteworthy features of the trends is their dependence on elevation (Fig. 4.7). Relative losses are largest at lower elevations, and trends are generally smaller at higher elevation. One can also bin the linear trends by mean midwinter temperature, where temperature is estimated for each site using algorithms related to the terrain (Hamlet and Lettenmaier 2005), for somewhat clearer results (Mote et al. 2005).

A strong dependence of relative trends on mean temperature has also been found in snowfall at weather stations in the western United States (Knowles et al. 2006), snow-cover duration in Switzerland (Scherrer et al. 2004), and mean March snow depth in Japan (Brown and Mote 2009). Absolute trends in SWE also depend on elevation

but in a more complicated way, being small at low elevation (owing to low mean SWE), most negative at middle elevation, and small again at high elevation (owing to small changes stemming from negligible dependence on temperature).

The generally smaller wintertime (Jan. 1 and Feb. 1) trends resemble the results for Northern Hemisphere snow-cover extent (Brown 2000, Lemke et al. 2007) and for mean snow depth in January and February at weather stations in Japan. In the case of snow cover extent, the smaller trends in winter than in spring can be understood by considering the energy balance and mass balance. During spring, lower solar zenith angle ensures stronger solar radiation, enhancing the snow-albedo feedback. In the case of western mountains, melt events during the accumulation season play a role in determining April 1 SWE at some sites (Mote et al. 2005), but the energy for winter melt events is primarily supplied by sensible heat flux, especially in the perennially cloudy Northwest. Enhanced spring losses of snowpack are related more to the earlier (by roughly ten days) arrival of the 0°C temperature at a given elevation owing to spring warming. These observations are consistent with the streamflow study of Stewart

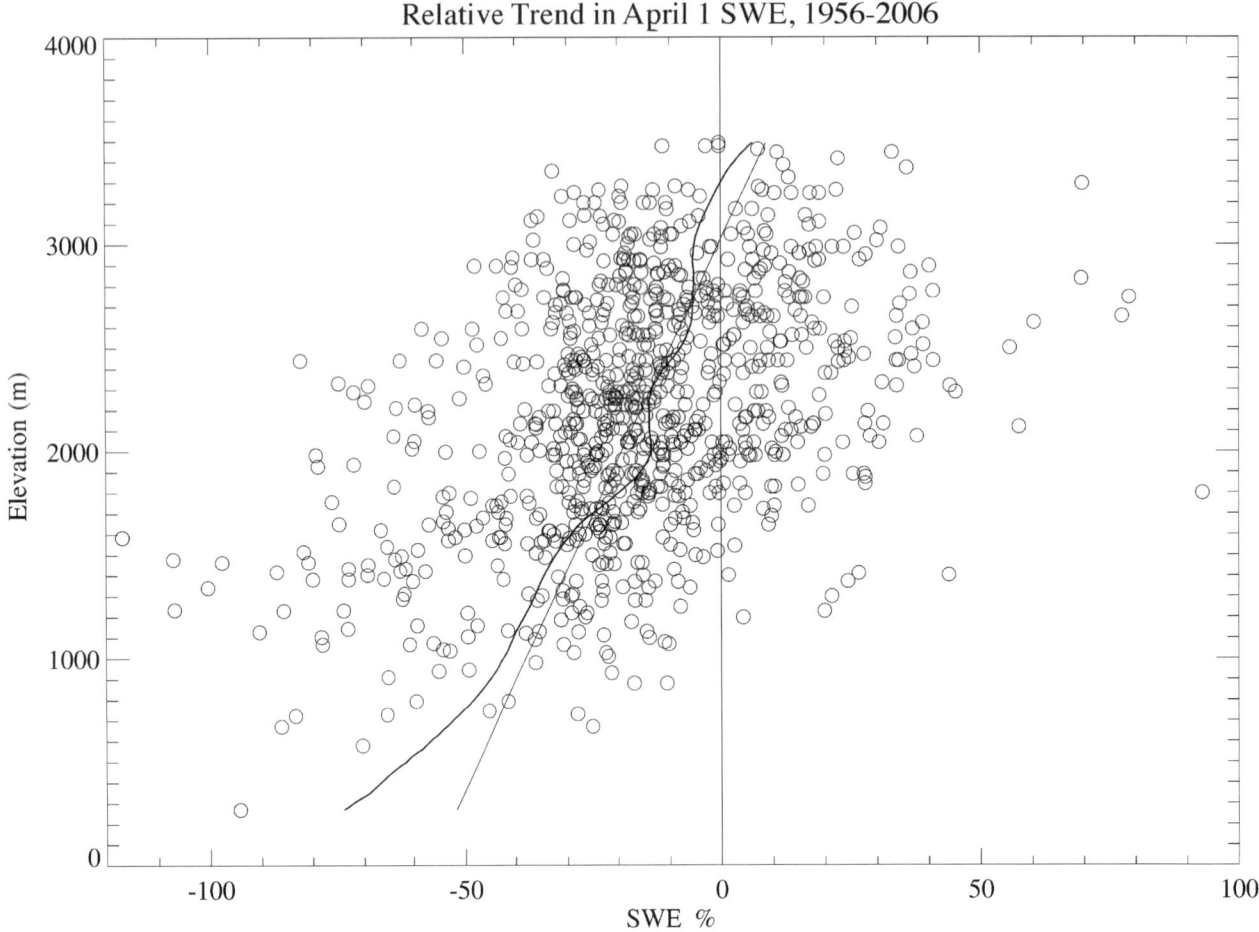

Figure 4.7. Relative trends from Fig. 4.5, plotted against elevation, along with linear and loess fits.

et al. (2005), who noted shifts in timing of peak spring snowmelt of 0–20 days in most of the West.

One way to illustrate the changing timing of spring with these monthly data is to plot the ratio of April 1 to March 1 SWE. I discarded a handful of instances for which the March 1 SWE was less than 5 cm, rendering the April/March ratio quite large. Fig. 4.8 shows the results: in a ratio remarkably similar to the fraction of negative trends in April 1 SWE, at 74 percent of the 748 locations the April/March ratio decreased. Some of the areas with positive trends in April/March ratio are the same places where April 1 SWE has increased, notably the Sierra Nevada, but many of the snow courses in the Sierra that had positive trends in Apr 1 SWE had negative trends in April/March SWE. Most of the easternmost snow courses along the front range of the Rockies, from Colorado to British Columbia, saw increases in the ratio, perhaps owing to a change in springtime precipitation in the Great Plains.

The largest decreases, 50 to 80 percent, occurred in a swath from Oregon through Nevada into Arizona.

Regional Time Series

For each of the regions outlined in Figure 4.3, I constructed a time series of regionally averaged SWE by determining when the region was adequately sampled and then combining normalized time series in each region.

As noted by Mote et al. (2007) for the Washington Cascades, early snow courses tended to have a higher mean elevation than the full set of snow courses that were active by 1950. Snow courses established after 1950 tended not to change the mean elevation. The starting year for each region is chosen as the earlier of (a) when the mean elevation reaches roughly its final value or (b) when half the snow courses are active. In most regions these starting years are only a few years apart and lead to a starting point

Trend in ratio of Apr 1/Mar 1 SWE, 1956-2006

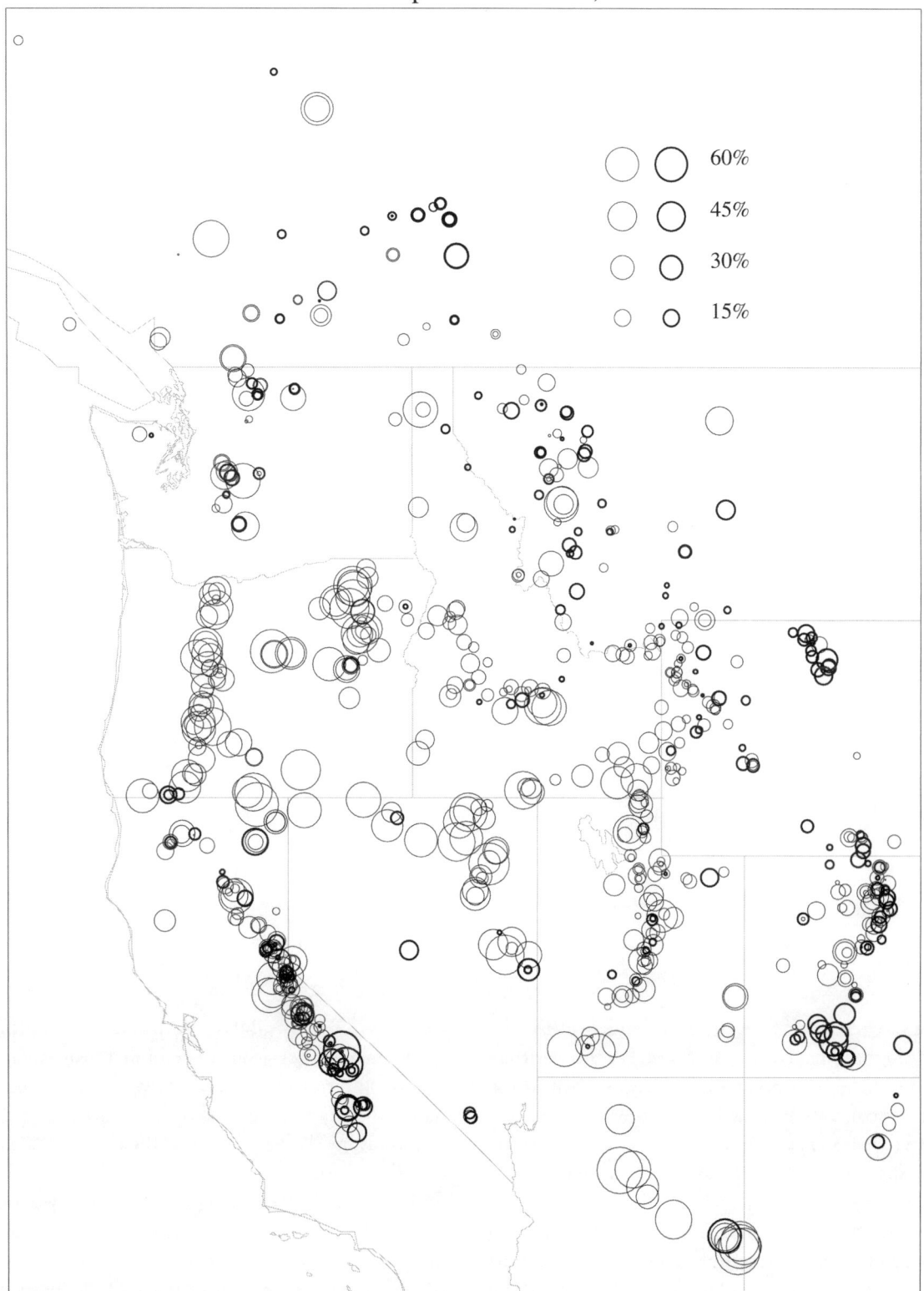

Figure 4.8. Map of trends in the ratio in April 1 to March 1 SWE, 1956–2006. Values of April 1 SWE greater than five times the March 1 value were omitted from the calculation.

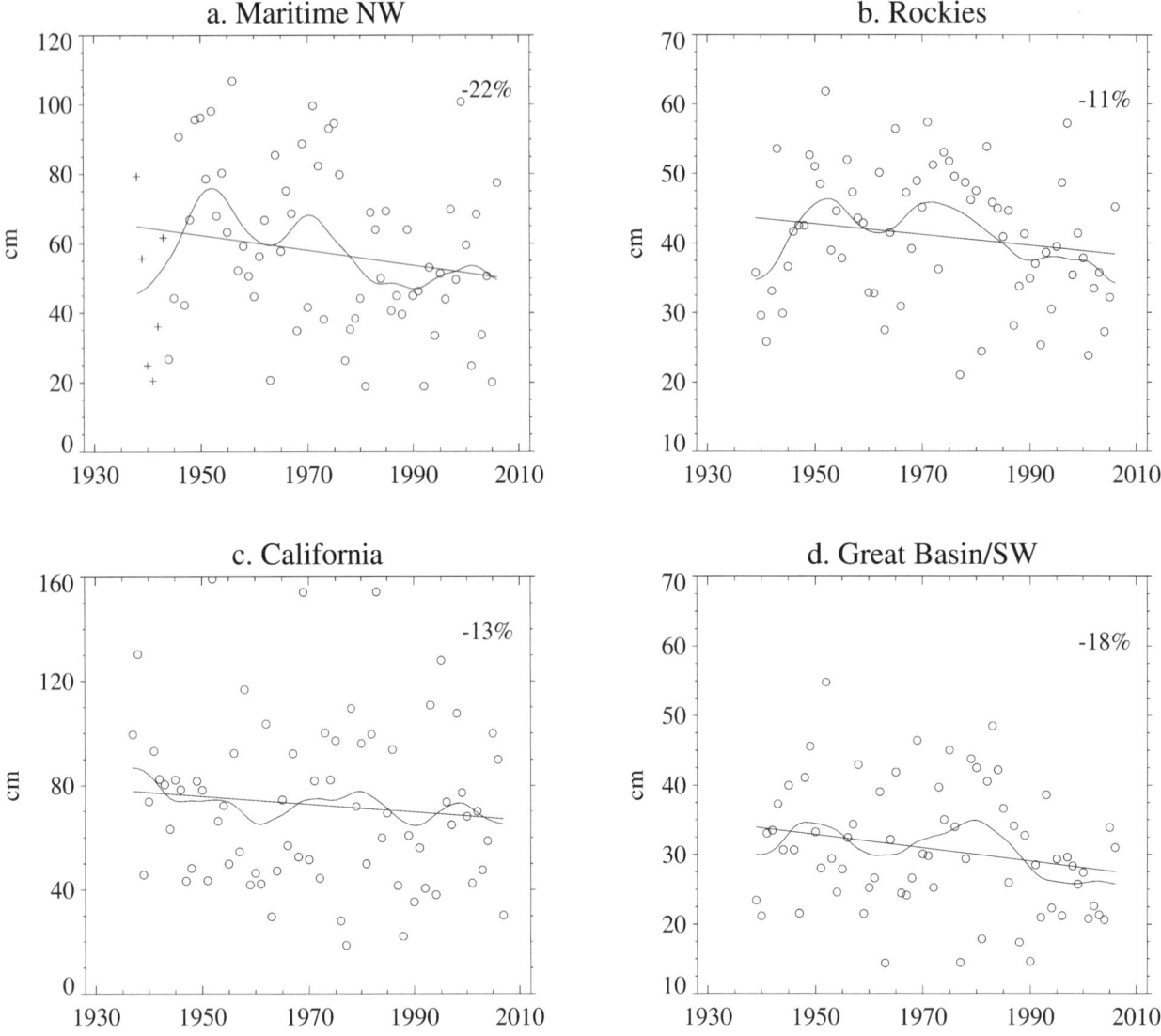

Figure 4.9. Time series of regionally averaged SWE for the four regions delineated in Figure 4.3a.

in the late 1930s: 1937 for California, and 1939 for the Rockies and the Great Basin/Southwest. For the maritime Northwest, the starting point is 1945; in order to make the time-series analysis comparable, additional values were estimated for 1938–1944 from the snow courses available for 1938–2006 (only about half as many), indicated in the figure by plus symbols.

The next criterion for inclusion in the regional averages was, again, 75 percent completeness, of which at least one value occurred at each end of the period in question. Finally, once a consistent set of snow courses was established, snow-course data were aggregated by first converting each reasonably complete time series of SWE to a time

series of z-scores by subtracting the mean and normalizing by the standard deviation, then for each year averaging the z-scores and converting back to SWE using the mean and standard deviation averaged over all the time series in the region (as in Clark et al. 2001). The time series (Fig. 4.9) were then smoothed.

In the maritime Northwest, three snowy periods are evident, in the early 1950s, early 1970s, and late 1990s, superposed on an overall decline. In fact, as Mote et al. (2007) showed for the Washington Cascades using hydrologic modeling, it is likely that the ending point of this smoothed time series is lower even than the drought of record in the late 1930s–early 1940s, despite considerably higher precipitation in recent years. For the maritime Northwest, if the objectively chosen starting point of 1945 is used, or even 1940, the decline is statistically significant ($p < 0.05$, two-sided t-test).

In the other regions the smoothed time series do not suggest such large declines, although in each case the most recent period is as low as the late 1930s–early 1940s. For none of these regions is the decline statistically significant, but in each region the declines since about 1938 have occurred despite increases in precipitation (Hamlet et al. 2007).

Discussion

The trends and variability in SWE (Figs. 4.5–4.8) can be seen approximately as the result of competition between fairly monotonic warming-driven declines at all but the highest altitudes, and more precipitation-driven increases and decreases (Mote 2006). In the Southwest and in spots elsewhere in the West, high precipitation in recent decades has increased snowfall, resulting in higher SWE despite higher temperatures. In the Cascades, however, very large declines resulted from a double blow of decreases in precipitation and large increases in temperature in a region where snow courses have high temperature sensitivity (Fig. 4.5).

These declines are consistent with advances in the timing of spring snowmelt (Cayan et al. 2001, Stewart et al. 2005) and with projections of future declines in snowpack related to global climate change (e.g., Hamlet and Lettenmaier 1999).

At other locations and at most places in winter, the median trends are closer to zero. During winter, melt events are rarer, and SWE is largely determined by total accumulation, which is less sensitive to winter temperature at most snow courses. In spring, as the competition between melt (which is sensitive to temperature) and accumulation (which becomes temperature-sensitive as the snow season wanes) intensifies, warming of the magnitude shown in Figure 4.1 can shift the balance and substantially reduce SWE, especially at low-elevation sites. The dependence of trends on season and elevation is thus consistent with the observed warming, which in turn is attributable on the scale of the West to rising greenhouse gases (Stott 2003).

Acknowledgments

I wish to thank the snow survey offices and hundreds of dedicated snow surveyors with the Natural Resources Conservation Service, the California Department of Water Resources, and the British Columbia Ministry of Environment for providing the snow-course data. This research was funded in part by the Joint Institute for the Study of the Atmosphere and Ocean under NOAA Cooperative Agreement No. NA17RJ1232, Contribution #1094.

References

Brown, R. D. 2000. Northern hemisphere snow cover variability and change, 1915–97. *J. Clim.* 13:2339–2355.

Brown, R. D., and P. W. Mote. 2009. The response of Northern Hemisphere snow cover to a changing climate. *J. Clim.*, doi: 10.1175/2008JCLI2665.1.

Cayan, D. R., S. A. Kammerdiener, M. D. Dettinger, J. M. Caprio, and D. H. Peterson. 2001. Changes in the onset of spring in the western United States. *Bull. Amer. Meteorol. Soc.* 82:399–415.

Clark, M. P., M. C. Serreze, and G. J. McCabe. 2001. The historical effect of El Niño and La Niña events on the seasonal evolution of the montane snowpack in the Columbia and Colorado River basins. *Water Resources Research* 37:741–757.

Hamlet, A. F., and D. P. Lettenmaier. 1999. Effects of climate change on hydrology and water resources objectives in the Columbia River Basin. *J. Amer. Water Resour. Assoc.* 35:1597–1623.

———. 2005. Production of temporally consistent gridded precipitation and temperature fields for the continental U.S. *J. Hydromet.* 6:330–336.

Hamlet, A. F., P. W. Mote, M. P. Clark, and D. P. Lettenmaier. 2007. Twentieth century trends in runoff, evapotranspiration, and soil moisture in the Western U.S. *J. Clim.* 20:1468–1486.

Knowles, N., M. D. Dettinger, and D. R. Cayan. 2006. Trends in snowfall versus rainfall for the western United States. *J. Clim.* 19:4545–4554.

Lemke, P., J. Ren, R. B. Alley, I. Allison, J. Carrasco, G. Flato, Y. Fujii, G. Kaser, P. Mote, R. H. Thomas, and T. Zhang. 2007. Observations: Changes in snow, ice and frozen ground. Pp. 337–383 in *Climate Change 2007: The Physical Science Basis.* Contribution of Working Group I to the Fourth Assessment Report of the Intergovernmental Panel on Climate Change (S. Solomon, D. Qin, M. Manning, Z. Chen, M. Marquis, K.B. Averyt, M. Tignor and H.L. Miller [eds.]). Cambridge Univ. Press, Cambridge, UK, and New York.

Mote, P. W. 2006. Climate-driven variability and trends in mountain snowpack in western North America. *J. Clim.* 19:6209–6220.

Mote, P. W., A. F. Hamlet, M. P. Clark, and D. P. Lettenmaier. 2005. Declining mountain snowpack in western North America. *Bull. Amer. Meteorol. Soc.* 86:39–49.

Mote, P. W., A. F. Hamlet, and E. P. Salathé, Jr. 2007. Has spring snowpack declined in the Washington Cascades? *Hydrology and Earth System Sciences Discussions* 4:2073–2110.

62 MOTE

Regonda, S. K., B. Rajagopalan, M. Clark, and J. Pitlick. 2005. Seasonal cycle shifts in hydroclimatology over the western US. *J. Clim.* 18:372–384.

Scherrer, S. C., C. Appenzeller, and M. Laternser. 2004. Trends in Swiss alpine snow days: The role of local and large scale climate variability. *Geophys. Res. Lett.* 31: doi 10.1029/2004GL020255.

Stewart, I. T., D. R. Cayan, and M. D. Dettinger. 2005. Changes towards earlier streamflow timing across western North America. *J. Clim.* 18:1136–1155.

Stott, P. A. 2003. Attribution of regional-scale temperature changes to anthropogenic and natural causes. *Geophys. Res. Letts.* 30: doi:10.1029/2003GL017324.

Trenberth, K. E., P. D. Jones, P. Ambenje, R. Bojariu, D. Easterling, A. Klein Tank, D. Parker, F. Rahimzadeh, J. A. Renwick, M. Rusticucci, B. Soden and P. Zhai. 2007. Observations: Surface and atmospheric climate change. Chapter 3 in *Climate Change 2007: The Physical Science Basis.* Contribution of Working Group I to the Fourth Assessment Report of the Intergovernmental Panel on Climate Change (S. Solomon, D. Qin, M. Manning, Z. Chen, M. Marquis, K.B. Averyt, M. Tignor and H.L. Miller [eds.]). Cambridge Univ. Press, Cambridge, UK, and New York.

5

Variability and Trends in Spring Runoff in the Western United States

Jessica D. Lundquist, Michael D. Dettinger, Iris T. Stewart, and Daniel R. Cayan

Abstract

In the western United States, over half of the human water supply is derived from mountain snowmelt, with the snow acting as a natural reservoir, delaying runoff until the spring and summer, when it is needed most. Interannual variability of both the magnitude and timing of spring runoff is tremendous, and western states have developed extensive reservoir systems to store water from wet years in order to weather droughts. In recent decades, however, important changes in snowpacks and runoff timing have been noted, with the fraction of annual streamflow that runs off during late spring and summer declining 10 to 25 percent. Warmer winters and springs have led to earlier snowmelt and a higher percentage of precipitation falling as rain rather than snow. Snowmelt runoff timing has advanced so that it arrives approximately one to three weeks earlier in 73 percent of mountainous catchments across western North America. Even conservative climate-change projections suggest that California could lose one-third of its present-day spring snowpack by the middle part of the twenty-first century, and that other western states will follow suit. Hydrologists have long known that snow is a key component of the water budget, and yet most operational runoff models are based on historically derived empirical relations, rather than on spatially and temporally distributed physical processes. In a changing climate, the empirical relations may not remain valid, so more observational attention to the high-altitude snowfields that supply so much of the region's water supplies is needed to provide a basis for predicting the magnitude and timing of snowmelt, sublimation, and runoff.

Introduction

In the western United States, over half of the human water supply is derived from mountain snowmelt, with the snow acting as a natural reservoir, delaying runoff and providing water in the spring and summer, when it is most needed. The total volume of spring runoff can vary by an order of magnitude from one year to the next, and the timing of spring runoff can vary by several weeks. Western water managers have invested heavily in reservoir and conveyance systems to store water from wet seasons and years for use in subsequent drought seasons and years.

Although populations and water demands in the region continue to grow, several recent studies have identified marked changes in springtime runoff. The fraction of annual streamflow that runs off during late spring and summer has declined by 10 to 25 percent since the 1950s (Roos 1991, Wahl 1992, Dettinger and Cayan 1995). Warmer winters and springs have led to earlier snowmelt (Cayan et al. 2001) and a higher percentage of precipitation falling as rain rather than snow (Knowles et al. 2006).

Snowmelt runoff timing has advanced by approximately one to three weeks in the large majority of mountainous catchments across western North America (Stewart et al. 2005, Regonda et al. 2005). These observed trends raise some urgent questions: (1) Are they connected to global warming, or can the shifts be explained by natural variability? (2) What can we expect in the future? (3) How can we prepare for future changes? This chapter summarizes observed changes in spring runoff in the western United States, offers plausible reasons for these changes and those expected in the future, and describes current efforts to cope with expected change.

Observed Trends in Spring Runoff

The examination of spring runoff timing began when Maury Roos (1987, 1991) noted that the fraction of total annual runoff occurring between April and July had declined in California since 1906 (Fig. 5.1). The April

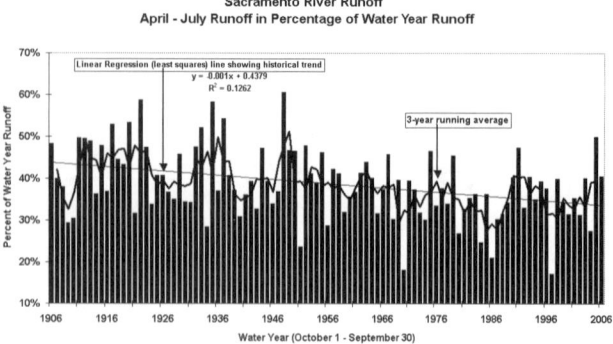

Figure 5.1. April–July fractional runoff for the Sacramento River in California. (Updated from Roos 1987.)

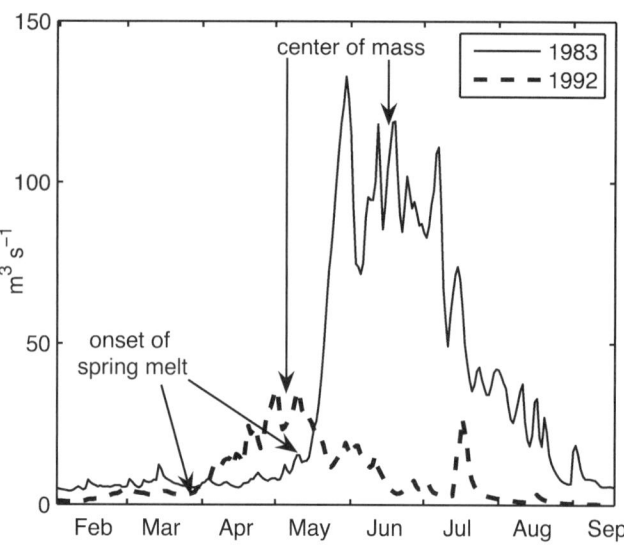

Figure 5.2. Daily streamflow discharges, Merced River at Happy Isles, 1983 and 1992. Both the timing and magnitude of snowfed rivers vary widely. Two methods of determining streamflow timing are the onset of spring melt and the center of mass of annual flow, both illustrated here.

through July (AMJJ) period encompasses most of the snowmelt season, and for most snowmelt-driven watersheds in the western United States, AMJJ flows are the most important contribution to the annual streamflow, comprising 50–80 percent of the annual total (Stewart et al. 2005). In contrast, mean AMJJ flow fractions are less than 30 percent at non-snowmelt-dominated gages, reflecting a lack of precipitation in much of the western United States during this period.

Furthermore, AMJJ flows follow the main winter storm season, during which many of the worst floods along the West Coast occur. During the winter storm season, reservoirs are only partially filled to allow space to mitigate floods. AMJJ flows are thus all the more important because this period is the most predictable portion of the annual runoff, allowing reservoirs and water supplies to be most confidently managed.

Following Roos's work, subsequent studies revealed that the California trends were not isolated and that the AMJJ fraction of runoff was declining in rivers throughout the western United States (Wahl 1992, Aguado et al. 1992, Pupacko 1993, Dettinger and Cayan 1995). The large regional coherence of these trends has been further supported by studies finding that the snow-fed rivers of the western United States rise and fall together each spring on time scales from days to year-to-year differences, along with the temperature patterns that control snowmelt (Peterson et al. 2000, Cayan et al. 1993).

Cayan et al. (2001) developed a second measure of spring runoff timing, the "spring pulse onset," which is the date of the beginning of the spring snowmelt-derived streamflow for snowmelt-dominated rivers. Their algorithm identifies the day when the cumulative departure from that year's mean flow is most negative, equivalent to finding the day after which most flows are greater than the

season's average (Fig. 5.2). In snowfed streams, the flows are typically an order of magnitude larger after the onset of snowmelt than before, making this measure as useful as it is simple. They found that these pulse dates are highly correlated with both regional temperatures and the dates of spring blooms of lilac (*Syringa vulgaris*) and honeysuckle (*Lonicera tatarica* and *L. korolkowii*); regionally, blooms occurred five to ten days earlier in the last half of the record (1976 to 2000) than in the first half (1950 to 1975).

Stewart et al. (2005) analyzed 302 stream gages in western North America for the period from 1948 to 2002 using both of the previous timing measures and a new measure, the timing of the center of mass of annual flow (CT) for each water year (Fig. 5.2), where $CT = \Sigma (t_i q_i)/ \Sigma q_i$, where t_i is time in days (or months) since the beginning of the water year (October 1), and q_i is the corresponding streamflow in that month or day. CT and the spring pulse onset date are strongly correlated ($r = 0.5$–0.8) for most gages and significantly correlated at 74 percent of the gages.

Regonda et al. (2005) used a measure similar to CT for a similar study period, calculating the date by which 50 percent of the water-year flow had passed a gage. CT is a robust measure of when most of the spring runoff arrived because, unlike the AMJJ fraction, it indicates whether the "missing" snowmelt-season water has shifted earlier or later. Also, unlike the spring pulse date, it is a measure

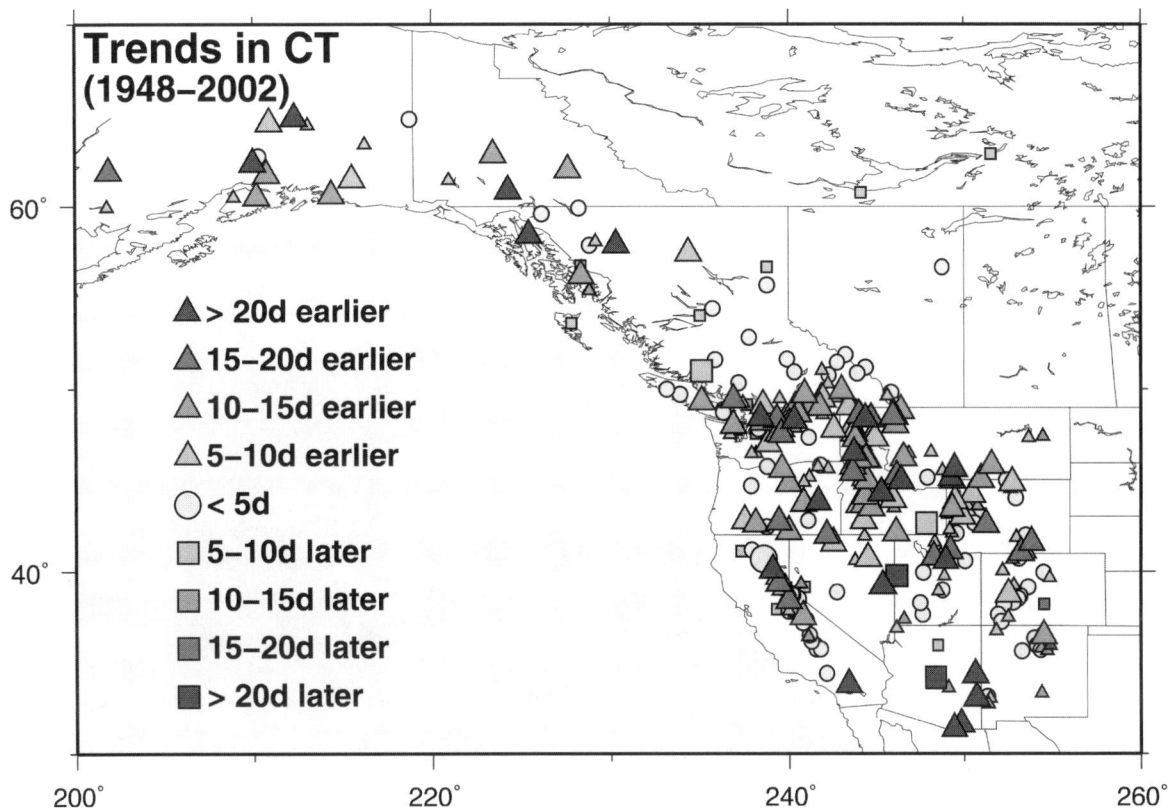

Figure 5.3. CT trends for 302 snowfed streamflow gages in western North America, 1948–2002. Larger symbols indicate significant trends at a 90 percent confidence interval. (From Stewart et al. 2005.)

at the seasonal/annual scale, rather than being controlled mainly by the events of a few brief weeks. Both Stewart et al. (2005) and Regonda et al. (2005) have reported statistically significant CT trends (Fig. 5.3).

Stewart et al. (2005) found that AMJJ fractional flows have declined for 81 percent of the gages, even while 50 percent have measured increased overall annual flows. In the Pacific Northwest, annual flows decreased overall during the 1948–2002 period as well. Spring pulse onset dates shifted 10 to 30 days earlier, with the largest changes observed in the Pacific Northwest and the Sierra Nevada. Of the 241 snowfed rivers, 66 percent (159) have experienced trends toward earlier spring pulse days by 3 days or more, and 54 percent (86) of these negative trends are significant at the 90 percent confidence level. Only 15–19 percent of the gages have shown no trends, or trends toward later spring pulse days in later years.

Results for CT were similar (Fig. 5.3). Trends toward earlier centers of mass have been measured at 73 percent (214) of the gages, and 49 percent of these earlier trends (105 gages) are significantly different from zero at 90 percent confidence levels. By contrast, CTs for most non-snowmelt-dominated gages, which usually are low-altitude coastal streams, have been trending in the opposite direction, toward later streamflow timing, with corresponding shifts of 5 to 25 days.

Most of these studies focused on the period of record after the mid-1940s because (1) this period has the most complete dataset available, and (2) the most significant contribution to trends took place during this period. Available data for the 1901–2002 period (not shown) reveal overall trends toward earlier streamflow timing for the coastal states, and smaller or no trends for gages in the interior (Stewart et al. 2005). The 1900–1948 period did not have consistently trending streamflow timings (Dettinger and Cayan 1995), suggesting that the observed trends are a late-twentieth-century phenomenon.

Causes

The 1948–2002 timing changes were caused primarily by rising temperatures that have been observed during the same period. This warming has been regional in extent (Cayan et al. 2001) and is believed to be part of global patterns of warming associated with increased greenhouse gases (Houghton et al. 2001, National Research Council

2001) in combination with warmth due to natural multi-decadal variations of the climate system (e.g., Mantua et al. 1997).

The timing of spring runoff is controlled by both temperature (Peterson et al. 2000) and precipitation (Aguado et al. 1992, Hamlet and Lettenmaier 1999, Stewart et al. 2005). Predicting future changes in western U.S. runoff requires an understanding of how each of these processes has influenced spring runoff in the past.

Influence of Temperature vs. Precipitation

Observational studies have shown clearly increasing trends in temperature and some less-certain increasing trends in precipitation over the West in the twentieth century (McCabe and Wolock 2002, Mote et al. 2003, Mote 2003b, Sheppard et al. 2002). Although warmer temperatures cause snow to melt earlier, increased winter precipitation results in larger snowpacks, which are correlated with later spring melt (Hamlet and Lettenmaier 1999, Stewart et al. 2005).

Additionally, the central timing of annual precipitation has tended to come about a week later over the past 50 years (Stewart et al. 2005), a trend reflected in the timing of non-snowfed rivers (which shifted to later streamflow) but masked in snowmelt-dominated rivers by the effects of warming temperatures. That is, temperature and precipitation forcings on streamflow have had opposing influences in snowfed streams in recent decades.

Hamlet et al. (2005) applied the variable infiltration capacity (VIC) hydrologic model (Liang et al. 1994, Cherkauer and Lettenmaier 2003) over the western United States to isolate the effects of precipitation and temperature from the period of 1916–1997. They found that although precipitation and temperature trends both contributed to changes in runoff timing, the combined effect of the temperature and precipitation trends on timing of peak snow accumulation and melt, and hence streamflow timing, was dominated by the temperature effects. Stewart et al. (2005) and Regonda et al. (2005) have shown the same temperature dominance using statistical analyses of historical flow timing and climate records.

Natural Variability vs. External Forcings

Atmospheric circulation patterns and water resources in the western United States are subject to modulations by multiyear climatic fluctuations such as the El Niño/Southern Oscillation (ENSO) air-sea interaction of the tropical Pacific (Allan et al. 1996), and its multidecadal counterpart, the Pacific Decadal Oscillation (PDO) (Mantua et al. 1997). ENSO and PDO are associated with recurring temperature and precipitation patterns across western North America (Ropelewski and Halpert 1986, Redmond and Koch 1991, Latif and Barnett 1994, Mantua et al. 1997, Cayan et al. 1999).

Through shifts in regional patterns of temperature and precipitation, large-scale atmospheric circulation patterns have strong influences on spring runoff patterns. For example, circulation-driven trends toward warmer winters since the late 1940s can explain much of the western trend in runoff timing (Dettinger and Cayan 1995). Warmer winters and springs in recent decades have been associated with deeper Aleutian lows and a southern displacement of the westerly wind fields over the central North Pacific, which brought warmer air masses over the West Coast of the United States (Fig. 5.4).

During the period from 1951 to 2000, the PDO shifted from a regime that favors cool conditions in the West to the opposite regime, which favors warmth, in a major step change from a cool to warm PDO in 1976–1977 (Mantua et al. 1997). The two multidecadal regimes of the PDO cover the period during which most records of streamflow timing were examined.

Analysis of the period from 1924 to 1976, when the PDO shifted from its warm/dry phase to its cool/wet phase, revealed trends toward larger magnitude and later timing of maximum snow accumulation, concentrated in the Pacific Northwest (Hamlet et al. 2005). Thus, the sign of the PDO and the timing of spring runoff are negatively correlated ($r = -0.2$ to -0.8, Stewart et al. 2005), such that the cool phase of the PDO corresponds to later melt, and the warm phase of the PDO corresponds to earlier melt.

Because of this negative correlation between timing and the PDO, and the large step toward warm PDO in the middle of the recent timing trends, the question arises of whether the timing trends were simply PDO responses. Toward this end, Stewart et al. (2005) estimated the trend that the PDO would be expected to impose on CTs and temperatures from interannual variations of the PDO within the separate PDO regimes from 1948 to 1976 and from 1977 to 1998. Applying the influences found before, and after, the 1976–1977 PDO shift to the PDO history as a whole demonstrated that the long-term PDO variations explained less than half of the observed timing trends at most snowmelt stations in the western United States (Fig. 5.5), with exceptions appearing in the Pacific Northwest, where PDO climatic influences are generally strongest (Mantua et al. 1997).

Furthermore, from 1977 to 1998, the PDO did not trend overall, and yet streamflow timing continued to trend toward earlier in the year. Other evidence also suggests that the shift in spring runoff timing cannot be entirely explained by the PDO. Stewart et al. (2005) used linear regression to remove the PDO component of the CTs and then correlated the residual with temperature alone. Although CT variations correlated with temperature shifts without the PDO component, when the technique was reversed, with the temperature component removed, the residual was not correlated with the PDO at most locations (Fig. 5.6). Hamlet et al. (2005) minimized the effects of decadal variability on streamflow timing by running the VIC model for back-to-back warm PDO epochs for the periods 1924–1946 and 1977–1995. In the absence of a PDO shift, large-scale warming still occurred, and the timing of snowmelt trended toward earlier in the spring. This result arose because the 1924–1946 "warm" PDO epoch was much cooler in the western United States than was the more recent warm phase.

The results of this model suggest that while decadal variability of the PDO probably does account for many trends in winter precipitation over shorter periods of the record, the temperature trends are greater than can be accounted for by the PDO. Because streamflow timing depends more on temperature than precipitation, these results suggest that the trends in spring streamflow are due to more than just the multidecadal natural variability of the PDO.

Warming temperatures affect the entire western United States in a similar manner, but the effects of ENSO vary between the north and south, due largely to north–south precipitation contrasts (Redmond and Koch 1991, Dettinger et al. 1998). Stewart et al. (2005) found that El Niño conditions are associated with warmer spring temperatures and lower winter precipitation in the northern portion of the western United States, and therefore with earlier snowmelt-derived streamflow, leading to positive correlations of ENSO indices with CT, ranging from r = 0.2 to r = 0.6.

In the southern portion of the study area, El Niño conditions are associated with higher-than-average winter precipitation, leading to a delay in snowmelt and negative correlations with CT ranging from r = −0.2 to r = −0.6. While interannual variations in streamflow timing can be at least partially explained by ENSO variations, the long-term trends toward earlier spring runoff that

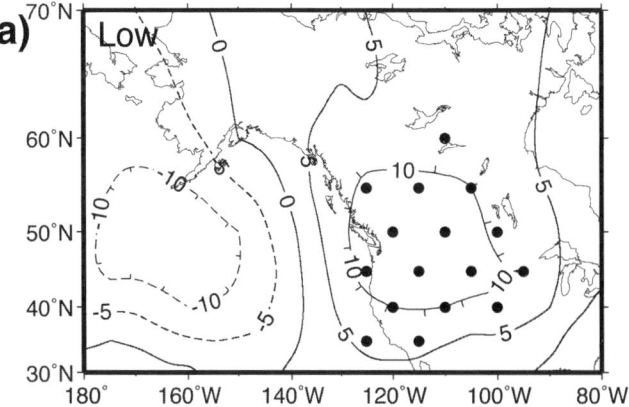

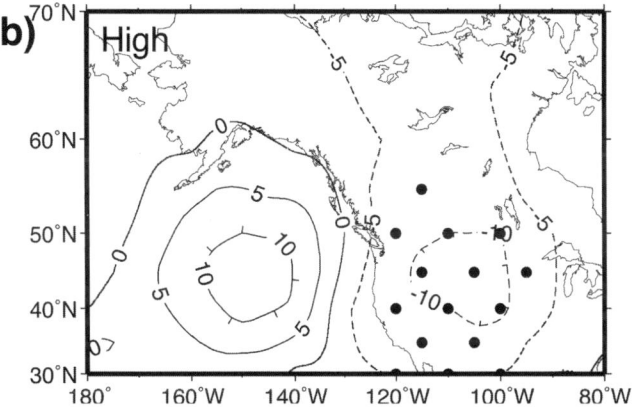

Figure 5.4. Composites of spring 700 mb height anomalies (in meters) for (a) low and (b) high values of the first principal component of western North American CTs from 1948 to 2002, associated with (a) earlier and (b) later timing of snowmelt-derived streamflow. Dashed lines indicate negative anomalies. Statistically significant correlations at the 95 percent confidence level are marked with small black circles. (From Stewart et al. 2005.)

have been observed throughout the northern and southern portions of the western United States cannot be explained by ENSO.

Regional Sensitivities

Changes in the timing of spring runoff vary primarily with elevation, such that areas with middle-elevation catchments, like the Pacific Northwest and the northern Sierra Nevada, have been the most sensitive to the observed warming trends (Mote 2003a, Regonda et al. 2005, Hamlet et al. 2005, Stewart et al. 2005). While shifts of CT to 10 to 20 days earlier are common in basins below 2,500 m, basins above 2,500 m exhibit little or no change. The higher basins receive winter precipitation at the coldest temperatures and then remain below freezing for

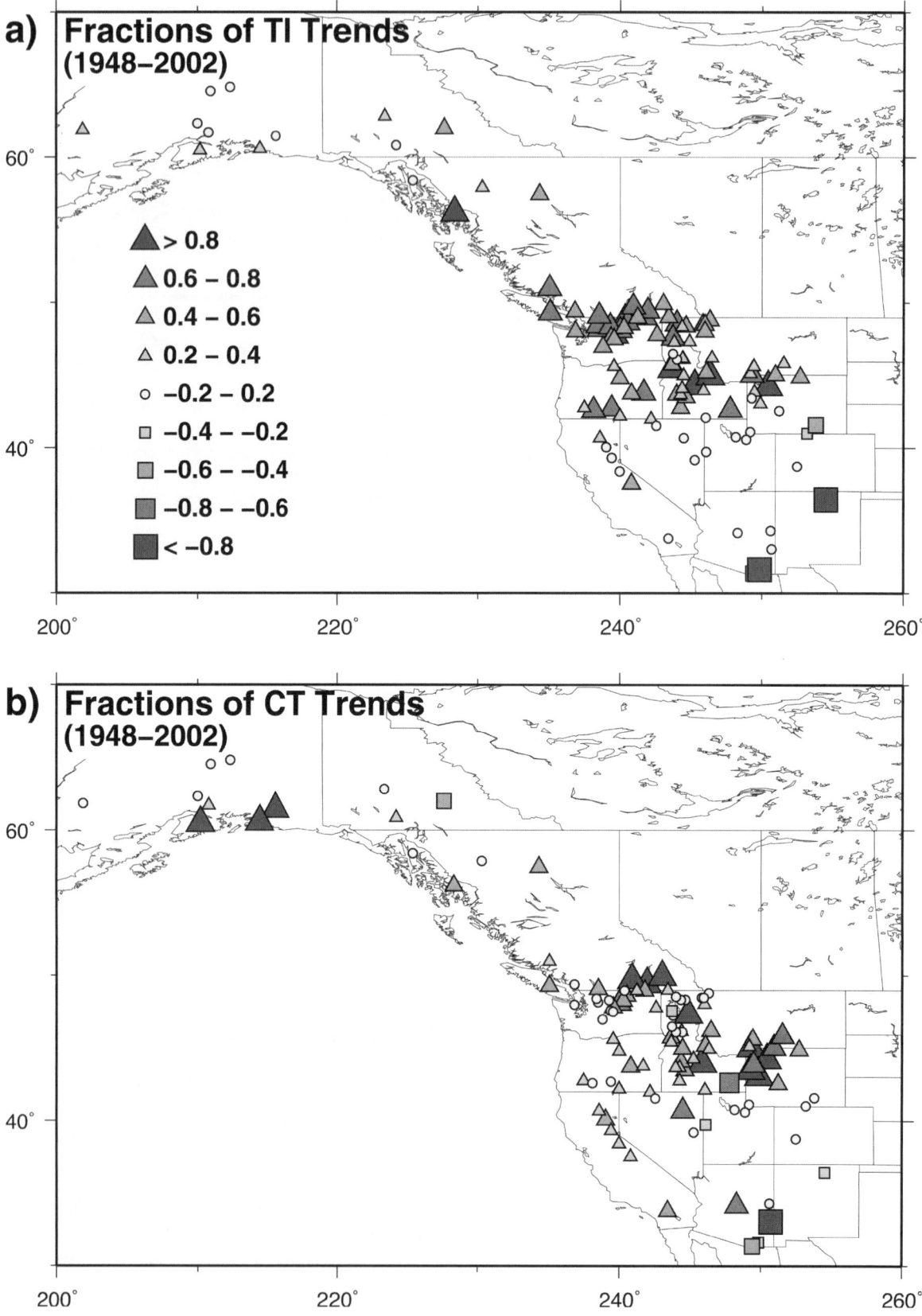

Figure 5.5. Fractions of (a) observed temperature (TI) and (b) observed CT trends that are explained by the PDO step-change "trend," using coefficients from regressions of CT and TI with PDO. Regression coefficients from the 1948–1976 and the 1977–1998 periods were averaged for use; fractions are shown only for gages with significant observed CT trends. (From Stewart et al. 2005.)

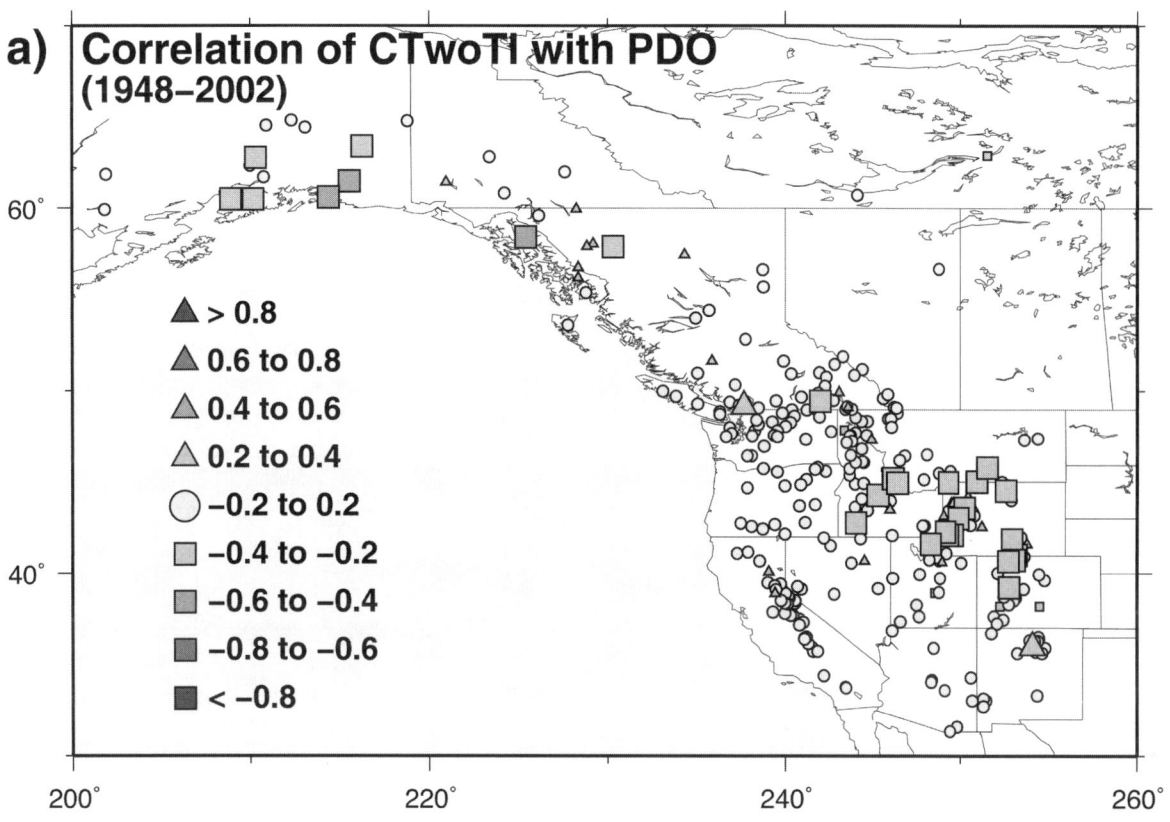

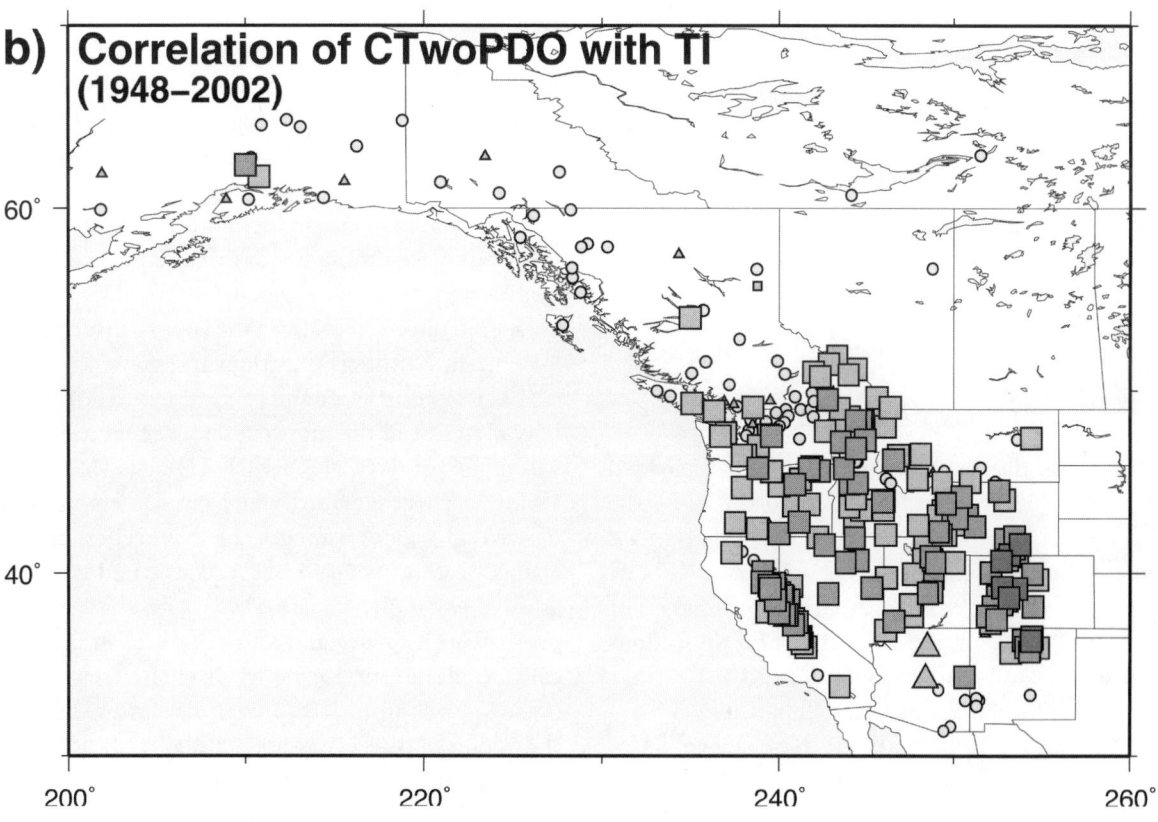

Figure 5.6. Correlation of (a) CT minus the portion of its variance attributable to the TI (CTwoTI), with PDO, and (b) CT minus the portion of its variance attributable to the PDO index (CTwoPDO), with TI. The larger symbols indicate statistically significant correlations at the 95 percent confidence level. (From Stewart et al. 2005.)

most of the spring (Regonda et al. 2005). Consequently, the high-elevation areas in the Rockies and southern Sierra have been relatively less affected by the statistically significant, but relatively modest temperature trends observed to date (Hamlet et al. 2005).

In contrast, the ratio of snow to winter rain is lower and more variable in the low-elevation Pacific Northwest (Serreze et al. 1999), and the shift of spring snows to spring rains enhances the trends toward earlier runoff. Long-term trends toward less snowfall and more rainfall in response to warming have occurred across most of the region (Knowles et al. 2006), and the rain-vs.-snow changes have been largest in the middle elevations. Because of these elevation effects, rivers in the warmer coastal states have been affected primarily by warming in the winter (Dettinger and Cayan 1995), whereas areas with a more continental climate are more sensitive to precipitation trends during the winter and to warming in late spring (Cayan et al. 2001, Hamlet et al. 2005, Knowles et al. 2006).

To provide a global context for the regional trends discussed in this chapter, an analysis of CTs in rivers around the globe with long-term (20-plus years) discharge records (described by Dettinger and Diaz 2000) is presented in Dettinger et al. 2001, Figure C.1. The results mapped there reveal that while some of the largest trends toward earlier spring melt have occurred in western North America, similar trends have occurred in rivers worldwide during the period from 1945 to 1993. Significant trends, using nonparametric Kendall's tau trend statistics, are found in rivers throughout eastern Europe and western Russia, across Canada, and, less robustly, in the Southern Hemisphere. As observed in the western United States, watersheds where cool season temperatures are often at or above freezing have exhibited the largest changes in streamflow timing.

Climate Projections

The studies described above demonstrate that snowfed runoff timing has been trending toward earlier in the year in western North America, and that this trend is controlled primarily by temperature and secondarily by precipitation (Stewart et al. 2005). Thus, projections of future streamflow characteristics depend on projections of future temperature and precipitation.

Projections from many different climate models have recently become available, including those from the U.S. PCM, Canadian CCCM, German ECHAM4, British HadCM3, Japanese NIES, and Australian CSIRO coupled ocean-atmosphere global climate models (as described by Dettinger 2006). A growing number of climate-change responses to plausible future greenhouse-gas emission scenarios have been simulated with such models—for example, in response to the A2, B2, and IS92a SRES emissions scenarios (Houghton et al. 2001), which represent projections of relatively rapid, moderate, and intermediate rates of twenty-first-century emissions increases, respectively.

Simulations by the six climate models listed above, each simulating responses to each of the three specified greenhouse-gas plus sulfate-aerosol emission scenario, indicate that temperatures in the western United States may increase between about +2.5 and +9°C by 2100, and that precipitation may increase, decrease, or remain the same (e.g., Fig. 5.7).

Dettinger (2006) resampled these projections to estimate probabilities for various levels of climate change in the western United States. Analyzed this way, the ensemble of projections shown in Figure 5.7 suggests, at typical grid cells in the interior West, that (1) temperatures warm in all projections, much as in Figure 5.7; (2) precipitation is not projected to change as much in the future; and (3) projection uncertainties increase as we look further into the future, more so for temperatures than for precipitation.

A resampling of the temperature projections across the conterminous United States (Fig. 5.8) illustrates that projected springtime warming in the interior West may be among the largest nationwide, and that uncertainties (as indicated by the growing widths of the 2050 distributions in Figure 5.8) are largest in the West. Thus the historical streamflow-timing trends discussed thus far can only be expected to continue and, indeed, to enter the interior West even more forcefully in the twenty-first century, unless current warming projections are askew.

Several efforts to model future streamflow scenarios have used one of the most conservative models of future warming, the NCAR Parallel Climate Model (PCM) (Washington et al. 2000), which gives ensemble estimates of +2°C to +3°C warming and ±10 percent precipitation change over the twenty-first century, assuming a business-as-usual emissions scenario. Even with this conservative projection, Knowles and Cayan (2002), using a physically based land-surface model, discovered that California could lose one-third of its present-day spring snowpack by the middle of the twenty-first century.

Using linear regression models, Stewart et al. (2004) projected that, at many gages, CT could shift to 30 to 40

PROJECTED CHANGES IN ANNUAL TEMPERATURE, NORTHERN CALIFORNIA

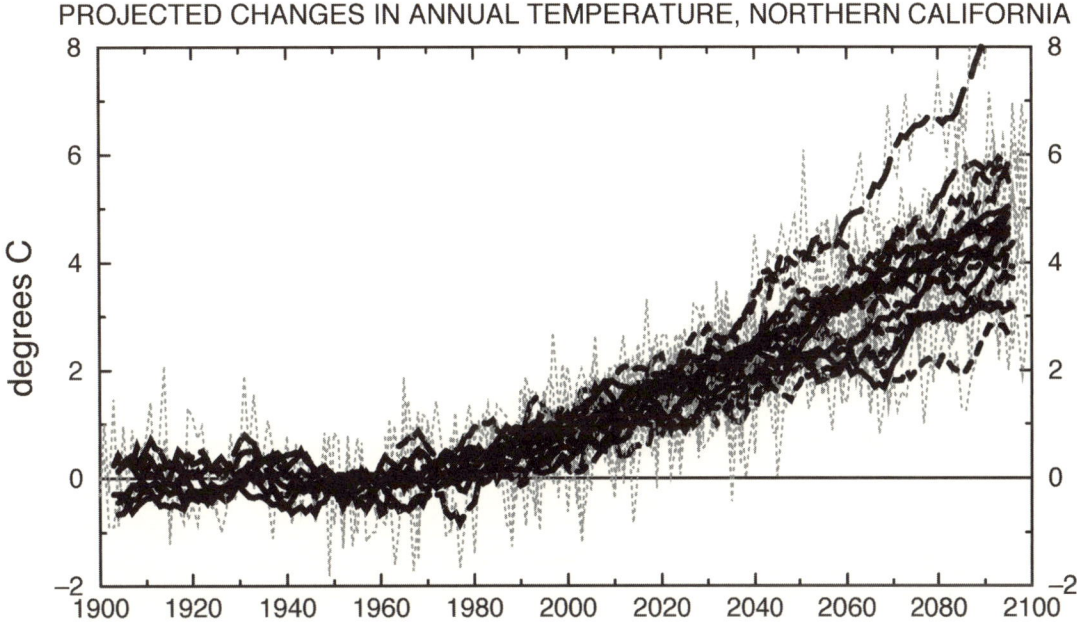

PROJECTED CHANGES IN ANNUAL PRECIPITATION, NORTHERN CALIFORNIA

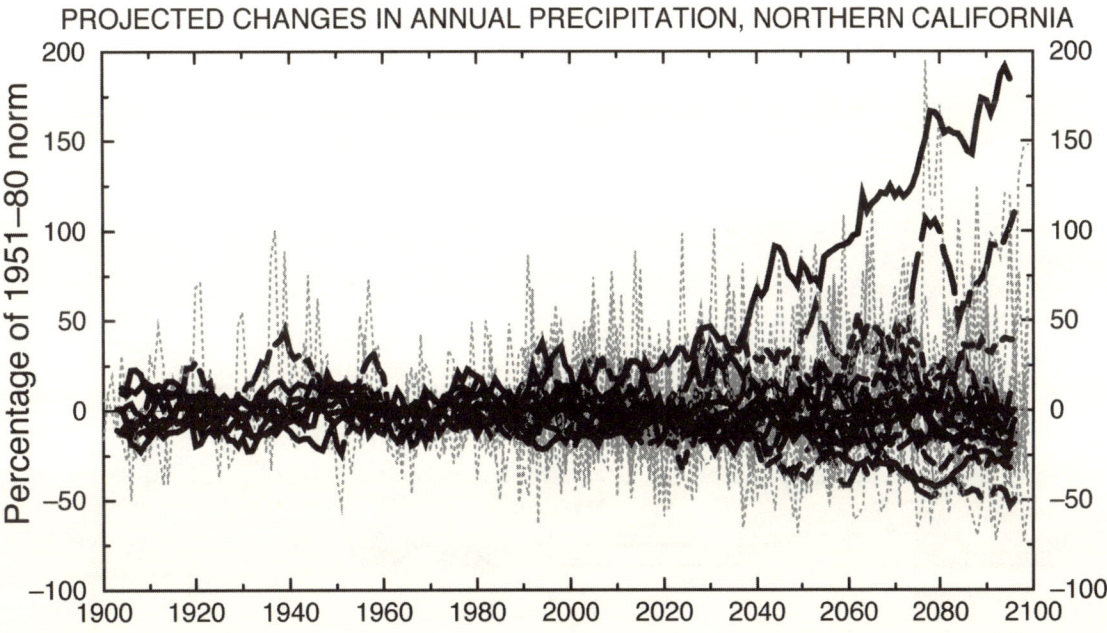

Figure 5.7. Ensembles of 23 simulations of historical and future temperature and precipitation changes projected by varying scenarios programmed into seven coupled, ocean-atmosphere general-circulation models. The scenarios range from one to five versions for each model, and extend from slight to marked increases over contemporary emissions in CO_2 and sulphate aerosols.

The models used are the Canadian CCCM climate model, Australian CSIRO model, European ECHAM4 model, British HadCM2 model, British HadCM3, Japanese NIES model, and US PCM model. (From Dettinger 2006).

DISTRIBUTIONS OF MONTHLY TEMPERATURE ANOMALIES RELATIVE TO 1950-99 MEANS

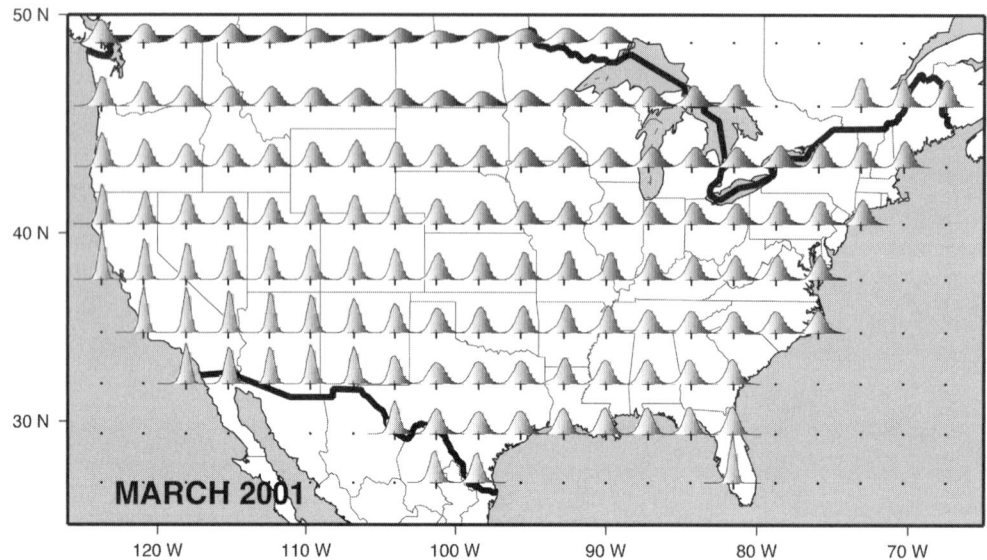

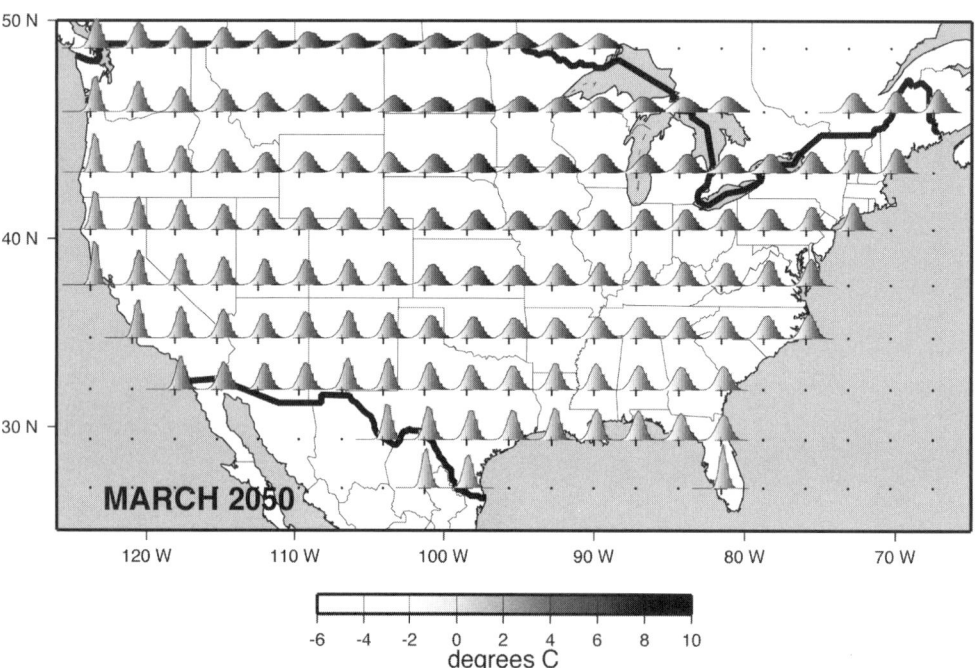

Figure 5.8. Probability density functions (pdfs) for projections of future temperature anomalies across the United States for (a) 2001 and (b) 2050, considering the ensemble of climate simulations shown in Figure 5.8. All pdfs are scaled the same, and all reflect the combination of natural temperature variability, uncertainties due to model differences, and uncertainties due to emission-scenario differences.

days earlier by the end of the twenty-first century, in good agreement with more detailed land-surface simulations, such as those by Wood et al. (2004) using the VIC model and by Dettinger et al. (2004) using the Precipitation-Runoff Modeling System (PRMS) (Leavesley et al. 1983).

These simulations and others (e.g., Jeton et al. 1996) agree with the observations that (1) increases in temperature advance the timing of spring runoff, (2) changes in precipitation change the magnitude of both spring and annual runoff, and (3) the most pronounced differences between how basins respond depend on what elevations they drain. Under a business-as-usual climate scenario, streamflow timing trends are projected to be continuations of the observed trends with no large-scale accelerations, unless emissions growth itself accelerates (Stewart et al. 2004, Dettinger et al. 2004). However, Lundquist and Flint (2006) recently have shown that once streamflow timing trends shift by more than about a month, the seasonal cycle of daily insolation may intervene to slow subsequent timing changes somewhat in many watersheds.

Preparing for the Future

Hydrologists have long known that snow is a key component of the water budget, and yet most operational runoff models are based on empirical relations between historical runoff and weather fluctuations in specific basins, rather than on invariant relations describing spatially and temporally distributed physical processes. In a changing climate, these empirical relations may become increasingly unreliable, and a historically based empirical model may not accurately represent changing snowcover, snowmelt amounts and contributions, and snowmelt rates. At the same time, recent work by Jain et al. (2005) suggests that the year-to-year variability in streamflow volumes has increased since 1972, resulting in more large floods and extreme droughts. Thus, predicting next year's flow timing may become even more difficult in the future.

Most of the observations and modeling efforts discussed above mention elevation as the primary factor affecting the magnitude of trends in spring runoff in various basins in the western United States. However, very few measurements exist at the highest elevations (e.g., Fig. 5.9, Lundquist et al. 2003), and surprisingly little is known about how temperature, precipitation, snowmelt, and runoff vary with elevation in different regions and weather conditions. Thus, more monitoring at high alti-tudes is needed to understand and track the anticipated changes in temperature and precipitation with elevation, and to predict the magnitude and timing of critical processes like snowmelt, sublimation, and runoff.

Recent increased hydroclimatic monitoring in Yosemite National Park has demonstrated that in some years, as in 2002, the onset of spring melt and runoff in California may be virtually independent of elevation (Fig. 5.10, Lundquist et al. 2004). This occurred when synoptic weather conditions resulted in over 13°C warming in less than a week, a temperature change large enough to overcome springtime lapse rates that typically hold the highest altitudes below freezing well after the lowest altitudes have thawed.

Understanding how surface- and free-air temperature lapse rates and precipitation lapse rates might change in a future climate is the subject of active research, with complex and often localized results (Aguado 1990, Barry 1990, Beniston et al. 1994, Beniston and Rebetez 1996, Dettinger et al. 2004, Pepin and Losleben 2002, Singh 1991, Williams et al. 1996). As snow disappears from lower-elevation areas, changes at high elevations will have a critical impact on western United States water supplies. Thus, further studies are needed (CIRMOUNT 2006).

Conclusions

During the past 50 years, the climate of the western United States has warmed. Warmer temperatures have led to earlier runoff and to more precipitation falling as rain rather than snow. These effects are greatest where temperatures are close to 0°C, as at middle elevations, and particularly in the coastal ranges of the Pacific Northwest. Climate projections agree that temperatures will continue to warm in the future.

Even the most benign of these projected climate changes are sufficient to significantly alter the landscape, hydrology, and land and water resources of the western United States, and those alterations are likely to become significant within roughly the next 25 years (Barnett et al. 2005, Dettinger et al. 2004, Van Rheenen et al. 2004). If even the modest observed warming in the last half of the twentieth century was sufficient to cause the hydrological changes reported in the literature, then prospects for additional climate-change impacts in the western United States should be taken seriously, despite significant remaining climate-change uncertainties.

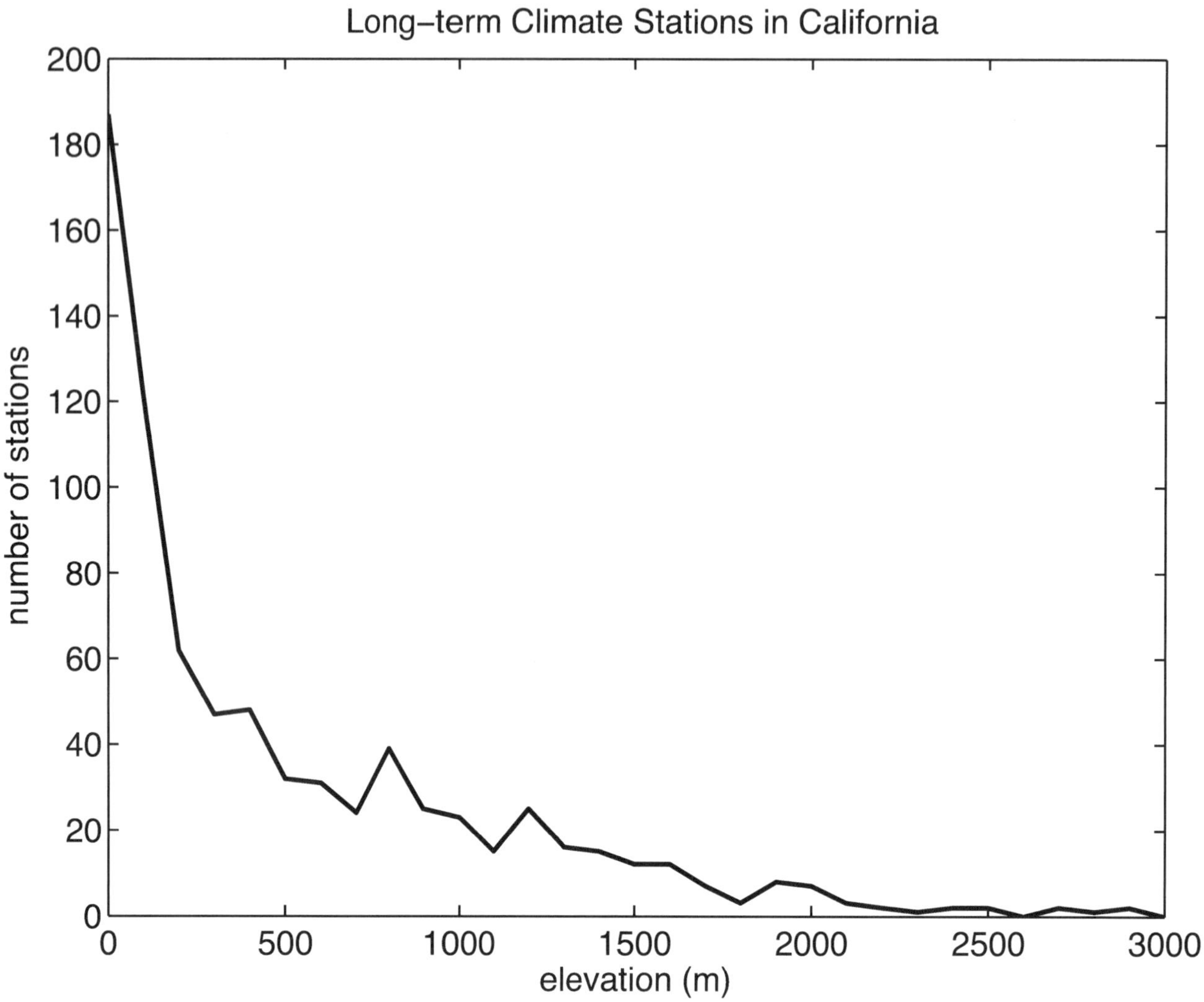

Long–term Climate Stations in California

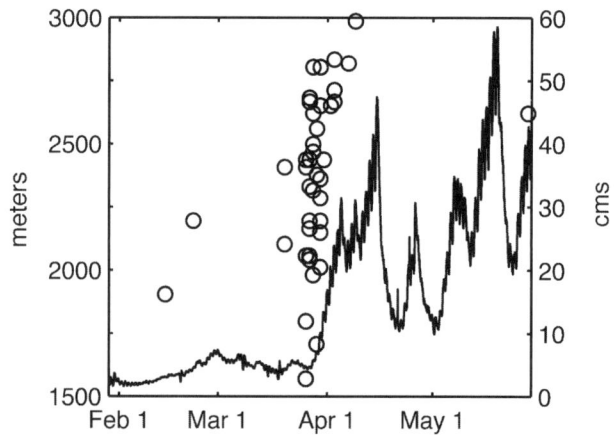

Figure 5.9. Most long-term climate stations in California (shown here) and throughout the West are located at low elevations. However, conditions at higher elevations are crucial for water resources and need to be more closely monitored. (From Lundquist et al. 2003)

Figure 5.10. In 2002, snow in the Sierra Nevada of California began melting simultaneously at all elevations (below about 9,000 ft) around the eighty-eighth day of the year (March 29), and streams rose dramatically. The line and the right axis show discharge at the Merced River at Happy Isles in Yosemite National Park, central Sierra Nevada. The circles show the date of maximum snow accumulation--that is, the date when melt began--at 45 snow pillows in the central Sierra Nevada, California, plotted against elevation, left axis. (From Lundquist et al. 2004.)

References

Aguado, E. 1990. Elevational and latitudinal patterns of snow accumulation departures from normal in the Sierra Nevada. *Theor. Appl. Climatol.* 42:177–185.

Aguado, E., D. Cayan, L. Riddle, M. Roos, 1992. Climatic fluctuations and the timing of West Coast streamflow. *J. Clim.* 5:1468–1483.

Allan, R. J., J. Lindesay, and D. Parker, 1996. El Niño, Southern Oscillation and Climate Variability. CSIRO Publication. Collingwood, Australia.

Barnett, T. P., J. C. Adam, and D. P. Lettenmaier. 2005. Potential impacts of a warming climate on water availability in snow-dominated regions. *Nature* 438:303–309. doi:10.1038/nature 04141

Barry, R. G. 1990. Changes in mountain climate and glacio-hydrological responses. *Mount. Res. Devel.* 10:161–170.

Beniston, M., and M. Rebetez, 1996. Regional behavior of minimum temperatures in Switzerland for the period 1979–1993. *Theor. Appl. Climatol.* 53:231–244.

Beniston M., M. Rebetez, F. Giorgi, and R. Marinucci, 1994. An analysis of regional climate change in Switzerland. *Theor. Appl. Climatol.* 49:135–159.

Cayan, D. R., S. A. Kammerdiener, M. D. Dettinger, J. M. Caprio, and D. H. Peterson. 2001. Changes in the onset of spring in the western United States. *Bull. Amer. Meteorol. Soc.* 82:399–415.

Cayan, D. R., K. T. Redmond, and L. G. Riddle. 1999. ENSO and hydrologic extremes in the western United States. *J. Clim.* 12:2881–2893.

Cayan, D. R., L. G. Riddle, and E. Aguado. 1993. The influence of precipitation and temperature on seasonal streamflow in California. *Water Resources Res.* 29:1127–1140.

Cherkauer, K. A., and D. P. Lettenmaier. 2003. Simulation of spatial variability in snow and frozen soil. *J. Geophys. Res.* 108:8858, doi:10.1029/2003JD003575.

CIRMOUNT Executive Committee. 2006. *Mapping New Terrain: Report of the Consortium for Integrated Mountain Climate Research in Western Mountains (CIRMOUNT)*. Pacific Southwest Research Station, USDA Forest Service, Misc. Pub. PSW-MISC-77. Albany, CA. 29 pp.

Dettinger, M. D. 2006. A component-resampling approach for estimating probability distributions from small forecast ensembles. *Clim. Change.* doi:10.1007/s10584-005-9001-6.

Dettinger, M. D., and D. R. Cayan. 1995. Large-scale atmospheric forcing of recent trends toward early snowmelt runoff in California. *J. Clim.* 8:606–623.

Dettinger, M. D., D. R. Cayan, H. F. Diaz, and D. M. Meko. 1998. North-south precipitation patterns in western North America on interannual-to-decadal timescales. *J. Clim.* 12:3095–3111.

Dettinger, M. D., D. R. Cayan, H. F. Diaz, and I. Stewart. 2001. Decadal variations and trends in snowmelt and streamflow timing: Global and North American patterns in the twentieth century. International HIGHEST II Conference: Climate Change at High Elevation Sites: Emerging Impacts, June 2001.

Dettinger, M. D., D. R. Cayan, M. K. Meyer, and A. E. Jeton. 2004. Simulated hydrologic responses to climate variations and change in the Merced, Carson, and American River Basins, Sierra Nevada, California, 1900–2099. *Clim. Change* 62:283–317.

Dettinger, M. D., and H. F. Diaz. 2000. Global characteristics of streamflow seasonality and variability. *J. Hydromet.* 1:289–310.

Hamlet, A. F., and D. P. Lettenmaier. 1999. Effects of climate change on hydrology and water resources in the Columbia River Basin. *J. AWRA* 35:1597–1623.

Hamlet, A. F., P. W. Mote, M. P. Clark, and D. P. Lettenmaier. 2005. Effects of temperature and precipitation variability on snowpack trends in the western U.S. *J. Clim.* 18:4545–4561.

Houghton, J. T., Y. Ding, D. J. Griggs, M. Noguer, P. J. van der Linden, X. Dai, K. Maskell, and C. A. Johnson, (eds.). 2001. *Climate Change 2001: The Scientific Basis.* Contribution of Working Group I to the Third Assessment Report of the Intergovernmental Panel on Climate Change. Cambridge Univ. Press, New York.

Jain, S., M. Hoerling, and J. Eischeid, 2005. Decreasing reliability and increasing synchroneity of western North American streamflow. *J. Clim.* 18:613–618.

Jeton, A. E., M. D. Dettinger, and J. L. Smith. 1996. *Potential Effects of Climate Change on Streamflow, Eastern and Western Slopes of the Sierra Nevada, California and Nevada.* USGS Water-Resources Investigations Report 95-4260.

Knowles, N., and D. Cayan. 2002. Potential effects of global warming on the Sacramento/San Joaquin watershed and the San Francisco estuary. *Geophys. Res. Lett.* 29:18–21.

Knowles, N., M. Dettinger, and D. Cayan. 2006. Trends in snowfall versus rainfall for the western United States. *J. Clim.* 19:4545–4559.

Latif, M., and T. P. Barnett. 1994. Causes of decadal climate variability over the North Pacific and North America. *Science* 266:634–637.

Leavesley, G. H., R. W. Lichty, B. M. Troutman, and L. G. Saindon. 1983. *Precipitation-Runoff Modeling System: User's Manual.* USGS Water-Resources Investigations Report 83–4238.

Liang X., D. P. Lettenmaier, E. F. Wood, and S. J. Burges. 1994. A simple hydrologically based model of land surface water and energy fluxes for general circulation models. *J. Geophys. Res.* 99:14, 415–14, 428.

Lundquist, J. D., D. R. Cayan, and M. D. Dettinger. 2003. Meteorology and hydrology in Yosemite National Park: A sensor network application. Pp. 518–528 in F. Zhao and L. Guibas (eds.), *Information Processing in Sensor Networks.* IPSN 2003, LNCS.

———. 2004. Spring onset in the Sierra Nevada: When is snowmelt independent of elevation? *J. Hydromet.* 5:325–340.

Lundquist, J., and A. Flint. 2006. Onset of snowmelt and streamflow in 2004 in the western United States: How shading may affect spring streamflow timing in a warmer world. *J. Hydromet.* 7:1199–1217.

Mantua, N. J., S. R. Hare, Y. Zhang, J. M. Wallace, and R. C. Francis. 1997. A Pacific interdecadal climate oscillation with impacts on salmon production. *Bull. Amer. Meteor. Soc.* 78:1069–1079.

McCabe, G. J., and D. M. Wolock. 2002. Trends and temperature sensitivity of moisture conditions in the conterminous United States. *Clim. Res.* 20:19–29.

Mote, P. W. 2003a. Trends in snow water equivalent in the Pacific Northwest and their climatic causes. *Geophys. Res. Lett.* 30:1601–1604.

———. 2003b. Trends in temperature and precipitation in the Pacific

Northwest during the twentieth century. *Northwest Sci.* 77: 271–282.

Mote, P. W., E. A. Parson, A. F. Hamlet, W. S. Keeton, D. P. Lettenmaier, N. J. Mantua, E. L. Miles, D. W. Peterson, D. L. Peterson, R. Slaughter, and A. K. Snover. 2003. Preparing for climatic change: The water, salmon, and forests of the Pacific Northwest. *Clim. Change* 61:45–88.

National Research Council. 2001. *Climate Change: An Analysis of Some Key Questions.* National Academy Press, Washington, D.C.

Pepin, N., and M. Losleben. 2002. Climate change in the Colorado Rocky Mountains: Free air versus surface temperature trends. *Int. J. Climatol.* 22:311–329.

Peterson, D., L. Smith, M. Dettinger, D. Cayan, and L. Riddle. 2000. An organized signal in snowmelt runoff over the western United States. *J. AWRA* 36:421–432.

Pupacko, A. 1993. Variations in northern Sierra Nevada streamflow: Implications of climate change. *Water Resour. Bull.* 29:283–290.

Redmond, K. T., and R. W. Koch. 1991. Surface climate and streamflow variability in the western United States and their relationship to large-scale circulation indexes. *Water Resour. Res.* 27:2381–2399.

Regonda S., B. Rajagopalan, M. Clark, and J. Pitlick. 2005. Seasonal cycle shifts in hydroclimatology over the western US. *J. Clim.* 18:372–384.

Roos, M. 1987. Possible changes in California snowmelt patterns. Pp. 22–31 in *Proc. 4th Pacific Climate Workshop.* Pacific Grove, CA.

———. 1991. A trend of decreasing snowmelt runoff in northern California. Pp. 29–36 in *Proc. 59th Western Snow Conference.* Juneau, AK.

Ropelewski, C. F., and M. S. Halpert. 1986. North American precipitation and temperature patterns associated with the El Niño/Southern Oscillation (ENSO). *Mon. Wea. Rev.* 114:2352–2362.

Serreze, M. C., M. P. Clark, R. L. Armstrong, D. A. McGinnis, R. S. Pulwarty. 1999. Characteristics of the western United States snowpack from snowpack telemetry (SNOTEL) data. *Water Resources Res.* 35:2145–2160.

Sheppard, P. R., A. C. Comrie, G. D. Packin, K. Angersbach, and M. K. Hughes. 2002. The climate of the US Southwest. *Clim. Res.* 21:219–238.

Singh, P. 1991. A temperature lapse rate study in western Himalayas. *Hydrology, Jour. Indian Assoc. of Hydrologists* 14:156–163.

Stewart, I. T., D. R. Cayan, and M. D. Dettinger. 2004. Changes in snowmelt runoff timing in western North America under a "business as usual" climate change scenario. *Clim. Change* 62:217–232.

———. 2005. Changes towards earlier streamflow timing across western North America. *J. Clim.* 18:1136–1155.

Van Rheenen, N. T., A. W. Wood, R. N. Palmer, and D. P. Lettenmaier. 2004. Potential implications of PCM climate change scenarios for Sacramento–San Joaquin river basin hydrology and water resources. *Clim. Change* 62:257–281.

Wahl, K. L. 1992. Evaluation of trends in runoff in the western United States. Pp. 701–710 *in Managing Water Resources During Global Change.* American Water Resources Assoc., Reno, NV.

Washington, W. M., J. W. Weatherly, G. A. Meehl, A. J. Semtner, T. W. Bettge, A. P. Craig, W. G. Strand, J. Arblaster, V. B. Wayland, R. James, and Y. Zhang. 2000. Parallel Climate Model (PCM) control and transient simulations. *Clim. Dyn.* 16:755–774.

Williams, M. W., M. Losleben, N. Caine, and D. Greenland. 1996. Changes in climate and hydrochemical responses in a high-elevation catchment in the Rocky Mountains, U.S.A. *Limnol. Oceanog.* 41:939–946.

Wood, A. W., L. R. Leung, V. Sridhar, and D. P. Lettenmaier. 2004. Hydrologic implications of dynamical and statistical approaches to downscaling climate model outputs. *Clim. Change* 62:189–216.

6

Climate Variability, Climate Change, and Wine Production in the Western United States

Michael A. White, Gregory V. Jones, and Noah S. Diffenbaugh

Abstract

Climate change in the western United States has altered and is likely to further alter the states and fluxes of water, carbon, and nutrients, with consequent impacts for biological and physical systems. In contrast to natural or quasi-natural systems in which many ecosystem constituents are geographically fixed and have limited ability to rapidly adapt to changing climates, agricultural systems are highly adaptable, with primary restrictions being land ownership and competing land use practices. This chapter explores the impacts of climate change on wine grape production in the coastal western United States (Washington, Oregon, and California). A review of prior literature demonstrates that long-term climate cycles such as the Pacific Decadal Oscillation (PDO), especially through interaction with the El Niño/Southern Oscillation (ENSO), have the potential to significantly affect wine grape production. In the late twentieth century, the warm phase of the PDO generally favored high-quality wine grape production, while poor vintages were associated with earlier occurrences of cool PDO coupled with neutral ENSO. Mid-to-late twentieth-century secular climate change in the western United States has tended to improve the thermal regime for most wine-producing regions, but if current trends continue, many of the current premium wine-producing areas, including the Napa and Sonoma valleys, may become too warm by the late twenty-first century. Calculation of a new multivariate index describing climate suitability for wine production and use of a high-resolution regional climate model show that by the late twenty-first century the counties most suitable for wine grape production will shift from northern and central coastal California to the coastal Pacific Northwest. Unfortunately, while the suggestion of a geographical shift may seem to imply merely a relocation of the wine grape

industry, evidence suggests that the counties likely to support a favorable thermal regime in the future climate will also experience a nearly twofold precipitation increase during the critical completion of the growing season and initiation of the ripening season—a factor likely to increase problems with diseases and pests.

Introduction

Worldwide, grapevine cultivation for the production of wine is of tremendous socioeconomic importance for specific countries and regions, yet it is also susceptible to the vagaries of short- and long-term climate variability. In this chapter we demonstrate that cyclic and secular climate variability and changes in the frequency of extreme temperatures are likely to create challenges for viticulture in the western United States, but also new geographic opportunities and potential alterations in wine styles.

Domesticated grapevine cultivation for the production of wine probably began during the Neolithic period, at least 7000 years BP (McGovern 2003). As one of the oldest cultivated plants, grapevines have since developed as an integral part of cultural history throughout much of Western civilization (Penning-Roswell 1989, Unwin 1991, Johnson 1994). Yet the grapevine's importance is belied by the narrow sliver of classic Mediterranean climate in which wines of the highest quality are produced. In contrast to more widely grown crops for which quality variations are not as critical for sensory perception, grapevines are thus at a greater risk from both short- and long-term climate variability and climate change.

Climate exerts both general and varietal-specific controls on wine grape production. In general, the overall wine style that a region produces is a result of the baseline climate, while climate variability determines vintage quality differences. Although localized exceptions are possible, perhaps due to extreme modifications such as deep mulching or use of polyethylene sleeves (Bowen et al. 2004), climate conditions must, to a very large extent, follow the Goldilocks paradigm: not too hot, not too cold. Assessed from general to specific, grapevine cultivation for the production of premium wine grapes occurs in climates characterized by the following three conditions.

First, average air temperature in the growing season ($TAVG_{GS}$, April to October in the Northern Hemisphere) must be between 12 and 22°C for the cultivation of premium *Vitis vinifera* wine grapes. This range may be divided into four separate climate/maturity zones (Jones et al. 2004): cool, 12 to 15°C, including ice wines, Müller-Thurgau, Pinot Gris, and Gewürztraminer; intermediate, 15 to 17°C, including Pinot Noir, Chardonnay, Sauvignon Blanc, and Semillon; warm, 17 to 19°C, including Tempranillo, Dolcetto, Merlot, Malbec, Viognier, Syrah (Shiraz), Cabernet Sauvignon, Sangiovese, Grenache, Carignane, and Zinfandel; and hot, 19 to 22°C, including some lesser-known indigenous varieties, table grapes, and raisins. Considerable overlap exists among groups, but the overall division into climate/maturity groups provides a useful level of aggregation.

Second, the summation of growing-degree days during the growing season (GDD_{GS}, summation of average daily temperatures greater than 10°C) must correspond to one of five Winkler regions (Amerine and Winkler 1944, Winkler et al. 1974): region I, with 1,111 to 1,390 GDD_{GS}; region II, with 1,391 to 1,670 GDD_{GS}; region III, with 1,671 to 1,950 GDD_{GS}; region IV, with 1,951 to 2,220; and region V, with 2,220 to 2,499 GDD_{GS}. While comestible wine may be produced in all regions, light- to medium-bodied dry wines of the best quality are produced in regions I and II, while region III produces full-bodied dry and sweet wines.

Third, daily temperatures should not be too extreme, and night minimum temperatures should not be too cold. Although cultural practices such as complete or partial burial of grapevines to prevent freeze/frost damage are possible, recurrence of extreme cold—defined as −12.2°C in winter (December through February) and −6.7°C in spring/fall (March through May, and September through November)—negatively impact wine grape production (White et al. 2006), often through mortality (Amerine and Winkler 1944, Winkler et al. 1974) and/or bud and wood damage (Watson 1998). Extreme heat, defined as temperatures greater than 35°C in either the growing sea-

son or ripening season (August 15 through October 15), negatively impacts wine grape production through inhibition of photosynthesis (Gladstones 1992) and reduction of color development (Kliewer and Torres 1972) and anthocyanin production (Mori et al. 2005).

Climate changes therefore have the potential to bring about changes in the frequency of short-term extreme temperatures influencing wine grape production, changes in regional wine styles, and spatial shifts in production suitability. Our goals in this chapter are, first, to present evidence linking climate variability with wine quality; second, to review how climate cycles and secular climate change may influence wine grape production in the western United States; and finally, to present a new climate analysis for Washington, Oregon, and California focused on a multifactorial county-specific analysis.

Climate and Wine Quality

The interrelationship between climate and wine quality and prices is especially well known for European wines for which long-term data are available (Ashenfelter et al. 1995, Jones and Storchmann 2001). As outlined above, favorable growing seasons produce higher-quality fruit, which, when appropriately processed, leads to higher ratings and higher market prices. Of the two possible measures of wine quality—price or ratings—vintage ratings are more useful given that long-term price data across regions and styles are rare and difficult to obtain. Vintage ratings, on the other hand, whether categorical or continuous scores, are broadly available from numerous sources (Jones et al. 2005) and may exist for single wines or as aggregations for entire regions and wine styles, such as California red wines.

Although vintage ratings have been criticized for subjectivity and possible "grade inflation" over time, they do, in fact, accurately reflect previously described weather conditions (Ashenfelter and Jones 2000, Jones and Davis 2000). Further, while rating systems differ in scoring techniques, the interannual anomalies of the various systems (e.g., *Sotheby's*, *Wine Enthusiast*, *Wine Spectator*, *Wine Advocate*, and others) are highly correlated ($r > 0.9$), indicating that this subjective measure of quality is a good quantitative representation of a vintage (Jones 1997, Reisman et al. 2003).

Based on the above research linking climate variability and wine quality, we assume for the remainder of this chapter that the two are related. Although case studies are possible for specific varieties, such as Cabernet Sauvignon

(*Wine Spectator* 2004) in Napa Valley, a comprehensive long-term data set for western North America does not exist. The focus of our discussion here, therefore, will be on variations in climate known to influence wine quality, not the relationship between vintage ratings and climate variability.

Cyclic Climate Variability

Climate cycles, particularly the El Niño/Southern Oscillation (ENSO), are an important influence on human and natural systems in western North America. The 1982 to 1983 El Niño, with events such as huge snowfall in the Intermountain West overwhelming Lake Powell Reservoir's holding capacity (White et al. 2005), introduced the term to a large, nonscientific audience. In the western United States, El Niño tends to increase winter precipitation in California while reducing it in the Pacific Northwest (Glantz 2001), but grape growers generally regard El Niño events as uniformly negative. This popular view of El Niño is unsupported: no research has shown a negative effect of ENSO events on wine grape production in the western United States (although such a relationship has been shown in Europe [Rodó and Comín 2000, Esteves and Orgaz 2001]).

The Pacific Decadal Oscillation (PDO), on the other hand, is a climate cycle with a probable influence on wine grape production in the western United States. Most evident in November through March, the PDO is related to variability in Pacific sea surface temperature (SST) (Frauenfeld et al. 2005) over a period of several decades and may be represented as the leading eigenvector of the mean monthly SSTs in the Pacific Ocean north of 20° latitude (Mantua et al. 1997). The PDO is usually considered to be in either a warm phase (above-normal SST for equatorial and west coast North America, and below-normal SST for the central and western North Pacific around 45°) or cool phase (opposite conditions). Although single-year values are possible for the PDO, multidecadal estimates are more often used (Gershunov and Barnett 1998, Gutzler et al. 2002), with warm phases from 1925 to 1946 and 1977 to 1998, and cold phases from 1890 to 1924 and 1947 to 1976 (Mantua and Hare 2002). Indications of a PDO shift to a new warm phase beginning in 1998 are, to date, equivocal.

Although it has pronounced influence on western North American climate, the PDO, in comparison to ENSO, is understudied: searches for "ENSO/El Niño" returned 9,856 articles from Web of Science in fall of 2007

and 1.3 million Google hits, whereas a search for "Pacific Decadal Oscillation" returned only 1,390 articles and 99,000 hits. Researchers have posited several theories as to the nature of the PDO, proposing that it is a result of nonlinear atmospheric teleconnections between tropical and extratropical systems (An et al. 2007), an ENSO signal with superimposed red noise (Gedalof et al. 2002, Newman et al. 2003), or SST anomalies forced by exogenous climate modes of variability, such as the Aleutian low, ENSO, and/or the Kuroshio-Oyashio Extension (Schneider and Cornuelle 2005). Further, and as shown below for viticulture, the use of ENSO as a predictive climate impact metric varies between the cool and warm phase of the PDO (Gershunov and Barnett 1998, McCabe and Dettinger 1999).

Prior research has shown at least two major impacts from the PDO on viticulture in the western United States, and California in particular. First, shifts in the PDO appear to have caused extensive climatic shifts generally beneficial to premium wine production (Nemani et al. 2001). Over the period from 1951 to 1997, which includes the famous 1976/1977 shift in the PDO, Pacific SSTs warmed by 0.72°C/47 years. Especially off the west coast of North America, SSTs are highly correlated with specific humidity (Nemani et al. 2001). Likely as a result of higher specific humidity over the Pacific coupled with an increase in southwesterly flows advecting warm, moist air over coastal land areas (Trenberth and Hurrell 1994), Pacific coastal SST and dewpoint temperatures on land are highly correlated (Nemani et al. 2001).

Probably influenced by these atmospheric teleconnections, climate changed asymmetrically in the Napa and Sonoma valleys (two regions capable of producing exceptional wines): maximum daily temperatures had no significant change, while minimum temperatures increased by 2.06°C/47 years. Similar results were found across all of the wine-growing regions of the western United States (Jones 2005). Increases were strongest since 1976, again suggesting an influence of the PDO. Most shockingly, over the 1951 to 1997 period, frost frequency declined by 20 days/47 years. GDD_{GS} summations also increased by 14 percent, native species spring phenology advanced by 18 days (from 1968 to 1994), and the date at which mature grapes could be harvested advanced by about three weeks. In short, nearly all climate changes over the study period were favorable for wine grape production. Of the suite of changes, the decline in frost was most related to detrended anomalies in north coast California vintage ratings, with an r^2 of 0.38.

Second, Jones et al. (2007) showed that while the PDO was important for Cabernet Sauvignon vintages in Napa Valley (Laube 1989, *Wine Spectator* 2004), an important interaction exists between PDO and ENSO such that especially bad vintages are predictable. Using 1948 to 2002 best-available daily weather records from the U.S. Historical Climatology Network, vintage-relevant climate metrics were not statistically different among ENSO modes (La Niña, El Niño, or neutral).

Unlike ENSO, PDO had statistically significant effects in most western United States wine regions on many wine-relevant climate metrics, including—during cold PDO—low temperatures in all seasons, increased frequency of spring frost, and later occurrence of the final frost during spring. In Napa Valley in particular, cold PDO was similarly associated with increased frequency of cold weather, but also with reduced GDD_{GS} accumulation and increased ripening period (August 15 through October 30) precipitation. The effects of PDO were increased during neutral ENSO (Jones and Goodrich 2007), such that the frost-free growing season declined by about one month, spring frost increased, and GDD_{GS} accumulation declined. Cold-phase PDO during La Niña, on the other hand, affected only minimum temperature.

In terms of vintage quality, Jones and Goodrich (2007) showed that ENSO phase had no statistically significant effect on Napa Cabernet Sauvignon vintage quality. PDO, on the other hand, resulted in a significant ($P < 0.01$) split between good vintages (warm PDO) and poor vintages (cold PDO). Most importantly, though, years with neutral ENSO and cold PDO produced the lowest-rated wines on record (average of 77.7 on a 0 to 100 scale). Of the 11 years of neutral ENSO-cold PDO, the highest rating was only 87, while the other five combinations of PDO-ENSO had highs ranging from 95 to 98.

Clearly, neutral ENSO/cold PDO is unlikely to produce high-quality vintages. In relating climate variability to vintage quality, Jones and Goodrich (2007) found that the leading component was composed of spring frost and cool growing seasons; using five principal components, 37 percent of variability in vintage ratings was explained, similar to earlier findings (Nemani et al. 2001). Thus, especially for California vintages, it is very likely that cyclic climate variations composed primarily of PDO, but with important ENSO interactions, will have a strong effect on wine production and vintage quality variation in the future.

Although not fully accepted, a renewed incidence of high frost frequency and colder temperatures may have been associated with unusually poor vintages in 1998, 2000, and 2002. To investigate the wider evidence of possible PDO-related shifts in climate, we investigated a climatic index of spring arrival using the spring indices (SI) model (Schwartz 2003). SI incorporates data from a network of lilac (*Syringa chinensis*) and honeysuckle (*Lonicera tatarica*, *L. korolkowii*) observation sites at which first leaf, 95 percent leaf, first bloom, full bloom, and end of bloom were recorded (Schwartz 1997).

A step-wise multiple regression model combines the phenology observation with climatic indices to predict an array of phenological stages, such as date when winter chill requirement is met, date of first leaf, and date of first bloom (most related to native species phenology). The main climate indices are accumulation of winter chill (degree days below 7.5°C) and heat accumulation (degree days above −0.6°C), which are further processed into a large number of derivative variables used in model development. Conceptually, the SI model represents the response of temperature-sensitive and well-watered plant species to seasonally integrated changes in temperature (Schwartz et al. 2006). For our purposes, the SI model is a generalized indicator of biological responses to variability in winter and spring temperatures, both of which are relevant for wine grape production.

We used daily maximum and minimum temperatures from the 1 km gridded daily Daymet meteorological database (Thornton et al. 1997) to simulate the SI for Washington, Oregon, and California (WOC). On average, the five years after the proposed PDO shift (1998 to 2003) had first bloom on average 7.6 days later than five years prior to the shift (1993 to 1997)—that is, spring appears to have begun one week later after the possible PDO shift, a simulation finding consistent with the lower temperatures found during cold PDO (Thornton et al. 1997; Fig. C.2).

For almost all of Washington, Oregon, and California, post-shift first bloom was later than pre-shift bloom. In some areas, especially on the west coast of the Pacific Northwest, spring differences approached three weeks. While not conclusive evidence, the simulations of first bloom suggest that even during a time of secular warming trends, climate cycles may introduce significant viticultural variations.

The importance of climate cycles is troubling because forecasting ENSO and PDO is as challenging, or more challenging, than forecasting secular climate change, the topic of the next section. Although understanding of ENSO and PDO continues to grow (Parker et al. 2007), fundamental predictability is still limited and faces "predictability barriers" (Wajsowicz 2007), the nature of which is only beginning to emerge (Mu et al. 2007).

Secular Changes in Average Temperatures

Long-term secular climate changes are occurring both worldwide (IPCC WG I AR4 2007) and in western North America. Western United States climate changes are discussed in more detail in other chapters, but briefly, for the purposes of viticulture, we outline the following critical theme: climate change will present both challenges and opportunities in the western United States, principally through the ability of grape growers in a given region to adapt to climate change by shifting grape varieties.

In a worldwide study of regions in which premium wines are produced, Jones et al. (2005) showed that for a given wine style, an optimal $TAVG_{GS}$ may be calculated. For a given region and wine style, the current climate may be below, near, or above the optimal range, with climate change either likely to increase or decrease a given region's suitability for a particular wine style. Similarly, shifts in GDD_{GS} accumulations may provide an opportunity for adaptation. Unless a region is already producing the wines of the warmest style, climate change may then encourage a shift in wine style rather than abandonment of wine grape growing entirely.

Over the 1948 to 2002 period, GDD_{GS} accumulations increased ($P < 0.01$) for all ten of the major western United States wine regions (Jones and Goodrich 2007). The coldest region, Puget Sound, had a mean of 972 GDD_{GS}, below the threshold for Winkler region I; the warmest region, the Central Valley of California, had a mean of 2618 GDD_{GS}, above the threshold for Winkler region V. For the late twentieth century, seven of ten regions were within the 1,111 to 1,950 GDD_{GS} range for Winkler regions I through III. If reported linear trends in GDD_{GS} are extrapolated, all seven regions will remain in Winkler regions I–III by 2050. However, by 2100, the Columbia Valley in Oregon, the Central Coast of California, and the California North Coast (including Napa and Sonoma valleys) would exceed the upper limit of 1950 GDD_{GS}. These projections are not based on general circulation model forecasts, which tend to indicate an accelerated pace of warming, but merely extrapolate historical 1948 to 2002 trends. A preferable alternative, and the subject of the next section, would be to incorporate regional climate model simulations.

Multivariate Climate Change Assessment

We used a hierarchical assessment focused on the central influential climate factors outlined above: $TAVG_{GS}$, GDD_{GS}, and extreme temperatures. For the baseline climate analyses, we again used the Daymet data set, here from 1980 to 2003. From these data, we calculated $TAVG_{GS}$, GDD_{GS}, the number of days with extreme heat in either the growing season or ripening season, and the number of days with extreme cold in either spring/fall or winter. For the future reference climate, we used 25 km simulations from the Abdus Salam Institute for Theoretical Physics regional climate model (RegCM3) (Giorgi et al. 1993a, Giorgi et al. 1993b, Pal et al. 2000), originally completed for Diffenbaugh et al. (2005) and White et al. (2006).

RegCM3 was run for two periods: the 1961 to 1989 reference climate (RF) and the 2071 to 2099 future climate under the A2 greenhouse gas emission scenario (A2) (IPCC WG I 2000). Boundary conditions were supplied by the National Aeronautics and Space Administration finite volume element model as described by Coppola and Giorgi (2005). We calculated $TAVG_{GS}$, GDD_{GS}, and days with extreme heat and cold for both RF and A2 integrations and then calculated the A2 minus RF difference, representing the change in climate metrics. We then reprojected and resampled the 25 km RegCM data to the 1 km Daymet grid and uniformly added the A2 minus RF differences to the daily Daymet record, producing a 24-year daily record representing the late twenty-first century. We term this climate to be the ΔDaymet climate. We conducted a three-part analysis: (1) wine maturity group based on $TAVG_{GS}$, (2) Winkler regions based on GDD_{GS}, and (3) a county-specific ranking system based on $TAVG_{GS}$, GDD_{GS} accumulation, extreme temperatures, and precipitation during the late growing and early ripening seasons.

Wine maturity groups in the cool category dominated the late-twentieth-century Daymet climate. In WOC (Washington, Oregon, and California) and Washington and Oregon separately, cool regions contained at least twice as much area as the next most common maturity group, intermediate (Fig. 6.1). For Oregon and Washington, warm groups were rare, and hot groups nonexistent.

California had a different pattern, with more equal coverage across maturity groups (Fig. 6.1), and the Central Valley accounting for most area in the hot group. In the ΔDaymet climate, the most consistent finding was a reduction in area for cool groups, which occurred in WOC as a whole and in all three states (Fig. 6.1). For the entire WOC, along with Oregon and California individually, the area in intermediate, warm, and hot groups increased although the relative magnitude varied. In California, there was little change in intermediate or warm groups, but the hot group declined slightly as a result of pixels becoming too hot.

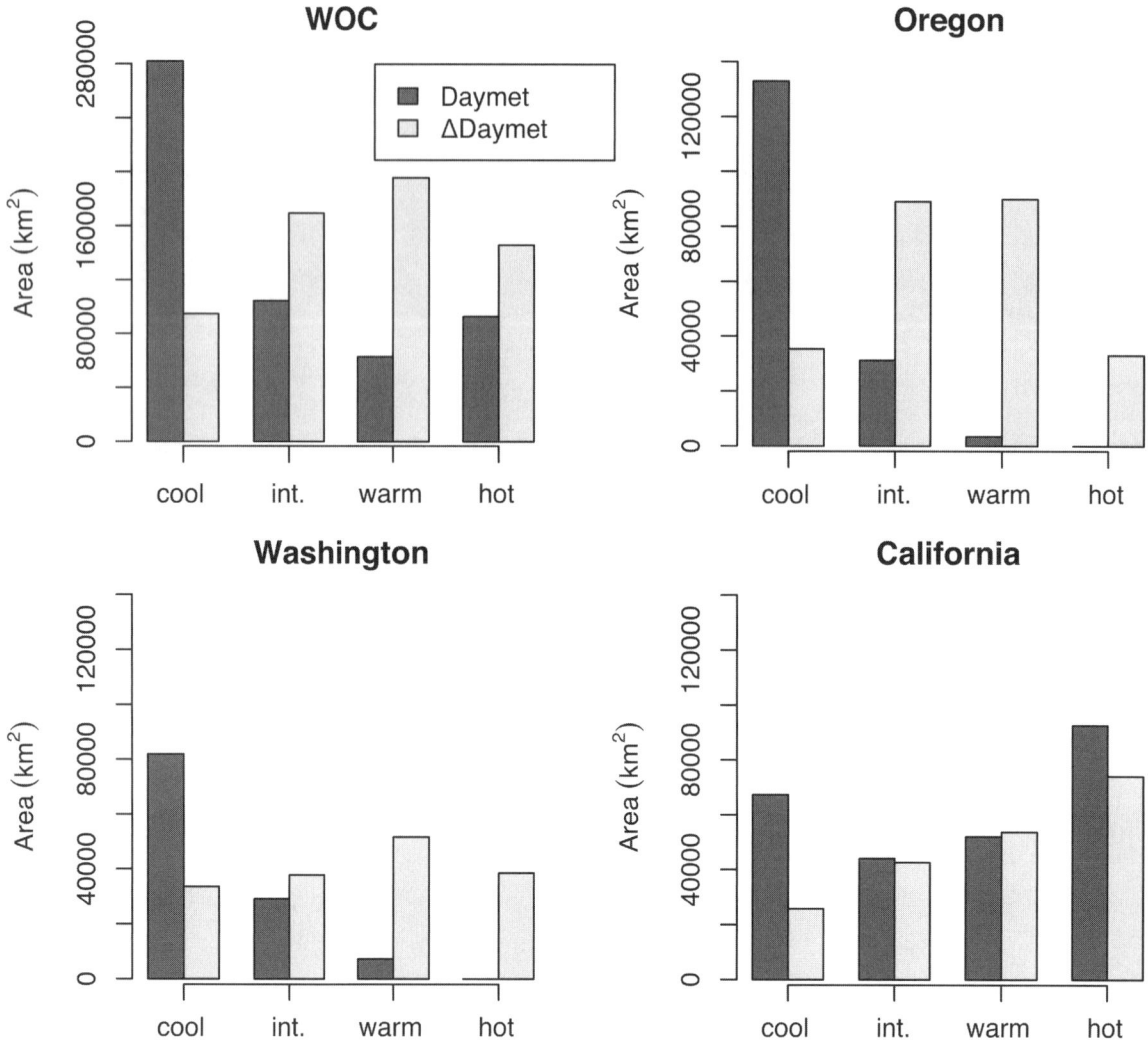

Figure 6.1. Area covered by cool, intermediate (int.), warm, and hot wine maturity groups for Washington, Oregon, and California (WOC) and individual states for the Daymet (late twentieth century) and ΔDaymet (late twenty-first century) climates. Maturity groups correspond to varieties capable of being ripened to high quality within a certain range of growing season (April through October) average temperature. See text for details.

Expressed as a percentage of the total area in WOC or of the individual states, differences showed a similar pattern but highlighted an overall improvement in conditions for Oregon and Washington, and a decline for California (Table 6.1). In the Daymet climate, WOC and the three states all had between 63 and 68 percent of total area in one of the four maturity groups. For Oregon and Washington, about 50 percent of area existed in the cool group. In the ΔDaymet climate, Oregon and Washington both increased their total area to over 90 percent, while less than half of California was suitable. Although these estimates seem to imply that a strikingly large percentage of total area exists in the western United States in one

of the four maturity groupings in both the Daymet and ΔDaymet climates, we stress that the full range of the four groups will tend to exaggerate the ability of a given region for the production of wines of the highest quality, as the extreme range of cold (12°C) and hot (22°C) are unlikely to consistently produce stellar vintages.

The GDD_{GS} analysis using Winkler region criteria suggests a general increase in wine-growing area for Washington and especially Oregon, with little overall change for California (Table 6.2). In the ΔDaymet climate, both Oregon and Washington are projected to experience increases in area in Winkler regions I to III, the regions most capable of supporting the production of high-quality

Table 6.1 Percentage of the coastal western United States (Washington, Oregon, and California [WOC]) and individual states in cold, intermediate, warm, and hot grape maturity groups

	Daymet Climate					ΔDaymet Climate				
	Cold	Int.	Warm	Hot	Total	Cold	Int.	Warm	Hot	Total
WOC	34	13	8	11	66	11	20	23	17	71
Oregon	53	12	1	0	66	14	35	36	13	98
Washington	47	17	4	0	68	19	22	30	22	93
California	16	11	13	23	63	6	10	13	18	47

Note: "Daymet Climate" represents the 1980 to 2003 record; "ΔDaymet Climate" represents the 2071 to 2099 climate simulated by the RegCM3 integration of the IPCC A2 GHG scenario.

Table 6.2 Percentage of the coastal western United States (Washington, Oregon, and California [WOC]) and individual states in Winkler regions I through V

	Daymet Climate						ΔDaymet Climate					
	I	II	III	IV	V	Total	I	II	III	IV	V	Total
WOC	12	7	5	4	6	34	16	20	15	9	6	66
Oregon	14	4	0	0	0	18	27	33	24	7	3	94
Washington	15	10	1	0	0	26	19	20	18	12	7	76
California	8	8	9	7	13	45	8	11	9	8	8	44

Note: "Daymet Climate" represents the 1980 to 2003 record; "ΔDaymet Climate" represents the 2071 to 2099 climate simulated by the RegCM3 integration of the IPCC A2 GHG scenario.

wines. In Oregon, expressed as a percentage of total area, region I increased from 14 to 27 percent, region II from 4 to 33 percent, and region III from 0 to 24 percent. Thus, while the analysis of maturity groups (Table 6.1) suggests a future shift away from cool-climate wines, the Winkler analysis indicates an increase in area for the coolest region I and II wines, particularly for the Pacific Northwest, thus highlighting contrasting results depending on the use of mean seasonal climate versus thermal accumulation.

Both the maturity group and Winkler region analysis, though, ignore the important and potentially negative influence of extreme heat or cold. To integrate $TAVG_{GS}$, GDD_{GS}, extreme heat, and extreme cold, we next calculated a simple index of climate quality for wine production with a mathematical range of -48 to 48 using the following four climate factors, each of which we scored as 1 (favorable) or -1 (unfavorable). The first factor considers $TAVG_{GS}$, the control of maturity groups:

$$TAVG_{GS} \text{ factor} = \begin{cases} 1 \text{ if } TAVG_{GS} > 12°C \\ \quad \text{and } TAVG_{GS} < 22°C \\ -1 \text{ if } TAVG_{GS} < 12°C \\ \quad \text{or } TAVG_{GS} < 22°C \end{cases} \quad (1)$$

The second factor considers GDD_{GS}, the control of Winkler regions:

$$GDD_{GS} \text{ factor} = \begin{cases} 1 \text{ if } GDD_{GS} > 1,111 \\ \quad \text{and } GDD_{GS} < 2,499 \\ -1 \text{ if } GDD_{GS} < 1,111 \\ \quad \text{or } GDD_{GS} < 2,499 \end{cases} \quad (2)$$

The third factor considers the presence of extreme heat, with consequent reduction of photosynthesis and degradation of acids, aromatic, and flavor compounds:

$$\text{heat factor} = \begin{cases} 1 \text{ if hot days} < 14 \\ -1 \text{ if hot days} \geq 14 \end{cases} \quad (3)$$

where a hot day is considered to be a day with maximum temperature exceeding 35°C. The final factor considers extreme cold, with consequent risk of bud, vine, or grape damage:

$$\text{cold factor} = \begin{cases} 1 \text{ if cold days} < 14 \\ -1 \text{ if cold days} \geq 14 \end{cases} \quad (4)$$

A cold day is considered to have a minimum temperature less than $-12.2°C$ in winter (December through February) and $-6.7°C$ in spring/fall (March through May and September through November). Thresholds are based on earlier work (White et al. 2006); the use of two weeks is not absolute but is based on a least restrictive consideration of extreme temperatures (White et al. 2006). The climate index is then:

$$\text{index} = \sum_{1980}^{2003} \begin{array}{l} \text{TAVG}_{GS} \text{ factor} + \text{GDD}_{GS} \text{ factor} \\ + \text{heat factor} + \text{cold factor} \end{array} \quad (5)$$

such that for any year the range is −4 to 4; summed over the 24-year record, the range is −48 for the worst possible wine climate to 48 for the best possible climate. The index thus jointly considers the influence of mean climate, thermal accumulation, extremes, and year-to-year climate variability. We calculated the climate index for all 1 km pixels in Washington, Oregon, and California in the Daymet and ΔDaymet climates.

As an alternative to presenting absolute scores, which are not translatable into a biophysically relevant value, we ranked the 133 counties in Washington, Oregon, and California based on the mean climate index from the upper quartile of the pixels within each county. We used the quartile approach because within any given county, a gradient of climates exists such that the climate index may have a low mean, while a smaller portion may be exceptionally well suited for wine grape production; we assumed that the top quartile was more representative of wine grape potential than the countywide mean.

In the Daymet climate, the highest-ranked counties existed along the northern and central California coast, especially in a cluster around the San Francisco Bay area. Areas of highly ranked counties also existed in southeastern Washington, the intermountain valleys of western Oregon, and counties along the foothills of the Sierra Nevada Mountains. Southeastern Oregon and the Central Valley and Sonoran Desert of California were the most poorly ranked regions.

While these Daymet climate index values are consistent with current areas of high-quality wine grape production, it is important to note that the rankings represent the upper quartile for a county and may thus overestimate or underestimate the distribution of wine grape potential. Napa County of California, for example, is ranked only the twenty-sixth best in the Daymet climate, when in reality Napa Valley consistently produces some of the highest-priced wines in the United States. This is due to the spatial representation of Napa Valley within Napa County: the valley itself is a relatively small area of the total county, and the valley pixels ranged from 46 to 48 in the climate index. The ranking map therefore should be used as an ap-

proximate, not absolute, indicator, with the changes from the Daymet to ΔDaymet climate being perhaps more important than the precise ranking in either climate.

In the ΔDaymet climate, the spatial distribution of county rankings shifted drastically, with a movement of highly ranked counties away from California and towards the coastal Pacific Northwest (Fig. C.3). The Central Valley and Sonoran Desert remained poor candidates for premium wine grape production, but most of coastal California declined from top-ten rankings to rankings typically from 40 to 60. Highly ranked counties existed from far northern California to the Canadian border. Counties highly ranked in eastern Washington disappeared in the ΔDaymet climate. Overall, of the top 20 counties in the Daymet climate, only two (Santa Cruz and San Mateo) remained in the ΔDaymet climate (Table 6.3). In the Daymet climate, 17 of the top 20 counties were in California, while in the ΔDaymet climate, 17 of the top 20 were in Oregon or Washington (Table 6.3).

The changes in wine grape–relevant climate suggested by Figures 6.1 and C.3 and Tables 6.1–6.3 may seem to suggest that climate change in the western United States should bring a shift in wine variety and the geography of wine grape production: in essence, while adaptation will be necessary, the overall industry will survive and indeed thrive. However, the geographic shift for wine production has been hypothesized to increase viticultural risks associated with excess moisture (White et al. 2006). To test this possibility, we calculated the county-average August through October average monthly precipitation (spanning the late growing season and most of the ripening season). As precipitation forecasts are far less certain than temperature forecasts, we used the same Daymet records for county ranking in the Daymet and ΔDaymet climates.

We found that the highly ranked counties in the current climate tended to have about 2 cm of precipitation per month in the Daymet climate, but in the ΔDaymet climate, the highly ranked counties experienced about 8 cm per month (Fig. 6.2). Consequently, the ΔDaymet minus Daymet differences for the top 40 counties tended to be about 5 cm per month: the best counties in the future, from a temperature point of view, are thus projected to be dramatically wetter during the critical late growing and early ripening seasons.

During growing and ripening seasons, excess moisture will tend to increase disease pressure (from mildew and rot), cause fruit flavor and composition dilution by water uptake, and lower the overall quality potential. If simulations of increased precipitation at higher latitudes are ac-

Table 6.3 The 20 counties with the highest climate index values in the Daymet and ΔDaymet climates

Rank	Daymet Climate County	State	ΔDaymet Climate County	State
1	Santa Cruz	California	Pacific	Washington
2	Santa Clara	California	Mason	Washington
3	Santa Barbara	California	Humboldt	California
4	San Francisco	California	Lewis	Washington
5	Marin	California	Santa Cruz	California
6	San Luis Obispo	California	Kitsap	Washington
7	Monterey	California	King	Washington
8	San Mateo	California	Island	Washington
9	Sonoma	California	Grays Harbor	Washington
10	Alameda	California	Lincoln	Oregon
11	San Benito	California	Polk	Oregon
12	Contra Costa	California	San Mateo	California
13	Marion	Oregon	Tillamook	Oregon
14	Multnomah	Oregon	Benton	Oregon
15	Mariposa	California	Wahkiakum	Washington
16	Washington	Oregon	Clatsop	Oregon
17	Ventura	California	Columbia	Oregon
18	Nevada	California	Snohomish	Washington
19	El Dorado	California	Coos	Oregon
20	Calaveras	California	Skagit	Washington

Note: Higher rank indicates a thermal climate more favorable for wine grape production.

curate (IPCC WG I AR4 2007), our results would tend to underestimate future shifts in the hydrologic regime, indicating a strong potential for an even more challenging future viticultural environment by the late twenty-first century.

Conclusions

The growing of wine grapes for producing premium wines in the western United States is influenced by climate in two central ways. First, climate determines where good-quality grapes may be grown and the resulting wine styles produced. Most wine grape–producing regions in the western United States experienced trends towards more favorable climates in the mid-to-late twentieth century. Our results suggest that by the late twenty-first century, favorable thermal climates for wine grape production will shift from coastal California to the coastal Pacific North-

west—a region our results show is likely to experience significant challenges associated with excess precipitation in the growing and ripening seasons.

Second, year-to-year climate variability at a given location seems to account for about 35 to 40 percent of variability in vintage ratings. Our reviews and research suggest that the largely unpredictable cycles associated with ENSO and PDO will continue to exert long- and short-term controls over vintage quality, and some evidence exists to support a contention of a shift to a cool PDO around the late twentieth century.

In summary, while the wine industry in the western United States is unlikely to be eliminated or even seriously degraded, significant geographic adaptation and continued research to increase viticultural and wine-making adaptive capacity, and to develop heat-tolerant and moisture- and disease-resistant varieties, will be required.

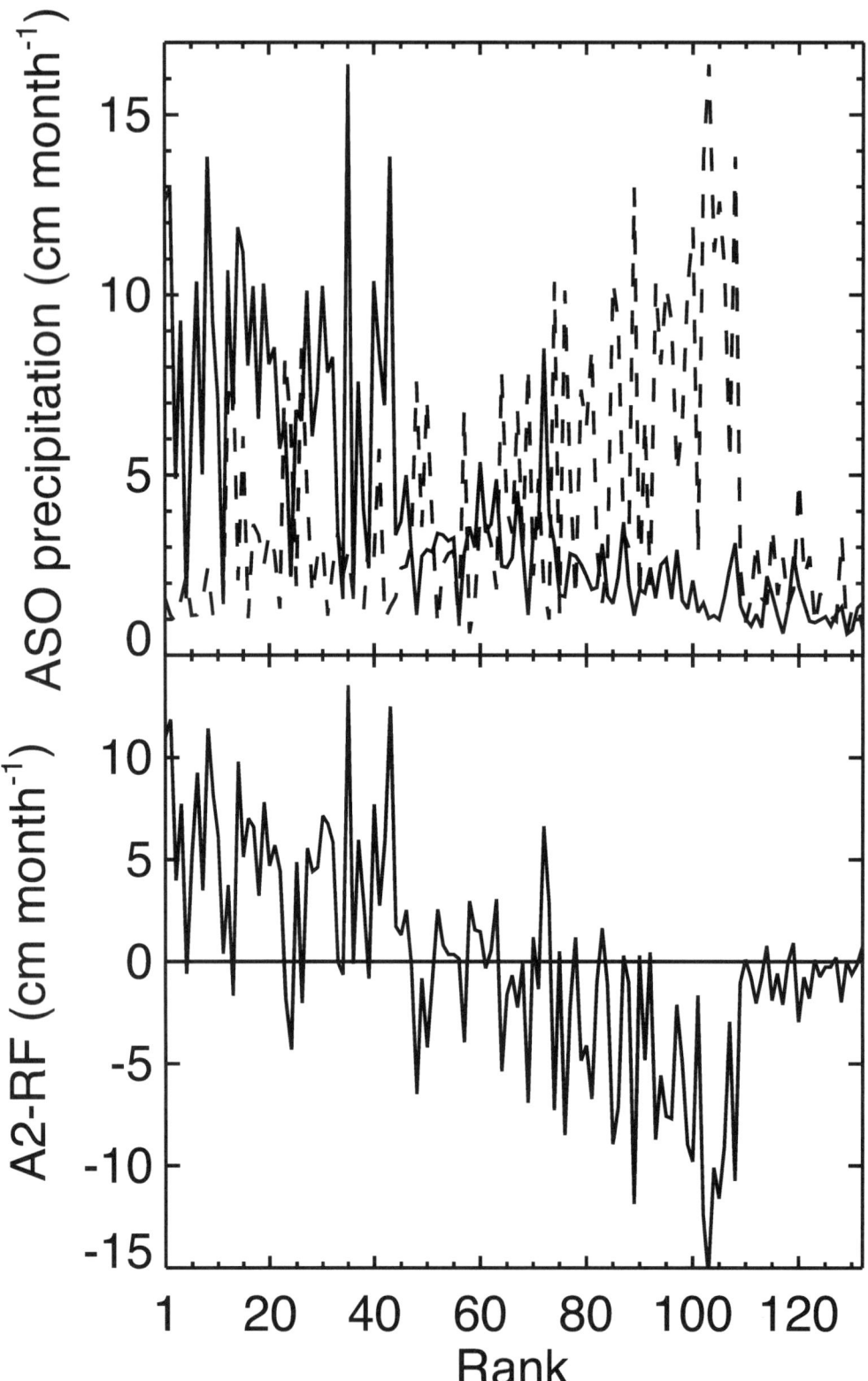

Figure 6.2. Mean monthly precipitation for August, September, and October (ASO) for all counties in Washington, Oregon, and California for the Daymet (late twentieth century) and ΔDaymet (late twenty-first century) climates. Both panels show the counties sorted left to right from best (1) to worst (133), based on the climate index (a thermal assessment). Top panel shows absolute values for Daymet (dashed line) and ΔDaymet (solid line) climates. Bottom panel shows the difference between the ΔDaymet and Daymet climates: the best-quality counties, from a thermal regime approach, are projected to be far wetter in the future climate.

Abbreviations

A2	RegCM3 integration of the IPCC A2 greenhouse gas emission scenario
ENSO	El Niño/Southern Oscillation
GDD_{GS}	Base 10°C growing degree summation during the growing season (April through October)
IPCC	Intergovernmental Panel on Climate Change
PDO	Pacific Decadal Oscillation
RegCM3	Abdus Salam Institute for Theoretical Physics regional climate model
RF	RegCM3 integration of the reference 1961 to 1989 climate
SI	Spring index (lilac and honeysuckle phenology model)
SST	Sea surface temperature
$TAVG_{GS}$	Average temperature during the growing season (April through October)
WOC	Washington, Oregon, and California

References

Amerine, M. A. and A. J. Winkler. 1944. Composition and quality of musts and wines of California grapes. *Hilgardia* 15:493–675.

An, S. I., J. S. Kug, A. Timmermann, I. S. Kang, and O. Timm. 2007. The influence of ENSO on the generation of decadal variability in the North Pacific. *J. Clim.* 20:667–680.

Ashenfelter, O., D. Ashmore, and R. Lalonde. 1995. Bordeaux wine quality and the weather. *Chance* 8:7–19.

Ashenfelter, O., and G. V. Jones. 2000. The demand for expert opinion: Bordeaux wine. In VDQS Annual Meeting, d'Ajaccio, Corsica, France. October 1998 Observatoire des Conjonctures Vinicoles Europeenes Cahiers Scientifique, Faculte des Sciences Economiques. Espace Richter, France.

Bowen, P. A., C. R. Bogdanoff, and B. Estergaard. 2004. Impacts of using polyethylene sleeves and wavelength-selective mulch in vineyards. I. Effects on air and soil temperatures and degree day accumulation. *Can. J. Plant Sci.* 84:545–553.

Coppola, E., and F. Giorgi. 2005. Climate change in tropical regions from high-resolution time-slice AGCM experiments. *Quart. J. Royal Meteorol. Soc.* 131:3123–3145.

Diffenbaugh, N. S., J. S. Pal, R. J. Trapp, and F. Giorgi. 2005. Fine-scale processes regulate the response of extreme events to global climate change. *Proc. Natl. Acad. Sci.* 102:15774–15778.

Esteves, M. A. and M. D. M. Orgaz. 2001. The influence of climatic variability on the quality of wine. *Int. J. Biometeorol.* 45:13–21.

Frauenfeld, O. W., R. E. Davis, and M. E. Mann. 2005. A distinctly interdecadal signal of Pacific Ocean-atmosphere interaction. *J. Clim.* 18:1709–1718.

Gedalof, Z., N. J. Mantua, and D.L. Peterson. 2002. A multi-century perspective of variability in the Pacific Decadal Oscillation: New insights from tree rings and coral. *Geophys. Res. Lett.* 29: Artn 2204.

Gershunov, A., and T. P. Barnett. 1998. Interdecadal modulation of ENSO teleconnections. *Bull. Amer. Meteorol. Soc.* 79:2715–2725.

Giorgi, F., M. R. Marinucci, and G. T. Bates. 1993. Development of a 2nd-generation regional climate model (Regcm2). 1. Boundary-Layer and radiative-transfer processes. *Mon. Weather Rev.* 121:2794–2813.

Giorgi, F., M. R. Marinucci, G. T. Bates, and G. Decanio. 1993. Development of a 2nd-generation regional climate model (Regcm2). 2. Convective processes and assimilation of lateral boundary-conditions. *Mon. Weather Rev.* 121:2814–2832.

Gladstones, J. 1992. *Viticulture and Environment*. Winetitles, Adelaide, Australia.

Glantz, M. 2001. *Currents of Change: Impacts of El Niño and La Niña on Climate and Society*. Cambridge Univ. Press, Cambridge, UK.

Gutzler, D. S., D. M. Kann, and C. Thornburgh. 2002. Modulation of ENSO-based long-lead outlooks of southwestern US winter precipitation by the Pacific decadal oscillation. *Weather Forecast* 17:1163–1172.

IPCC WG I. 2000. *Special Report on Emissions Scenarios*. Cambridge Univ. Press, Cambridge, UK.

IPCC WG I AR4. 2007. *Climate Change 2007: The Physical Science Basis*. Cambridge Univ. Press, Cambridge, UK.

Johnson, H. 1994. *The World Atlas of Wines*. 4th ed. Simon and Schuster, New York.

Jones, G. 1997. A synoptic climatological assessment of viticultural phenology. Master's thesis. Univ. of Virginia, Charlottesville.

———. 2005. Climate change in the western United States grape growing regions. *Acta Horticult.* 689:41–60.

Jones, G. V., and R. E. Davis. 2000. Climate influences on grapevine phenology, grape composition, and wine production and quality for Bordeaux, France. *Am. J. Enol. Viticult.* 51:249–261.

Jones, G. V. and G. B. Goodrich. 2007. Influence of climate variability on the western U.S. wine regions and wine quality in the Napa Valley. *Clim. Res.* In press.

Jones, G. V., N. Snead, and P. Nelson. 2004. Geology and wine. 8. Modeling viticultural landscapes: A GIS analysis of the terroir potential in the Umpqua Valley of Oregon. *Geosci. Canada* 31:167–178.

Jones, G. V., and K. H. Storchmann. 2001. Wine market prices and investment under uncertainty: An econometric model for Bordeaux Crus Classes. *Agr. Econ.* 26:115–133.

Jones, G. V., M. A. White, O. R. Cooper, and K. Storchmann. 2005. Climate change and global wine quality. *Clim. Change* 73:319–343.

Kliewer, W. M., and R. E. Torres. 1972. Effect of controlled day and night temperatures on grape coloration. *Am. J. Enol. Viticult.* 23:71–77.

Laube, J. 1989. *California's Great Cabernets: The Wine Spectator's Ultimate Guide for Consumers, Collectors, and Investors*. Wine Spectator Press, New York.

Mantua, N. J., and S. R. Hare. 2002. The Pacific Decadal Oscillation. *J. Oceanog.* 58:35–44.

Mantua, N. J., S. R. Hare, Y. Zhang, J. M. Wallace, and R. C. Francis. 1997. A Pacific interdecadal climate oscillation with impacts on salmon production. *Bull. Amer. Meteorol. Soc.* 78:1069–1079.

McCabe, G. J., and M. D. Dettinger. 1999. Decadal variations in the strength of ENSO teleconnections with precipitation in the western United States. *Int. J. Climatol.* 19:1399–1410.

McGovern, P. E. 2003. *Ancient Wine: The Search for the Origins of Viniculture*. Princeton Univ. Press, Princeton, NJ.

Mori, K., S. Sugaya, and H. Gemma. 2005. Decreased anthocyanin biosynthesis in grape berries grown under elevated night temperature condition. *Scientia Horticult.* 105:319–330.

Mu, M., W. S. Duan, and B. Wang. 2007. Season-dependent dynamics of nonlinear optimal error growth and El Nino–Southern Oscillation predictability in a theoretical model. *J. Geophys. Res.-Atmos.* 112.

Nemani, R. R., M. A. White, D. R. Cayan, G. V. Jones, S. W. Running, and J. C. Coughlan. 2001. Asymmetric warming along coastal California and its impacts on the California premium wine industry. *Clim. Res.* 19:25–34.

Newman, M., G.P. Compo, and M. A. Alexander. 2003. ENSO-forced variability of the Pacific Decadal Oscillation. *J. Clim.* 16:3853–3857.

Pal, J. S., E. E. Small, and E. A. B. Eltahir. 2000. Simulation of regional-scale water and energy budgets: Representation of subgrid cloud and precipitation processes within RegCM. *J. Geophys. Res.-Atmos.* 105:29579–29594.

Parker, D., C. Folland, A. Scaife, J. Knight, A. Colman, P. Baines, and B. W. Dong. 2007. Decadal to multidecadal variability and the climate change background. *J. Geophys. Res.-Atmos.* 112.

Penning-Roswell, E. 1989. *Wines of Bordeaux*. 6th ed. Penguin Books, London and New York.

Reisman, C., M. Matthews, and D. Block. 2003. Analysis of the effects of weather of California cabernet sauvignon vintage quality. In ASEV 53rd Annual Meeting, Portland, Oregon. *Am. J. Enol. Viticult.* (sponsor).

Rodó, X., and F. A. Comín. 2000. Links between large-scale anomalies, rainfall and wine quality in the Iberian Peninsula during the last three decades. *Global Ch. Biol.* 6:267–273.

Schneider, N., and B. D. Cornuelle. 2005. The forcing of the Pacific decadal oscillation. *J. Clim.* 18:4355–4373.

Schwartz, M. D. 1997. Spring index models: An approach to connecting satellite and surface phenology. In H. Leith and M. D. Schwartz (eds.), *Phenology of Seasonal Climates*. Backhuys, Netherlands.

———. 2003. Phenoclimatic Measures. Pp. 331–343 in M. Schwartz (ed.), *Phenology: An Integrative Environmental Science*. Kluwer Academic Pub., New York.

Schwartz, M. D., R. Ahas, and A. Aasa. 2006. Onset of spring starting earlier across the Northern Hemisphere. *Global Ch. Biol.* 12:343–351.

Thornton, P. E., S. W. Running, and M. A. White. 1997. Generating surfaces of daily meteorological variables over large regions of complex terrain. *J. Hydrol.* 190:214–251.

Trenberth, K. E., and J. W. Hurrell. 1994. Decadal atmosphere-ocean variations in the Pacific. *Clim. Dyn.* 9:303–319.

Unwin, T. 1991. *Wine and the Vine: An Historical Geography of Viticulture and the Wine Trade*. Routledge, London and New York.

Wajsowicz, R. C. 2007. Seasonal-to-interannual forecasting of tropical Indian Ocean sea surface temperature anomalies: Potential predictability and barriers. *J. Clim.* 20:3320–3343.

Watson, J. 1998. Washington Viticulture: The Basics. In J. Watson (ed.), *Growing Grapes in Eastern Washington*. Washington State Univ., Pullman.

White, M. A., J. C. Schmidt, and D. J. Topping. 2005. Application of wavelet analysis for monitoring the hydrologic effects of dam operation: Glen Canyon Dam and the Colorado River at Lees Ferry, Arizona. *River Res. Appl.* 21:551–565.

White, M. A., N. S. Diffenbaugh, G. V. Jones, J. S. Pal, and F. Giorgi. 2006. Extreme heat reduces and shifts United States premium wine production in the twenty-first century. *Proc. Natl. Acad. Sci.* 103:11217–11222.

Wine Spectator. 2004. *Wine Spectator's Ultimate Buying Guide*. 8th ed. Running Press, Philadelphia.

Winkler, A., J. Cook, W. Kliewere, and L. Lider. 1974. *General Viticulture*. 4th ed. Univ. of Cal. Press, Berkeley.

7

Credibility and Confidence

Assessing Climate-Change Impacts
on Biological Systems in North America

John H. Matthews and Camille Parmesan

Abstract

Three distinct approaches to climate-change impacts research (CCIR) on ecological systems have emerged: mechanistic, nonmechanistic correlational, and meta-analytic. While mechanistic studies largely pursue a deductive reasoning and experimental hypothesis-testing paradigm (strong evidence), the latter two types normally rely on inductive reasoning and observational data (weak evidence), especially when exploring indirect effects of climate shifts. Nonmechanistic correlational studies are the most common type of CCIR and will likely remain so for the future, but they face skepticism about the veracity of their conclusions and methods. Conservation ecologists exploring climate-change impacts via nonmechanistic correlational studies can often increase the strength of weak evidence through comparative methodologies such as sign-switching observations and explicit temporal and spatial comparisons, which will increase their studies' applicability and usefulness for resource managers and policy makers. We present evidence from one example each of range shifts, phenological shifts, and community-abundance shifts that have been inferred to result from climate change.

Introduction

Research on the biological impacts of anthropogenic climate change have lagged behind studies of abiotic effects. Many habitat categories and taxa have received scant attention from biologists searching for the fingerprints of climate shifts, and few synthetic studies have been conducted. These delays reflect the youth of this topic, which only began to coalesce by the mid-1990s (reviewed in Parmesan 2006). The deficits of climate-change impact research (CCIR) are starkly clear when comparing these studies with other conservation ecology areas such as biological impacts from land-use change, habitat fragmentation, and industrial pollution. Although climate science suggests that ecological shifts seen over the past century are relatively minor compared to those projected to occur by 2100, numerous studies already document significant species responses to an array of altered climate variables. By implication, coming decades should see more species affected, as well as more dramatically affected species.

As a result, research on biological impacts is expanding from population- and species-level effects to multitrophic, community, and ecosystem impacts. Ideally, research will also move from the study of realized organismal impacts to realistic projections for impacts given existing climate trends. Policy makers and resource managers will also need to shift to a broader scale to create and implement climate-aware policy and management plans, which are largely absent from the long-term strategies of current governmental and nongovernmental organizations. Both changes will depend on the emergence of a more mindful and climate-aware ecological epistemology.

The outlines of this epistemology are now emerging. This review is meant to describe these outlines for biologists who are conducting CCIR, as well as for consumers of climate-impacts research, particularly policy makers and resource managers. These thoughts are also intended to guide those who must explain climate-impact science to a public that may not intuitively appreciate its methods or conclusions, and to the climate scientists producing the abiotic data driving this area of ecological research. Because this volume grows out of research into climate impacts on the western United States, the examples used will be taken from that region.

This chapter focuses on two topics. We begin by briefly describing the kinds of ecological research that typify CCIR and the major categories of impacts that have been observed to date. Few of these are unique to climate-impacts research, but the pressing need for sound science and clear trends demands that conservation ecologists conducting climate-aware research embrace approaches that have not yet been widely absorbed in the discipline. Second, we briefly provide some considerations for those designing, funding, and reviewing CCIR.

The Scope of Climate-Change Impacts Research: Constraints and Opportunities

Conservation biology as a discipline relies heavily on the history, methodology, and approaches of academic ecology and, at least in the United States, government funding; funding of nongovernmental organizations (NGOs) is much less important than state and federal research money in North America. Given that conservation research often operates with tight schedules and restricted budgets, individual projects are often narrow and limited in wider applicability by necessity. Moreover, addressing larger questions also entails challenges built into the funding infrastructure of research. Long-term and large-scale ecological studies, which would be the most useful types of projects for policy makers and resource managers, are very rare. As Pimm (1991:1) has written more generally about community ecology,

> Look at any current ecology journal and you will see that our studies are very brief: ten years is a long-term study. In the United States, such a study would require a minimum of four consecutive National Science Foundation grants, and such continued support is enjoyed only by a very few.... So where is the ecology of tens to hundreds of species over decades to centuries, across hectares to thousands of square kilometers? Should we care that this region of ecological research is rather sparsely populated...? The most pressing ecological problems involve many species and their fate across decades to centuries, over large geographical areas.

Ten years is a tiny snapshot of ecological change, particularly given that such a period could hardly register as a meaningful climate "unit." As a result, most time series mine data from previous research—work often designed to study issues predating concern for anthropogenic climate change—or from proxies never designed for research purposes, such as trapping records. Likewise, basic life history data are often lacking. For most taxa, range maps are not sufficiently documented or spatially resolved to track historically significant changes in abundance or extent. Information on range limits from previous decades or centuries is even more scarce (Delcourt and Delcourt 1987).

Europe and North America have a larger body of data to draw on for historical and spatial analyses, but for many regions of the world, especially marine and freshwater systems and most tropical habitats, the science of impact research faces severe challenges. Notable institutional responses include the National Phenological Network in the United States and efforts by the U.S. National Science Foundation to fund and maintain long-term ecological research and the National Ecological Observatory Network. However, these and other efforts are largely building, young institutions, with benefits to be paid in the future. As a result of the dearth of good temporal and spatial data, ecologists must often rely on indirect and inferential measures of changes through time to provide a more complete picture of climate-change impacts.

Classifying Climate-Change Impacts Research by Approach and Impact

Given these restrictions, an impressive number of researchers have studied the effects of climate shifts since the mid-1990s, most of which have been published since 2002 (Parmesan 2006). These fall into three basic approaches.

Mechanistic studies are arguably the most persuasive means of linking climate change with shifts in behavior. Even when they include inferential and inductive analyses, these results are tested via positive experiments to implicate particular drivers of change, such as mismatches in temperature tolerances or phenologies. Mechanistic studies answer *why* change is occurring in a population or community, usually through direct physiological effects. Through the use of deductive reasoning and classical hypothesis testing, such research is firmly in the tradition of experimental science that produces "strong evidence," even though some replicates may be a result of "natural experiments" (such as opportunistic factorial studies resulting from variation between habitat patches).

Parmesan (1996), for instance, used records gathered by butterfly collectors from western North American over the twentieth century to trace changes in the extent of *Euphydryas editha* populations. She found that these populations had moved higher in altitude and farther north, and the extent of the mean movement corresponded to a 0.7°C warming described by climate scientists over the same period (Karl et al. 1996). She hypothesized that the basis for the range shift was due to a phenological mismatch between *E. editha*'s host plant species and the developmental rate of the caterpillars. Experimental work in the field confirmed that a 2°C rise caused the *E. editha* larvae to starve.

This study was unusual because *E. editha* is a relatively well known species, with an extensive literature describing the relationship between the species and temperature (e.g., Parmesan 2003) and because amateur collectors had sampled populations for more than a century over the whole species' range and were willing to share their data. Very few species meet any of these requirements. Although mechanistic projects are powerful case studies of how climate shifts influence biological systems, they are likely to remain relatively rare in number and limited to taxa whose behavior has been an object of research (such as model organisms; e.g., Rodriguez and Rodriguez-Trelles 1998) or long-standing interest (as with economically important plants; e.g., Rodó and Comín 2000).

Butterflies and birds also overpopulate such work, as well as North American and European species more generally. These taxa and regions have been a reliable source of natural history records on scales ranging from decades to centuries. Groups such as coral or woody plants that produce tissues that record climate variation are also well represented (Masson-Delmotte et al. 2003). However, the very richness of detail in mechanistic studies may also limit their applicability. The story of a single species or population may not be that relevant to other taxa sharing the same habitat, and even congenerics may demonstrate quite different responses to the same climate phenomena (e.g., Parmesan et al. 2005), presumably as a result of differential mechanistic processes.

Nonmechanistic correlational studies, another approach to climate-impacts research, make up the bulk of the current CCIR literature, a trend that is likely to continue. These studies look for organismal responses in association with changes in one or more climate variables, usually via regression, covariance, and/or correlation analyses. Most data used in such research could be classified as observational. Although researchers may suggest mechanisms that are driving these shifts, these hypotheses are not tested directly and remain speculative. Thus, nonmechanistic correlational studies employ inductive reasoning. Often the ability to develop mechanistic tests requires extensive analysis beyond the scope of existing

natural history knowledge about the species or community of interest (e.g., thermal tolerance).

Nonmechanistic correlational studies typically represent opportunistic comparisons, as when data collected in the past (often without regard to future studies) are used as a baseline for looking at changes over time. Dobzhansky's work with seasonal variations in *Drosophila* genotypes was not designed to address our concerns with climate impacts, for example, but his work serves as a powerful means of examining genetic evolution as a response to decadal-scale warming trends (Rodriguez and Rodriguez-Trelles 1998). Such comparisons contrast organismal responses between two or more distinct climate periods or, in the case of spatial studies, two or more climate regions.

A few researchers have even sought to find more unusual temporal-contrast data sets found in the oral traditions of indigenous cultures (e.g., Sagarin and Micheli 2001). While often limited in detail, such sources are at least a qualitative means of making temporal comparisons.

The range of possible questions may also be highly restricted by the scope or type of previously collected data. Thus, these studies tend to focus on quite different groups than those used for mechanistic studies. Moreover, nonmechanistic correlational studies face credibility challenges, given their dependence on inductively argued "weak" evidence that does not test and isolate mechanisms driving the patterns of interest (Harvey 1991, Brooks and McLennan 1991).

Meta-analyses, the third approach to CCIR, by definition synthesize trends from multiple studies. Like nonmechanistic correlational studies, these patterns are unlikely to result in classically defined "strong" evidence. However, their analytical power derives from their ability to inferentially glean large-scale processes and patterns from studies that were focused on more narrow impacts (Gurevitch and Hedges 2001). Such studies have only recently begun to emerge for climate-impacts research, but they are serving as excellent barometers for the extent and types of effects realized to date. Ideally, meta-analyses should also act as a basis for setting policy relevant to impacts.

Mechanistic, nonmechanistic correlational, and meta-analysis studies have been used to explore both direct and indirect impacts from anthropogenic climate change. Direct impacts assume an unmediated response (often physiological) by individual organisms to meteorological and/or atmospheric variables, such as elevated CO_2 levels or thermal tolerance. Often direct-impact questions are best framed as controlled manipulations (i.e., mechanistic studies).

Questions about indirect impacts can include all three research approaches. To date, most studies of indirect responses have focused on shifts in range limits, phenology (i.e., time-sensitive behavior), and abundance. A small but growing number of researchers have gone further by attempting to classify these responses as ecological or evolutionary reactions to climate change (Thomas et al. 2001, Parmesan 2006).

A Crisis of Inference

Indirect-impact questions are most likely to employ inductive means of analysis, and most have been addressed by nonmechanistic correlational studies. Is inference a less useful or important mode of research? For studies of physiological responses to altered atmospheric composition, hydrology, or temperature, mechanistic approaches are appropriate. However, the use of reciprocal transplants to test range or phenology shifts may be impossible or even, in some cases, unethical. Thus, Dunn and Winkler (1999) found a significant advance in mean egg-laying date for tree swallows (*Tachycineta bicolor*) that matched widespread trends for warmer May temperatures. While a direct effect on egg development or phenological shifts in behaviors such as mating or egg laying cannot be ruled out, neither can indirect causes, such as changes in the emergence phenology of prey species. Given the nature of the data, direct and indirect climate impacts cannot be distinguished.

Unfortunately, other hypotheses cannot be ruled out of such studies either. Indeed, factors unrelated to climate shifts may be influencing the object of nonmechanistic correlational studies. Given that few clear management responses have been developed to respond to realized and anticipated climate change, this is not a trivial issue. A misattribution could lead to bad conservation policy or, even worse, a reduction in the credibility of conservation science with the public or policy makers. Does this issue imply that nonmechanistic correlational studies should be avoided or valued less than mechanistic studies?

We believe such studies must continue to play their important role in CCIR. Dunn and Winkler (1999), as well as studies described in this volume, exemplify the processes of building a chain of evidence attributing organismal changes to anthropogenic climate change. However, not all studies are so well constructed. Ecologists carry a significant responsibility to address the issues and concerns of the public, policy makers, and resource managers by demonstrating why climate shifts matter to organisms and ecosystems, even when this responsibility may also expose them to criticism. Given the existing state of political and public debate regarding the impact of human influence on world climate systems, biological research into climate-change impacts requires more care when proposing, conducting, or reporting work than with more traditional forms of conservation-relevant research.

Indeed, because the weighting of evidence and proof is closely scrutinized by skeptics of the anthropogenic role in modern climate change, CCIR rhetoric in the twenty-first century demands a careful reevaluation of how climate models are applied and described with reference to biological data. Until the glass ceiling of public consensus on the human role in shifting global climate is broken, CCIR ecologists must be prepared to work harder to explain the logic of their research and how conclusions have been reached. Nonmechanistic correlational studies must be as convincing as possible to avoid being dismissed as "mere correlational research" or "soft science" (Lomborg 2002). Given that nonmechanistic correlational studies of indirect impacts are likely to continue to dominate CCIR for the foreseeable future, how can we improve their scientific authority and inductive power? In other words, can we make weak evidence stronger?

The traditional image of scientific methodology imagines a process of hypothesis testing via experimental manipulation of single variables. Unfortunately, this image does not match ecology as a discipline very well given the scale and complexity of most ecological systems. As Hilborn and Mangel (1997:13) have noted, "[Ecological] experiments may be difficult to perform, so [ecologists] rely on observation, inference, good thinking, and models to guide our understanding of the world."

The use of inferential hypothesis testing rather than deductively reasoned controlled experiments has been unapologetically established in sciences such as astronomy, physics, and geology (Bowler 1993). Indeed, dependence on weak evidence permeates climate science (Weart 2003). The use of accurate and precise thermometers extends back only to the late nineteenth century, and for only a few locations (IPCC 2001a, 2001b). Consequently, most climate data are obtained instead from proxies such as boreholes, ice cores, oceanic sediment data, and tree-ring growth rates. In fact, many of the scientific debates that have raged over climate trends are critiques of the chains of evidence used to tie precipitation, atmospheric carbon concentrations, sea-surface temperature, or other variables to a proposed indicator (IPCC 2001a, 2001b).

In process if not substance, these are arguments about evidence that seems weak, incomplete, inaccurate, and/or inadequate.

Inferential hypothesis testing has a long history in population biology; for example, Darwin's *On the Origin of Species by Natural Selection* (1859) is largely based on such arguments. The methodology has only recently developed both more formal methodologies and wider credibility, particularly with the sophisticated comparative approaches now widely used in phylogenetic systematics (Harvey 1991, Brooks and McLennan 1991) and the development of iterative Monte Carlo/Markov chain model-testing, which has penetrated ecology through hierarchical Bayesian methodologies (Clark and Lavine 2001, Dose and Menzel 2004, Ellison 2004). Resource managers and policy makers, especially those trained in a more deductive paradigm, may be uncomfortable with inferential science, even when sympathetic with the overall message that human-induced climate change is occurring and having a detectable biological impact (Parmesan and Yohe 2003).

How Strong Is Your Weak Evidence?

A Pew survey of biological impacts from anthropogenic climate change in the United States acknowledged the importance for high-quality, nonmechanistic correlational studies by suggesting a series of questions to evaluate the quality of evidence of alleged impacts (Parmesan and Galbraith 2004). Some of the issues raised would be relevant to any ecological study (e.g., reducing sampling bias, increasing statistical power), but three are specifically relevant to nonmechanistic correlational studies:

1. Is the time span considered appropriate for the impact? Do the data span an ecologically and climatologically meaningful period of time?
2. Is the geographic scale appropriate for the species or community and the alleged impact?
3. How many lines of support exist for the conclusions? Are there multiple attributions for a cause-effect relationship between the impact and observed shifts in climate?

The Intergovernmental Panel on Climate Change has recognized similar issues with climate-impacts research (Ahmed et al. 2001). These questions also reflect concerns about applicability, generality, and the need for more large-scale and long-term studies, as discussed by Pimm (1991). Certainly with respect to CCIR, weak evidence can be employed in powerful and meaningful ways.

The use of weak evidence for claims of climate-change impacts in no way prevents a study from being sound, insightful, or widely relevant. Weak evidence need not be unpersuasive. Instead, it requires more careful qualification and analysis than strong evidence, particularly given the current political sensitivity to global-warming issues.

Excellent studies exist of the judicious application of weak evidence by itself, and two examples will be presented discussing western North American impacts on phenology and range shifts at the population and species levels, in turn, as well as a third that examines the impacts on marine communities in the western United States.

Range Shift: American Pikas in the Great Basin

Beever et al. (2003) used a variety of sources to compile records of the locations of American pika (*Ochotona princeps*) populations in the U.S. Great Basin dating back to 1913. Seven of the 25 early populations studied appeared to have gone extinct by the end of the twentieth century (Fig. 7.1). Although the authors compared many potentially causative factors—including habitat area, elevation extent, relative latitude, and proximity to roads—none of the factors proved common to all of the extinct populations except for the presence of grazing livestock.

The authors developed several hypotheses to explain the extinctions, including biogeographic and climate-change impact theories, but none were tested experimentally, so the mechanism of extinctions remained speculative. However, a factorial analysis suggested that the extinct populations were located at relatively lower altitudes compared to extant populations, were on mountain slopes that were comparatively isolated (and thus difficult to recolonize), and had smaller mean habitats. The authors inferred that these extinctions were fueled in part by climate-change impacts on plant community distributions.

Phenology Shift: North American Tree Swallows

Dunn and Winkler (1999) examined more than 21,000 breeding records for tree swallows (*Tachycineta bicolor*) collected between 1959 and 1991. A statistically significant relationship was found between egg-laying date and mean May temperature, with the mean egg-laying date advancing nine days over the years surveyed (Fig. 7.2). The authors did not believe that there was a direct link between the onset of breeding or egg-laying and temperature per se, but speculated that there might be a connection between temperature and the timing of poikilothermic prey abundance. In other words, the median emergence date for a prey insect species might be advancing as a result of an

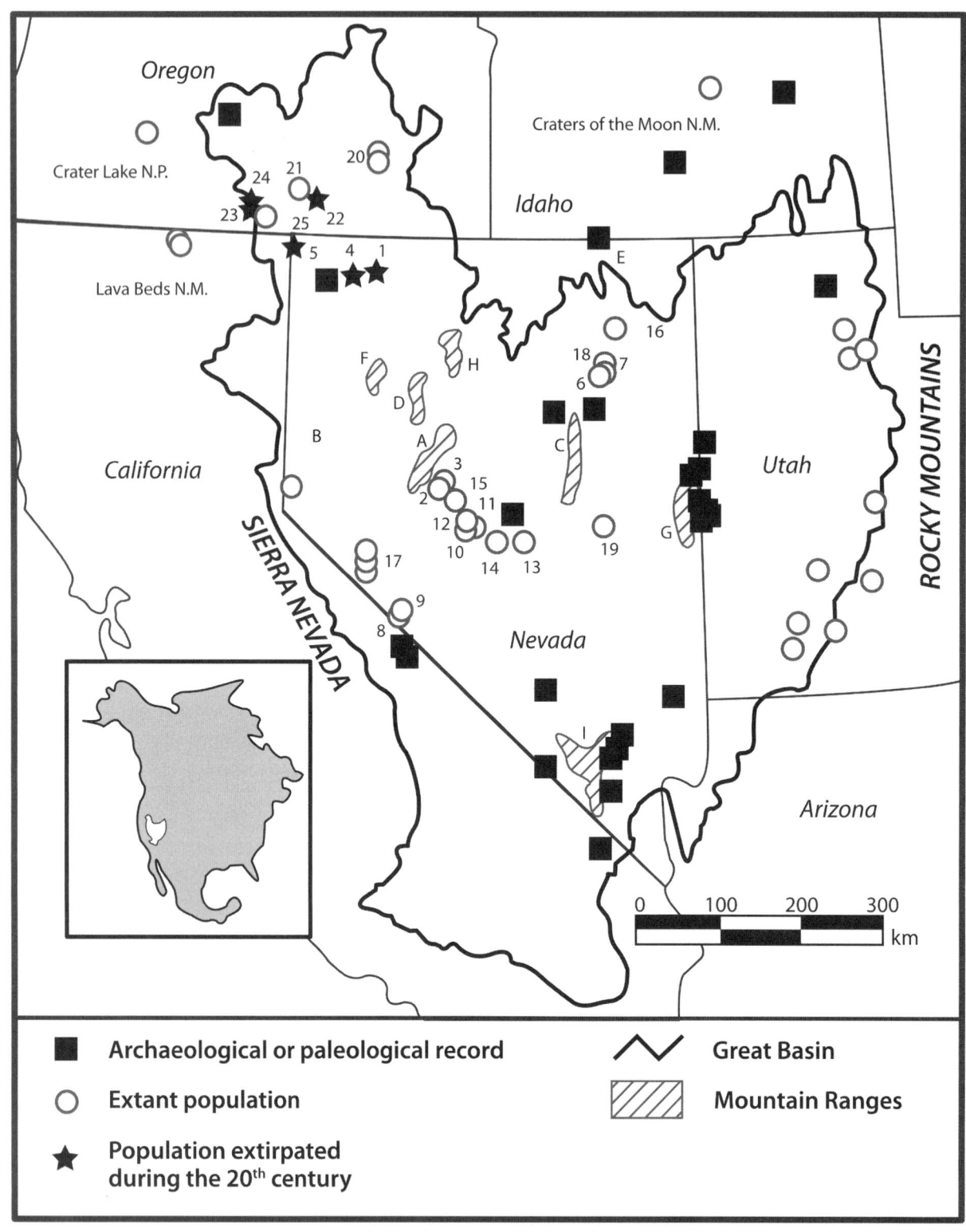

Figure 7.1. Current and historical sites of American pikas (*Ochotona princeps*) in the Great Basin region. Since 1913, low-altitude populations have been much more likely to go extinct than those at high altitudes. (Source: Beever et al. 2003.)

Legend:

■ Archaeological or paleological record

○ Extant population

★ Population extirpated during the 20th century

⋀ Great Basin

▨ Mountain Ranges

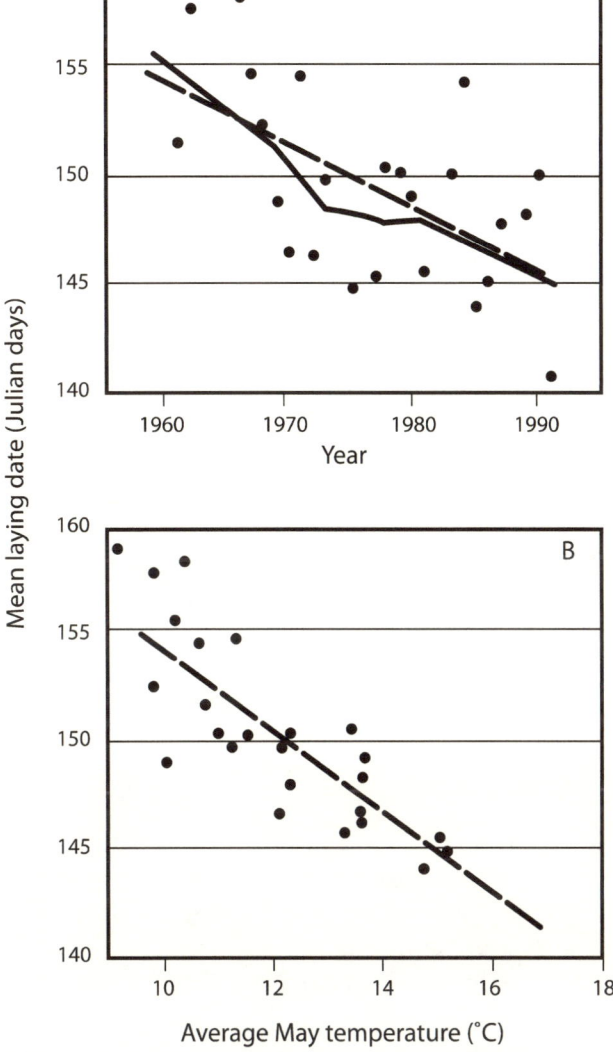

Figure 7.2. Mean egg-laying dates for North American tree swallows (*Tachycineta bicolor*) have advanced with warming temperatures. (Source: Dunn and Winkler 1999.)

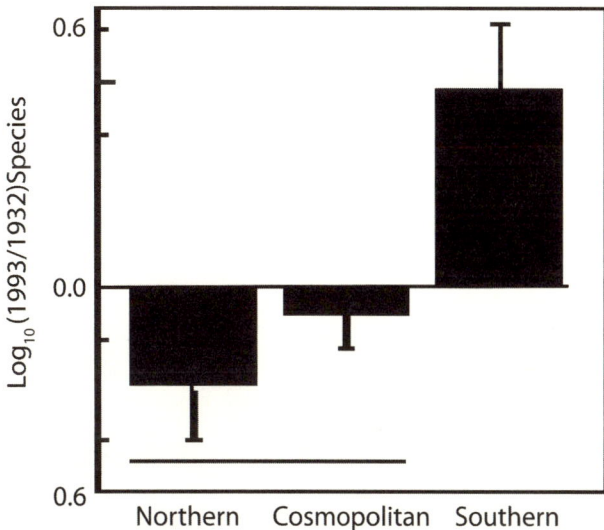

Figure 7.3. Over a 60-year period, the relative abundance of coastal marine species in California has shifted. Species near the northern end of their ranges have tended to increase in abundance, whereas species near the southern end of the ranges have tended to decline. (Source: Sagarin et al. 1999.)

earlier onset of spring. As with Beever et al. (2003), Dunn and Winkler left this hypothesis untested, but the wealth of records showing an advance in tree swallow egg-laying is impressive in itself.

Community-Abundance Shifts:
Shallow Marine Communities on the California Coast
Sagarin et al. (1999) compared the results of a transect of abundance and makeup of shallow marine invertebrate communities conducted in 1931–1933 in Monterey Bay, California, with an identical survey conducted at the same locale between 1993 and 1996. Mean shoreline water temperature had increased 0.79°C over the period between the two studies. Individual species were described

as northern, southern, or cosmopolitan, based on their distribution along the Pacific Coast. Sagarin et al. found that northern (and presumably more cold-adapted) and cosmopolitan species had generally declined in abundance at the research site, whereas southern species had generally increased in abundance (see Fig. 7.3). Very few contrary species trends were found, and no mechanistic hypotheses were tested.

None of these studies addressed mechanistic issues behind observed changes; the authors' speculations remain untested and thus weak. Indeed, in many cases it may be found that range shifts follow from phenology shifts or changes in community makeup, that these categories are not exclusive or deterministic, and that they are likely complex and multifactorial. However, these studies do present sufficiently convincing evidence that correlations between population-level shifts have resulted from climate changes over the research period by spanning ecologically and climatically significant time spans with meaningful sample sizes.

Evidential Strength Training: Transforming Weak Patterns

The studies by Dunn and Winkler (1999) and Beever et al. (2003) are similar to other climate-impacts research in their focus on a subset of populations of a single species.

Indeed, the strength behind the conclusions of Sagarin et al. is the similarity in response across species. Thus, a simple principle applicable to many studies is that multiple supporting correlations (across species or across behaviors in a single species) are more convincing than single-population or single-species projects (Parmesan and Yohe 2003).

The comparative method is based on the assumption that statistical power is derived from multiple species or population comparisons (Harvey 1991). Weak evidence can be strengthened to improve the soundness of its conclusions through multiple correlations; there is strength in numbers, even when experimental data are lacking. Of course, meta-analyses take this approach by definition (e.g., Root et al. 2003, Thomas et al. 2004). However, like an Internet search engine, the quality of a meta-analysis's results is largely a reflection of the content it has to work with.

An extension of this approach was employed by Parmesan and Yohe (2003) in the concept of sign-switching, which is the tracking of biological responses to swings in some climate-related variable. In this case, the twentieth century exhibited well-documented decadal swings in temperature. Looking over a broad range of studies, Parmesan and Yohe asked whether northern range boundaries had shifted higher in latitude/altitude during warming periods and then "reversed sign" to shift to lower latitudes/altitudes during cooling periods.

Their analysis showed that 100 percent of species for which such long-term data were available did indeed exhibit sign-switching. Of course, little or no experimental manipulation is possible with sign-switching studies, at least in the context of a retrospective or observational study. Nonetheless, the probability that sign-switching is due to chance alone decreases rapidly with finer resolution, longer tracking periods, and evidence of sign-shifts for additional species. Indeed, these shifts are the temporal equivalent of natural experiments so widely used in ecological and evolutionary research. Each sign-shift is effectively a separate data point.

We believe that some of these approaches are applicable to current and future climate-impact research. Our recommendations include:

1. Collecting multipopulation and/or multisite data.
2. Using data spanning at least 20 years, even if these data are discontinuous.
3. Employing multiple species or examining multiple behaviors in a single species whenever possible to avoid criticisms of a positive publishing bias.
4. Examining data for signs of sign-switching across species/behaviors.
5. Identifying possible mechanisms and, if possible, testing these mechanisms or describing how they might be tested.
6. Explicitly exploring hypotheses for explanations that are not dependent on climate change.

The widespread skepticism about anthropogenic impacts on climate requires that researchers studying biological impacts become more aware that our work also has a nonscientific audience. It is our hope that researchers can become more sensitive to this larger community by improving the quality of nonmechanistic correlational evidence of climate impacts. The increased credibility thereby gained should allow climate-impact research to provide more powerful, persuasive, and ultimately more useful recommendations for policy makers and resource managers.

Acknowledgments

We gratefully appreciate the comments offered by two anonymous reviewers, and we are particularly thankful to Fred Wagner, who organized our symposium and fostered our fruitful interactions.

References

Ahmad, Q. K., R. A. Warrick, T. E. Downing, S. Nishioka, K. S. Parikh et al. 2001. Methods and tools. See IPCC 2001a, pp. 105–143.

Beever, E. A., P. F. Brussard, and J. Berger. 2003. Patterns of apparent extirpation among isolated populations of pikas (*Ochotona princeps*) in the Great Basin. *J. Mammal.* 84:37–54.

Bowler, P. J. 1993. *The Norton History of the Environmental Sciences.* W. W. Norton, New York.

Brooks, D. R., and D. A. McLennan. 1991. *Phylogeny, Ecology, and Behavior: A Research Program in Comparative Biology.* Univ. of Chicago Press, Chicago.

Clark, J. S., and M. Lavine. 2001. Bayesian statistics: Estimating plant demographic parameters. In S. M. Scheiner and J. Gurevitch (eds.), *Design and Analysis of Ecology Experiments,* 2nd ed. Oxford Univ. Press, New York.

Darwin, C. 1859. *On the Origin of Species by Natural Selection.* Murray, London.

Delcourt, P. A., and H. R. Delcourt. 1987. Late-quaternary dynamics of temperate forests: Applications of paleoecology to issues of global environmental change. *Quat. Sci. Rev.* 6: 129–146.

Dose, V., and A. Menzel. 2004. Bayesian analysis of climate change impacts in phenology. *Global Change Biol.* 10: 259–272.

Dunn, P. O., and D. W. Winkler. 1999. Climate change has affected the breeding date of tree swallows throughout North America. *Proc. R. Soc. London Ser. B* 266:2487–2490.

Ellison, A. M. 2004. Bayesian inference in ecology. *Ecol. Lett.* 7:509–520.

Gurevitch, J., and L. K. Hedges. 2001. Meta-analysis: Combining the results of independent experiments. *In* S. M. Scheiner and J. Gurevitch (eds.), *Design and Analysis of Ecology Experiments,* 2nd ed. Oxford Univ. Press, New York.

Harvey, P. H. 1991. *The Comparative Method in Evolutionary Biology.* Oxford Univ. Press, New York.

Hilborn, R., and M. Mangel. 1997. *The Ecological Detective.* Princeton Univ. Press, Princeton, NJ.

IPCC (Intergovernmental Panel on Climate Change). 2001a. *Climate Change 2001: The Science of Climate Change.* Contribution of Working Group I to the Intergovernmental Panel on Climate Change Third Assessment Report. J. T. Houghton, Y. Ding, D. J. Griggs, M. Noguer, P. J. van der Linden, X. Dai, K. Maskell, and C. A. Johnson (eds.). Cambridge Univ. Press, Cambridge, UK.

———. 2001b. *Climate Change 2001: Impacts, Adaptation, and Vulnerability.* Contribution of Working Group II to the Intergovernmental Panel on Climate Change Third Assessment Report. J. J. McCarthy, O. F. Canziani, N. A. Leary, D. J. Dokken, K. S. White (eds.). Cambridge Univ. Press, Cambridge, UK.

Karl, T. R., R. W. Knight, D. R. Easterling, and R. G. Quayle. 1996. Indices of climate change for the United States. *Bull. Amer. Meteorol. Soc.* 77:279–292.

Lomborg, B. 2002. *The Skeptical Environmentalist: Measuring the Real State of the World.* Cambridge Univ. Press, Cambridge, UK.

Masson-Delmotte, V., G. Raffali-Delerce, P. A. Danis, P. Yiou, M. Stievenard, F. Guibal, O. Mestre, V. Bernard, H. Goosse, G. Hoffman, and J. Jouzel. 2003. Changes in European precipitation seasonality and in drought frequencies revealed by a four-century-long tree-ring isotopic record from Brittany, western France. *Clim. Dyn.* 24:57–69.

Parmesan, C. 1996. Climate and species' range. *Nature* 382:765–766 .

———. 2003. Butterflies as bio-indicators of climate change impacts. In C. L Boggs, W. B. Watt, and P. R. Ehrlich (eds.), *Evolution and Ecology Taking Flight: Butterflies as Model Systems.* Univ. of Chicago Press, Chicago.

———. 2006. Ecological and evolutionary responses to recent climate change. *Ann. Rev. Ecol., Evol., and System.* 37:637–669.

Parmesan, C., S. Gaines, L. Gonzalez, D. M. Kaufman, J. Kingsolver, A. T. Peterson, and R. Sagarin. 2005. Empirical perspectives on species borders: From traditional biogeography to global change. *Oikos* 108:58–75.

Parmesan, C., and H. Galbraith. 2004. *Observed Impacts of Global Climate Change in the U.S.* Pew Center on Global Climate Change, Arlington, VA.

Parmesan, C., and G. Yohe. 2003. A globally coherent fingerprint of climate change impacts across natural systems. *Nature* 421: 37–42.

Pimm, S. 1991. *The Balance of Nature? Ecological Issues in the Conservation of Species and Communities.* Univ. of Chicago Press, Chicago.

Rodó, Xavier, and F. A. Comín. 2000. Links between large-scale anomalies, rainfall and wine quality in the Iberian peninsula during the last three centuries. *Global Change Biol.* 6:267–273.

Rodriguez, M. A., and F. Rodriguez-Trelles. 1998. Rapid microevolution and loss of chromosomal diversity in *Drosophila* in response to climate warming. *Evol. Ecol.* 12:829–838.

Root, T. L., J. T. Price, K. R. Hall, S. H. Schneider, C. Rosenzweig, and J. A. Pounds. 2003. Fingerprints of global warming on wild animals and plants. *Nature* 421:57–60.

Sagarin, R. D., J. P. Barry, S. E. Gilman, and C. H. Baxter. 1999. Climate-related change in an intertidal community over short and long time scales. *Ecol. Monog.* 69:465–490.

Sagarin, R., and F. Micheli. 2001. Climate change in nontraditional data sets. *Science* 294:811.

Thomas, C. D., E. J. Bodsworth, R. J. Wilson, A. D. Simmons, Z. G. Davies, M. Musche, and L. Conradt. 2001. Ecological and evolutionary processes at expanding range margins. *Nature* 411:577–581.

Thomas, C. D., A. Cameron, R. E. Green, M. Bakkenes, L. J. Beaumont, Y. C. Collingham, B. F. N. Ferreirade Siquelra, A. Grainger, L. Hannah, L. Hughes, B. Huntley, A. S. van Jaarsveld, G. F. Midgley, L. Miles, O. L. Phillips, and S. E. Williams. 2004. Extinction risk from climate change. *Nature* 427:145–148.

Weart, S. R. 2003. *The Discovery of Global Warming.* Harvard Univ. Press, Cambridge, MA.

8

Ecological Consequences of Forest Insect Disturbance Altered by Climate Change

Jesse A. Logan and James A. Powell

Abstract

Unprecedented outbreaks of native bark beetles are occurring in forests throughout the mountains of western North America. Any one of these events would be unusual; their simultaneous occurrence is nothing short of remarkable. Significant biogeographical events are occurring at a continental scale, and a warming climate is the one commonality across all of these spectacular outbreak events. Mountain pine beetle (*Dendroctonus ponderosae* Hopkins) populations are responsible for three of the more impressive of these events, and in this chapter we describe recent case histories that illustrate the unique attributes of the current situation. These case histories involve outbreaks within the current range of mountain pine beetle in areas with microclimates that were previously too cold; unusually large and intense outbreaks in high-elevation pines; and range expansion north and east beyond the historical distribution in Canada. After describing each of these three situations, we briefly discuss their importance with respect to global climate warming.

Introduction

Native insects are the greatest forces of natural change in forested ecosystems of North America. In aggregate, insect and pathogen disturbances affect an area almost 45 times larger than that affected by fire, resulting in an economic impact that is nearly 5 times as great (Dale et al. 2001). Of these natural agents of ecosystem disturbance and change, the bark beetles are the most obvious in their impact. Of these, the mountain pine beetle (*Dendroctonus ponderosae* Hopkins) has the greatest economic impact (Samman and Logan 2000), primarily because it is one of a handful of bark beetles that are true predators,

meaning they must kill their host to successfully reproduce, and they often do so in truly spectacular numbers (see Fig. C.4).

Although the mountain pine beetle is an aggressive tree killer, it is a native component of natural ecosystems; in this sense, the pine forests of western North America have coevolved (or at least coadapted) in ways that incorporate mountain pine beetle disturbance in the natural cycle of forest growth and regeneration. Such a relationship in which insect disturbance is "part and parcel of the normal plant biology" has been termed a "normative" outbreak (Mattson 1996).

This normative relationship between native bark beetles and their host forests is currently undergoing an apparent shift, exemplified by an unusual sequence of outbreak events. Massive outbreaks of spruce beetle have recently occurred in western North America from Alaska to southern Utah (Berg 2003, Ross et al. 2001). A complex of bark beetles is killing ponderosa pine in the southwestern United States at levels not previously experienced during the period of European settlement. Piñon pines are being killed across the entire range of the piñon/juniper ecotype in the West, effectively removing a keystone species in many locations. Mountain pine beetle outbreaks are occurring not only at greater intensity, but in locations where they have not previously occurred (Logan and Powell 2001, Carroll and Safranyik 2004).

Any one of these events would be unusual, but their simultaneous occurrence is remarkable. In many of these instances the outbreaks are anything but normative; they are occurring in novel habitats, with potentially devastating ecological consequences.

What is going on here? The root of these unprecedented outbreaks appears to be directly related to an unusual weather pattern. Although drought, particularly in the Southwest, is playing an important role in some of these outbreaks, the dominant and ubiquitous factor at the continental scale is the sequence of abnormally warm years that began sometime during the mid-1980s (Berg 2003). Regardless of the underlying causes, the impact of warming temperatures on bark beetle outbreaks has resulted in a renewed research interest focused on understanding and responding to the economic and ecological threat of native insects functioning as an invasive pest. Because of its ecological importance and economic impact, the mountain pine beetle is the subject of much of this research. Development of predictive models is an important component of this research effort (Logan and Powell 2004).

In this chapter, we describe application of a mathematical model that was designed to represent the basic mountain pine beetle ecological response to weather (temperature) and climate. This model has been used to analyze three ongoing mountain pine beetle outbreaks. The first is occurring in the Sawtooth Valley of central Idaho, an area with a microclimate that has previously been too cold for regular outbreaks. The second outbreak involves the expansion of mountain pine beetle activity into high-elevation habitats that were previously too cold to sustain outbreak populations. The final interaction of interest is the alarming range expansion of mountain pine beetles north and eastward in Canada.

Model Description

As poikilothermic organisms, mountain pine beetles are profoundly affected by environmental temperature, and both weather and climate play critical roles (Amman 1973, Safranyik 1978). Seasonal temperature patterns in large part determine both the latitudinal and elevational distribution, or range, and are also key in the initiation and collapse of outbreaks. Historically, the geographic range of this beetle has been limited to areas with enough thermal energy to complete the life cycle in one year (univoltinism). Additionally, as discussed further below, synchrony of adult emergence is critical for the beetle to overcome the substantial host response that defends trees from attack. The term "adaptive seasonality" (Logan and Powell 2001) has been used to describe the seasonal temperature pattern that results in these two conditions. In order to more clearly understand the role of adaptive seasonality in mountain pine beetle ecology, we have maintained a long-term investment in developing and analyz-

ing mathematical models relating seasonal temperatures to adaptive seasonality.

A mathematical framework is extremely useful for evaluating adaptive seasonality because of the temporal complexity and noise of the seasonal temperature cycle. Measures of seasonal temperature are essentially unlimited; however, without some conceptual framework to evaluate this complex cycle in terms of the individual organism's ecology, it is impossible to tell which measures are important. A mathematical model provides the necessary conceptual framework for this evaluation (Logan et al. 2003).

As with any successful organism, the relationship of the mountain beetle to its physical environment is multifaceted and complex. However, the basic constraints on population performance are temperature dependent. The key aspects of mountain pine beetle ecology that were considered in model development are:

1. The mountain pine beetle is aggressive and must kill its host to successfully reproduce.
2. Host trees have (co)evolved effective defensive compounds in their resin that must be overcome by attacking beetles.
3. Beetles overcome tree defenses by a mass attack strategy that numerically overwhelms tree defenses—that is, the number of essentially simultaneous attacks depletes both the constitutive and induced resin capacity of the tree (Raffa and Berryman 1987).
4. There is no compelling evidence of diapause or another physiological timing mechanism that synchronizes the lifecycle, including emergence.

Our model determines adaptive seasonality based on three criteria: (1) Is the population univoltine? (2) Is adult emergence synchronous? (3) Is emergence at an appropriate time of year? If the answers to all three questions are affirmative, then the annual temperature cycle is determined to be adaptive; if any condition is violated, then the seasonality is not adaptive (Logan and Powell 2001). Note that in the applications of the model discussed below, we are interested in the warming of a maladapted cold habitat. The situation of overwarming an already suitable thermal habitat would require slightly different criteria (i.e., evaluation of synchrony loss by moving into a band of fractional voltinism that is between uni- and bivoltine).

The temperature-dependent model that we developed is typical of many insect phenology models in that the structure is a cascading sequence of developmental stages in which the completion of one life stage initiates the beginning of the next. The rate of progression through an individual life stage is determined by integrating the specific developmental rate of each life stage at an hourly time step. The high temporal resolution is necessary because the temperature-dependent developmental rate is typically nonlinear, with both low- and high-temperature thresholds (Logan et al. 1976, Logan and Powell 2001, Powell and Logan 2005). Initial analysis of our model results was empirically based (i.e., the model was run for a given period of time and then results were examined).

These investigations resulted in interesting ecological predictions, but left unresolved questions regarding model veracity because predictions were dynamically complex and somewhat counterintuitive. Subsequent research (Powell et al. 2000, Jenkins et al. 2001) has expanded our understanding of the model's dynamic properties and has reinforced the credibility of earlier model predictions (Powell and Logan 2005). In the following three sections, we discuss the predictions from our model and their ecological implications.

Sawtooth Valley

The Stanley Basin in central Idaho is an area of remarkable scenic beauty (see Fig. C.5). The Salmon River has its headwaters here, and the basin includes the heart of the Sawtooth National Recreation Area. This area also provides a unique opportunity for research involving mountain pine beetle dynamics, and the USDA Forest Service Western Bark Beetle Project has maintained an active research program there for more than 15 years.

The Stanley Basin is well within the geographic distribution of the mountain pine beetle and contains ample forests of susceptible lodgepole pine. However, the steep high mountains surrounding the narrow valley produce a cold air sink that results in year-round low temperatures. Due to this mesoscale climate, the thermal habitat has historically been only marginally suited for mountain pine beetles. This has had two important results: first, instead of the dramatic boom-and-bust outbreak cycles of more favorable climates, historical populations have tended to be maintained at sub-outbreak levels for prolonged periods. This has allowed for detailed population dynamics research at one site over a long time period.

Second, climate marginality means that slight variations in annual weather patterns result in immediate and measurable population responses. (See Logan and Bentz [1999] for a more detailed description of the Stanley Basin

in relation to mountain pine beetle ecology and outbreak activity.) Despite the historical marginality of the climate from the mountain pine beetle's perspective, the Stanley Basin has experienced a full-scale outbreak during the past five years (Fig. C.6).

The original purpose of our work in the Stanley Basin was to validate a life-system model of the mountain pine beetle (B. J. Bentz, pers. comm.). To accomplish this end, weather monitoring equipment was installed in trees that had been successfully attacked, and life-history field samples were taken periodically throughout the ensuing year. During the original years of the study, research locations were moved every year as small spot infestations moved around thermally favorable locations, primarily on lower-elevation, west-facing aspects of the valley. The infestation at this time was characteristic of the historical interpretation for the Stanley Basin of the mountain pine beetle operating as an opportunistic species in a thermally marginal habitat.

Populations, however, were beginning to build in the early 1990s. This incipient outbreak was immediately shut down by the record-breaking cold summer of 1993. Populations remained at extremely low levels for the next several years. However, by the late 1990s, populations were building rapidly, finally culminating in an outbreak of major proportions. What we witnessed appears to be a shift from a marginal thermal habitat to a highly favorable habitat.

During the time period of our work in the Stanley Basin, the overall temperature was warming. Figure 8.1 shows the general temperature trend in the National Climate Data Center (NCDC) Idaho Division 4, which includes the Stanley Basin, over the period 1895–2002. There are three take-home lessons from this plot. First, the general trend line began a steady increase starting in the mid-1980s. Second, the mean annual temperature for 1993 was cold, but the 1996–1997 winter was not remarkably so. However, the temperature during the 1997 summer (June–August) was the coldest on record. The observation that cool summer temperatures, not minimum winter temperature, was responsible for shutting down an incipient outbreak is counter to the conventional wisdom that minimum winter temperature is the primary limiting factor on mountain pine beetle population success (Safranyik 1978). Finally, the cold pulse of 1993 (possibly related to the worldwide effect of the Pinatubo eruption) was short-lived, and temperature the following year, and every year since, has maintained a steady upward trend.

Model analysis of yearly temperatures based on

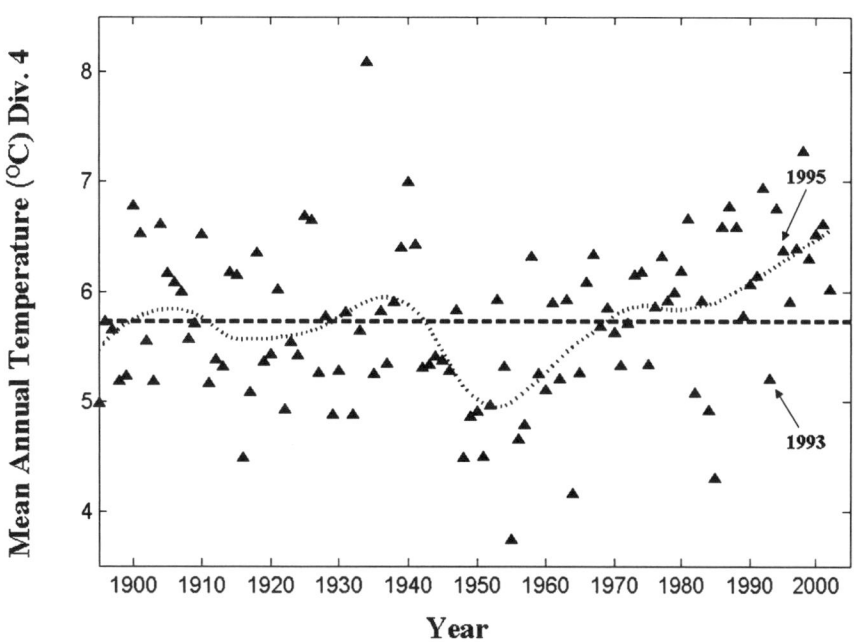

Figure 8.1. Annual mean temperatures for Idaho National Climate Date Center Division 4, which includes the Stanley Basin, smoothed by a robust Lowess quadratic fit that includes a bracket of one-third of the data for each estimated smoothed point. Note the strong increasing trend from 1980 on, and the relative position of the 1993 and 1995 years, well below and at the trend line, respectively.

phloem temperatures at individual study sites has indicated that in the first half of the 1990s, populations were predicted to have asynchronous emergence with fractional voltinism (four generations in five years). Subsequent to 1995, seasonality has been predicted to be synchronous and univoltine. For reasons previously discussed, fractional voltinism is not adaptive, whereas synchronous univoltinism is. Figure 8.2 shows the switch in voltinism plotted concurrently with aerial detection survey (ADS) data collected by flying over a region in a light aircraft and noting areas with red-topped lodgepole pine on a digital sketch pad with a mapping data base. Polygons formed must enclose at least ten impacted trees but may also include regions with complete mortality. The sketch maps are therefore difficult to convert into actual numbers of attacked trees, but give a good indication of total area impacted by mountain pine beetle.

The switch from maladaptive to adaptive seasonality occurred concurrently with the shift from a sub-outbreak population phase to an exponentially growing population. Comparison of Figures 8.1 and 8.2 illustrates the necessity of considering the full annual microhabitat temperature cycle. Mean annual temperatures are valid for indicating general trends, but the full seasonal temperature cycle is necessary for evaluating the impact on population dynamics. In effect, the model provides a lens to focus the complex daily temperature cycle into a single prediction of population response.

Several important general conclusions can be drawn from our experience in the Stanley Basin. Both the population response to the cold summer of 1993 and the speed of recovery were immediate, indicating the value of mountain pine beetle populations as an indicator species for climate change. It is important, however, to note that this response is expressed as a threshold (Logan and Bentz 1999). The steady trend-line increase shown in Figure 8.1 is transformed into the threshold response of Figure 8.2. The resulting effect of a warming climate in the Stanley Basin (assuming the continuing trend in Fig. 8.1) will be a shift from a marginal thermal habitat to a thermally favorable habitat. Management implications are obvious, and both federal (USDA Forest Service) and state (Idaho Division of Forestry) officials have taken an active interest in formulating a response to this new challenge.

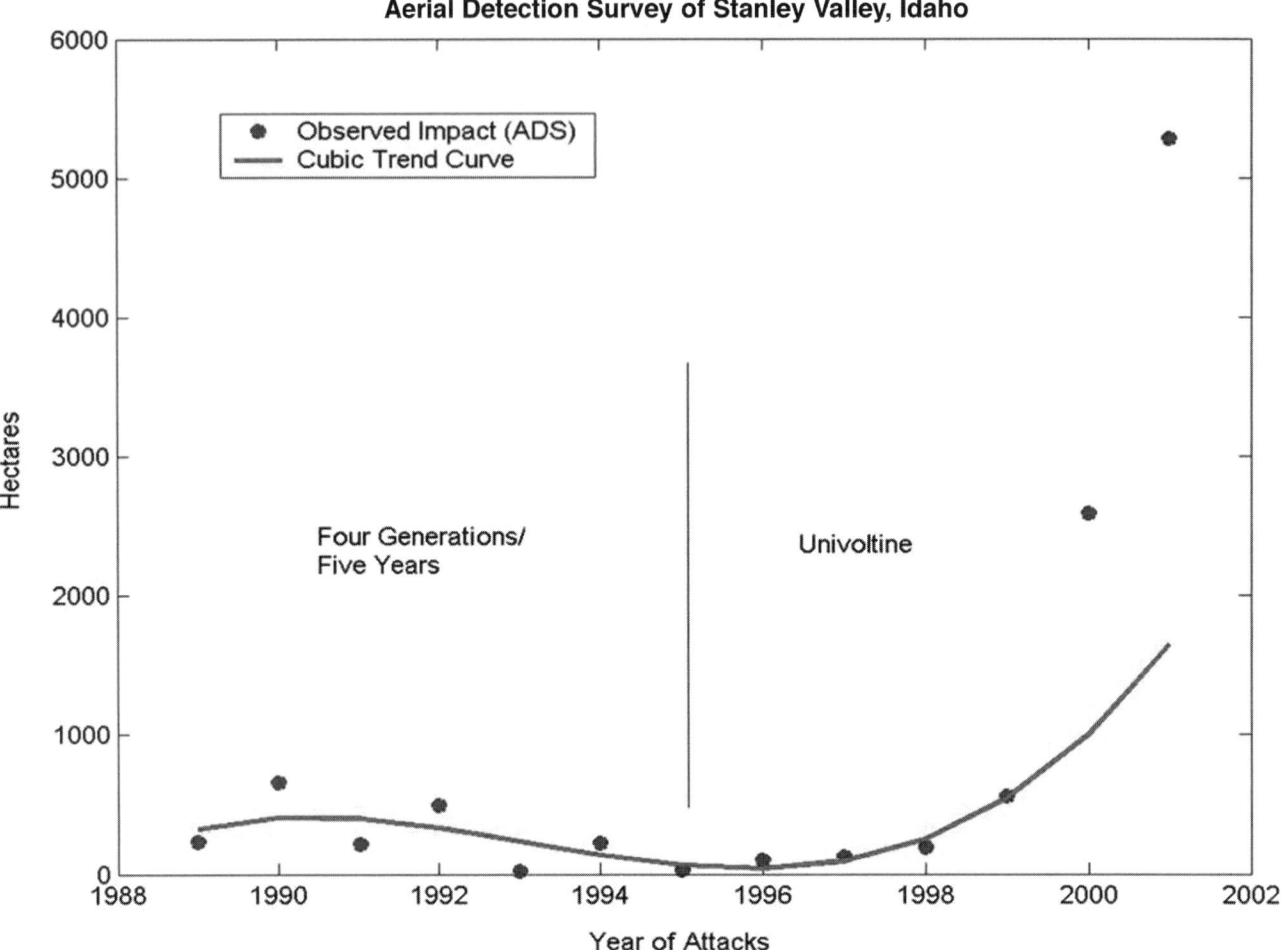

Figure 8.2. Comparison of aerial detection survey (ADS) data and model-predicted voltinism from field population sampling sites (B. J. Bentz, pers. comm.). Simulation runs prior to 1995 resulted in maladapted seasonality (fractional voltinism and resulting complex cycles). Simulation results for every year after 1995 indicated adaptive seasonality with synchronous univoltine populations.

Invasion of High-Elevation Pine Ecosystems

Concurrent with our work in the Stanley Basin, we became interested in high-elevation pines—specifically whitebark pine (*Pinus albicaulis* Engelmann). High-elevation five-needle pines, including whitebark pine, are charismatic species that occupy some of the harshest habitats on the North American continent. In central Idaho, climax forests of whitebark pine begin as an upright growth form in the upper subalpine zone at approximately 8,500 ft (2,500 m) and continue into the alpine zone as a krummhotz growth form (Arno 2001). The high-elevation five-needle pines are typically long-lived species that are not particularly adapted to insect outbreak disturbances (in contrast to lodgepole pines, which incorporate such disturbances in their reproductive strategy [Amman and Schmitz 1988]). In fact, one explanation

for the high-elevation distribution of five-needle pines is an escape strategy from the host of insect pests and pathogens that occur in more benign lower-elevation environments.

There are many interesting and unique aspects to the relationship between elevation, habitat, and mountain pine beetle ecology, but the primary point of interest with respect to climate change is the relationship between voltinism and elevation. The steep elevational gradient of the Rocky Mountains typically results in related habitat zones. Associated with these zones, the conifer habitat is partitioned into lower-elevation pines (e.g., lodgepole or ponderosa), a mid-elevation zone of non-host mixed conifers, and at higher elevations, five-needle pines.

Accompanying the change in forest type with increasing elevation is the response of mountain pine beetle pop-

Figure 8.3. Photo taken near the south site, Railroad Ridge, in the White Cloud Mountains of central Idaho. The climax whitebark pine forests on Railroad Ridge have a healthy balance of age classes and exhibit significant recruitment. The oldest whitebark pine on record (Perkins and Swetnam 1996) is within a kilometer or two of this location, and the area has, so far, not been impacted by white pine blister rust.

ulations to concurrently decreasing temperatures. In the lower-elevation pine forests, mountain pine beetle populations are typically univoltine. In the non-host mixed conifer zone, populations exhibit fractional voltinism, varying between one and two generations per year. In high-elevation pine forests, there is typically not enough thermal energy to complete the life cycle in less than two years, and populations are semivoltine at best. As noted earlier, outbreaks have been restricted to synchronous univoltine populations. Despite this fact, significant mountain pine beetle mortality occurred in whitebark pine during the 1930s (Perkins and Swetnam 1996). Motivated by this mortality event, and also because of the magnitude of predicted warming with likely greenhouse gas emissions (IPCC 1990), we initiated a long-term study site on Railroad Ridge in the White Cloud Mountains of central Idaho.

At approximately 10,000 ft (3,000 m), Railroad Ridge is superb whitebark pine habitat (Fig. 8.3) (Logan and Powell 2001). Beginning in 1995, we established four weather-monitoring sites: one on top of the ridge, and three on aspects corresponding to the cardinal directions south, east, and north. One of our early computer simulation experiments (Logan and Bentz 1999) was made by performing a bifurcation analysis of mountain pine beetle seasonality for microhabitat phloem temperatures recorded at the north site (Fig. 8.4). This analysis was obtained by initiating oviposition on some reasonable but arbitrary date (in this case August 1), allowing the model to run for 50 transient generations and then plotting the last 20 ovipositional dates reduced by modulo 365.

The bifurcation procedure was followed for temperatures obtained by adding ΔT to each observed hourly temperature of the original 1996 temperature series. Plotting a single point (e.g., Regions A and C in Fig. C.6)

indicates an attracting fixed point where all 20 final ovipositional dates are the same value. A set of m output values plotted at a given ΔT indicates an orbit of m ovipositional dates over an unknown period of years. Each of these dates is attracting in a small range of ovipositional dates, fragmenting the population over time and resulting in asynchronous emergence. As noted earlier, an attracting fixed point is analogous to synchronous emergence, while an orbit of ovipositional dates indicates lack of synchrony. As shown in Figure C.6, regions of asynchronous emergence (e.g., region b) separate regions of synchronous emergence (e.g., regions a and c).

Exploring the adaptive seasonality of the observed temperature cycle ($\Delta T = 0$), we note that one of the necessary conditions, synchrony, has been met. However, two years are required to complete the life cycle; that is, the life cycle in this thermal environment is semivoltine (two years per generation). This result for high-elevation environments is consistent with the conventional wisdom that populations are semivoltine at high elevations. As described earlier, semivoltine populations invariably result in high winter mortality due to vulnerable life stages entering winter in one or both years. Thus we concluded that mountain pine beetle populations were semivoltine in thermal regimes corresponding to the 1996 Railroad Ridge temperatures, and that they were unable to achieve outbreak densities, though they may have been able to maintain endemic populations functioning as a fugitive species in snow-felled or otherwise weakened trees (J. C. Vandygriff, personal observation).

As temperature was systematically increased, the model predicted loss of synchrony and complex cycles resulting from the fractional voltinism (more than one generation every two years, but less than one generation per year). As described earlier, different initial oviposition

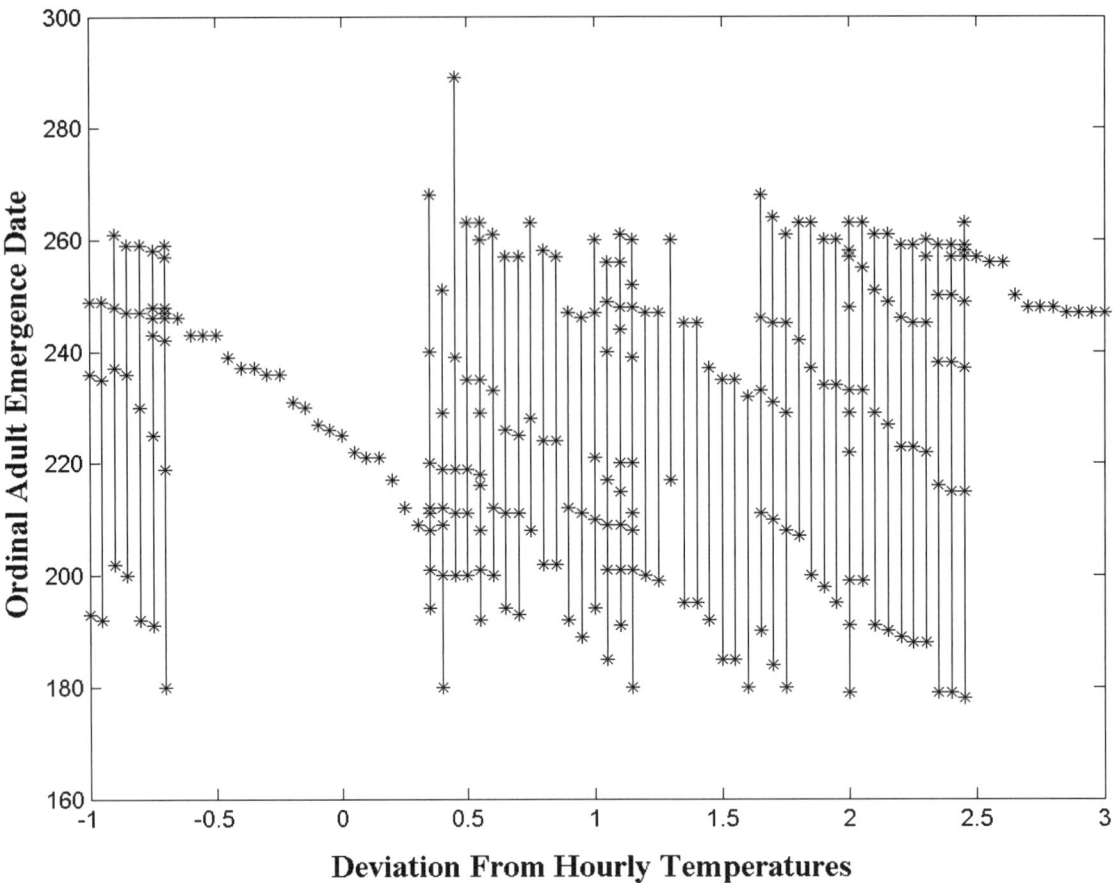

Figure 8.4. Bifurcation plot for the 1996 temperatures at the Railroad Ridge south site. Julian day for emergence appears on the vertical axis, while the horizontal axis indicates degrees C added to each hourly temperature (raising the mean annual temperature by the indicated amount) (see Powell and Logan 2005). More than one plotted point indicates maladaptive seasonality, while a single point indicates synchronous univoltine populations. The observed annual temperature cycle results in prediction of seasonality at the beginning of a synchronous semi-voltine band. Increasing temperature by approximately 2.5°C would result in a shift to an adaptive thermal regime of synchronous univoltine populations.

dates will be attracted to different points on the orbit, resulting in a lack of synchrony. In other words, warming of the observed maladapted semivoltine temperature cycle of 1995 would actually *reduce* the suitability of the habitat until a threshold for univoltinism was reached at approximately 3°C.

Increasing mean annual temperature by 3°C resulted in synchronous univoltine populations, transforming a maladapted thermal habitat into an adaptive one. An increase of this magnitude is well within the predictions for global warming anticipated to occur by midcentury, with obvious implications for the high-elevation five-needle pines. Similarly, a look back in time discloses that the major 1930s outbreak in whitebark pine occurred during the only time in the previous century that mean summer

temperatures were more than 2.5°C above average (Logan and Powell 2001).

During 2001 we became aware of an ongoing mountain pine beetle outbreak in whitebark pine on Snow Bank Mountain in extreme west-central Idaho. This site was interesting because mortality was almost exclusively restricted to whitebark pine, even though a mixed-conifer forest immediately below the whitebark pine zone contained many host trees (lodgepole and ponderosa pine). Snow Bank is approximately 2,000 ft lower in elevation than Railroad Ridge. An earlier shift to synchronous univoltine populations and subsequent outbreak potential at lower/warmer locations would therefore be consistent with climate-warming predictions.

We also observed increasing mountain pine beetle

activity in lower-elevation host pines north of Railroad Ridge. We first observed significant mortality at our highest-elevation site on Railroad Ridge during the fall of 2003, with significant mortality (outbreak populations) during the summer of 2004 (Fig. C.7). The outbreak activity in whitebark pine beginning during the summer of 2003 was widespread and spectacular (Fig. C.8). In fact, the extent and intensity of the outbreak was surprising, even though we had been predicting the possibility of such an event for the previous ten years (Logan et al. 1995).

In a recent article, Bradley et al. (2004) reported that model predictions (the average of seven prominent coupled global circulation models) with CO_2 doubling scenarios indicated particularly intense summer warming at high elevations (above 3,000 m) and latitudes between 35° and 55°N. This corresponds almost exactly with the northern U.S. Rocky Mountains, and the predicted warming occurs at the most biologically sensitive time of year. Ongoing analysis of our temperature data from both Railroad Ridge and Snow Bank Mountain will provide data for evaluating the accelerated impact of climate warming at these elevations and latitudes.

Other factors may also play a role in the apparent sensitivity of whitebark pine ecosystems. As already noted, the mountain pine beetle acts in these situations as an invasive native species, perhaps outstripping the dispersive potential of associated native biological controls. Whitebark pine may also not be as well chemically defended as the more common host pines. Finally, the thick phloem of whitebark pine (relative to lodgepole pine) is a superior food resource for mountain pine beetles (Amman 1984). This is perhaps why even small whitebark pines (< 8 in. DBH) produce significant numbers of beetles. We hope that an ongoing, active research program will ferret out which combination of factors is responsible for the apparent sensitivity of whitebark pine to mountain pine beetle outbreaks.

In summary, our observations at Railroad Ridge and Snow Bank Mountain indicate the fragility of whitebark pine ecosystems, along with all the social and ecological services they provide, to climatic release of mountain pine beetle populations and resulting outbreaks. Clearly, there have been previous warm periods in the evolutionary history of whitebark pine, and the ecosystem has survived. Today, however, with the predicted high rate of anthropogenic warming combined with introduced pathogens and other global and human use changes (such a fragmentation of habitat and impacts of land use), the system is more susceptible than in previous extended warm periods. There is reason to be concerned about the loss of the biological "rooftop" of the Rockies as a result of climate warming.

Range Expansion to the North

High-elevation pine habitats are not the only at-risk ecosystems for mountain pine beetle invasions. In an article we published several years ago detailing the invasive potential for high-elevation systems(Logan and Powell 2001), we noted, almost as an aside, the analogy between high-elevation and northerly range expansion. Our model simulation results further predicted that range expansion north of the historical distribution of mountain pine beetle into jack pine (*Pinus banksiana* Lamb.) would be likely with a CO_2 doubling scenario. The map we published struck a responsive chord with our Canadian colleagues, primarily because of the explosive outbreaks that were occurring in Canada at that time (Fig. C.4b). If mountain pine beetle is successful in colonizing jack pine, there is a continuous connection of suitable host species across central Canada, down the east coast of the United States, and all the way to Texas (Fig. 8.5). What we are describing here is a potential biogeographic event of continental scale with unknown, but potentially devastating, ecological consequences.

Although a detailed description of the Canadian situation is beyond the scope of this chapter, several recent developments are worth noting. The first is the truly astounding scale of the outbreak in Canadian lodgepole pine (8.7 million ha as of summer 2005). More importantly, this outbreak has expanded far beyond the mountain pine beetle's historical range, which was largely restricted to British Columbia west of the Continental Divide. The recent outbreak event has breeched the divide through a low pass to the north and is now progressing eastward, roughly following the Peace River drainage. Isolated outbreak populations are currently within 50 km of the distribution overlap between lodgepole and jack pine.

Concurrent with the northward and eastward expansion of mountain pine beetle in British Columbia is the northward expansion of the southern pine beetle (*Dendroctonus frontalis* Zimmermann) along the east coast of the United States. Infestations were observed as far north as New Jersey in 2001, the first since an isolated event recorded in 1939. In the years since 2001, the initial outbreak area has expanded, doubling in size from 1,270 acres in 2002 to 2,508 acres in 2003 according to the Web page of the New Jersey Department of Environmental Protection

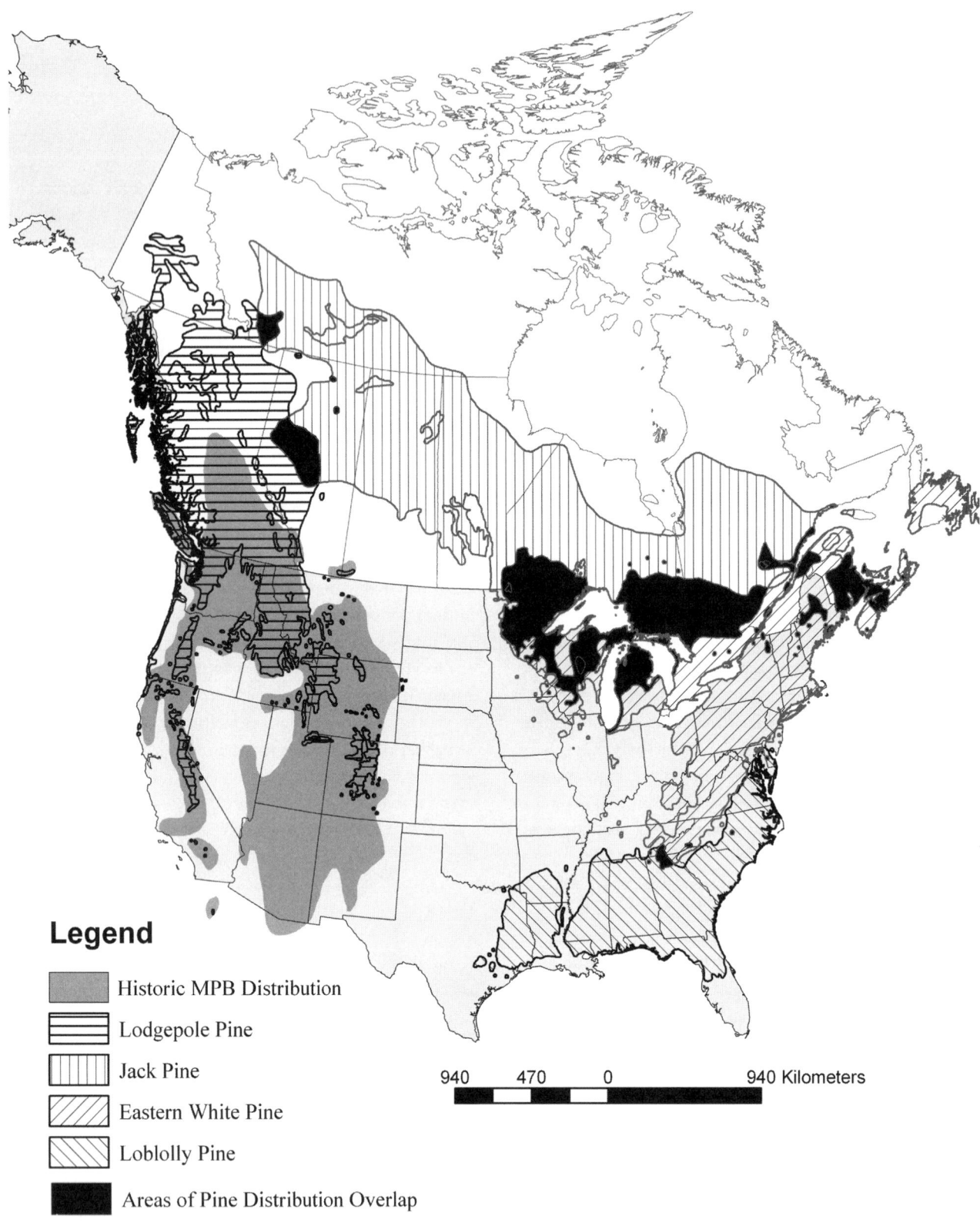

Legend

▨	Historic MPB Distribution
▤	Lodgepole Pine
▥	Jack Pine
▨	Eastern White Pine
▨	Loblolly Pine
■	Areas of Pine Distribution Overlap

940   470   0   940 Kilometers

Figure 8.5. The historical distribution of mountain pine beetle along with several widely distributed pine species. These distributions show the connection between the distribution of mountain pine beetle and essentially all North American pines. See text for a discussion of the significance of this connection in relationship to climate warming.

Division of Parks and Forestry (http://www.state.nj.us/dep/parksandforests/forest/njfs_spb .html). It is entirely conceivable that in the not-too-distant future we will see competition between the two most important bark beetles in North America in terms of economic impact. This hypothetical situation would indeed raise interesting ecological and economic issues.

Summary and Conclusions

The outbreaks of mountain pine beetle described above are unusual in that they have exceeded any previously observed during the period of historical record (European settlement). These observations can be summarized as:

1. Intensification of impact within the range of historic distribution. The outbreak in the Stanley Basin is an empirical example of this prediction. Although an outbreak occurred in this area during the warm period of the 1930s, this prior event resulted from a short pulse of several warm years. The current situation—if, in fact, it is the initiation of a predicted warming climate due to greenhouse gas accumulation—is fundamentally different. Outbreaks are a consequence of simultaneously occurring events, and as such, the probability of an outbreak is the product of the individual probabilities for these independent events. By increasing the probability of one of these (adaptive seasonality) to essentially one, the joint probability of occurrence is substantially increased. As a result, outbreaks of increased frequency and intensity are to be expected across the current distribution of mountain pine beetle. An exception to this general prediction may be in the southern limits of distribution. Model predictions are that excessively warm temperatures will lead to a breakdown in synchrony and loss of adaptive seasonality. Unfortunately, a number of indigenous species (including Mexican bark beetle, roundheaded bark beetle, southern pine beetle) are capable of occupying any niche vacated by mountain pine beetle.

2. A general northerly shift in population distribution. Again, models predict new areas of occupation in Canada as well as in historically occupied areas in the south that become too warm (Allan Carroll and Jacques Régnière, pers. comm.). This prediction also needs to be tempered by the fact that mountain pine beetle thermal ecology exhibits a significant amount of genetic and phenotypic plasticity (Bentz and Mullins 1999), and the capacity for northern populations to genetically adapt to a warming climate is unknown.

3. Encountering a lodgepole/jack pine overlap. Logan and Powell (2001) noted that the geographic distance between the historical northern distribution of mountain pine beetle in lodgepole pine and the distribution of unoccupied jack pine was less than the predicted range expansion under a CO_2 doubling scenario. The current situation in British Columbia appears to be following this prediction.

4. Continental-scale invasion of jack pine. If the mountain pine beetle successfully invades jack pine, the boreal distribution of jack pine across the North American continent will lead to consequences that are, at the least, economically devastating. The ecological consequences can only be guessed, but the history of invasive species impacts provides little optimism.

5. Subsequent invasion of pines southward along the eastern United States. As indicated in Figure 8.5, there is a continuous connection of pine distributions all the way from the western Rocky Mountains to the southeast United States. During historical times, the Great Plains has served as an insurmountable barrier to the eastward migration of mountain pine beetles. Once this barrier is breached to the north, as it has been in the Peace River valley of British Columbia, the previously inaccessible habitat of boreal and eastern pines will become vulnerable. The ecological impact of mountain pine beetle in these new habitats can only be surmised, but there is reason to expect the impact of a native invasive species will be no less than an exotic invasive species. The potential economic impact is even less ambiguous. There is every reason to expect major disruption of timber resources, particularly along the invading front. Intensified research is needed to predict rates of invasion, capacity for beetle genotypic and phenotypic adaptation to changing climate, and development of management strategies using potential terrain and habitat features as barriers to invasion.

6. Range expansion into high-elevation five-needle pines, with subsequent loss of biodiversity. The extent and intensity of mortality that occurred in whitebark pine during the summer of 2003 was a surprise, even though we had predicted potential problems with a warming climate (Logan and Bentz 1999). If our current observations are an indication of climate warming, then the question becomes "Can we formulate an effective management response?" Management of these systems is difficult because they are remote, inaccessible, spatially discontinuous, and extensive.

Although we do have direct beetle population management options—including effective chemical pesticides, pheromone strategies, and vegetation management—these approaches are labor intensive and expensive. That fact, combined with the nature of the habitat, precludes their application over any meaningful proportion of whitebark pine distribution. However, giving up hope is not an option; the resource is simply too valuable. Perhaps we can gain insights from the adaptive strategies that have allowed whitebark pine to survive past prolonged warm periods. Through a combination of these insights with advanced technologies of computer modeling, remote sensing, and landscape ecology, there is hope of formulating a viable management response.

As noted in the introduction, the situation with mountain pine beetle is only one of several current unprecedented outbreaks involving a variety of bark beetles. One commonality across time and space for all of these events is increasing temperatures that began in the western United States during the mid-1970s and continue to the present. The time has passed for simply being aware of the potential consequences of a warming climate, and it is now time to begin formulating viable responses. Insights from paleoecology may be helpful in doing so; however, other aspects of global change complicate the current situation. The accelerating pace of habitat fragmentation may foreclose important adaptive strategies of the past. Fire suppression and other management practices have altered community relationships. Finally, introduced pests and pathogens (e.g., white pine blister rust) are further stressing systems, rendering them increasingly vulnerable to other disturbances.

Predicting community-level effects of a warming climate is a complex problem because of inherent nonlinearity and resulting threshold events. We have some understanding of the threshold response that mountain pine beetle exhibits to warming temperature (Logan and Bentz 1999, Logan and Powell 2004, Powell and Logan 2005), but this situation is unusual and has resulted only because of the long-term research commitment to this insect. We know much less about other western bark beetle species. About the only definitive thing that can be said is that the impacts of a warming climate will be expressed through disruption of coadapted community interactions long before the direct effects become obvious (Malanson 2001, Logan et al. 2003)—and that the unexpected is to be expected. Formulation of effective management responses in the face of such uncertainty presents serious challenges, and the only hope lies in the effective integration of advanced technology with in-depth ecological knowledge.

Acknowledgments

The technical support and ecological insights provided by Jim Vandygriff are gratefully acknowledged. Without his dedication and ability to maintain delicate weather instrumentation under some of the most severe conditions in North America, none of our high-elevation work would have been possible.

Research reported herein was supported in part by the National Science Foundation of the United States under grant number DMS 0077663.

References

Amman, G. D. 1973. Population changes of the mountain pine beetle in relation to elevation. *Environ. Entomol.* 2:541–547.

———. 1984. Mountain pine beetle (Coleoptera: Scolytidae) mortality in three types of infestations. *Environ. Entomol.* 13(1): 184–191.

Amman, G. D., and R. F. Schmitz. 1988. Mountain pine beetle–lodgepole pine interactions and strategies for reducing tree losses. *AMBIO* 17:62–68.

Arno, S. F. 2001. Community types and natural disturbance processes. Pp. 74–88 in D. F. Tomback, S. F. Arno, and R. E. Keane (eds.). *Whitebark Pine Communities: Ecology and Restoration.* Island Press, Washington, D.C.

Bentz, B. J., J. A. Logan, and J. C. Vandygriff. 2001. Regional life history variation in *Dendroctonus ponderosae. Can. Entomol.* 133:375–387.

Bentz, B. J., and D. E. Mullins. 1999. Ecology of mountain pine beetle (Coleoptera: Scolytidae) cold hardening in the intermountain west. *Environ. Entomol.* 28(4):577–587.

Berg, E. E. 2003. Fire and spruce bark beetle disturbance regimes on the Kenai Peninsula, Alaska. *Proc., 2nd Intern. Wildland Fire Ecology and Fire Management Cong.,* Orlando, FL.

Bradley, R. S., F. T. Keimig, and H. F. Diaz. 2004. Projected temperature changes along the American cordillera and the planned GCOS network. *Geophys. Res. Lett.* 31:L16210.

Carroll, A. L., and L. Safranyik. 2004. The bionomics of the mountain pine beetle in lodgepole pine forests: Establishing a context. Pp. 19–30 in T. L. Shore, J. E. Brooks, and J. E. Stone (eds.), *Proc., Mountain Pine Beetle Symposium: Challenges and Solutions.* October 30–31, 2003, Kelowna, British Columbia, Canada.

Dale, V. H., L. A. Joyce, S. McNulty, R. P. Neilson, M. P. Ayres, M. D. Flannigan, P. J. Hanson, L. C. Irland, A. E. Lugo, C. J. Peterson, D. Simberloff, F. J. Swanson, B. J. Stocks, M. B. Wotton. 2001. Climate change and forest disturbance. *BioScience* 51:723–734.

IPCC (Intergovernmental Panel on Climate Change). 1990. *Climate Change: The IPCC Scientific Assessment.* Report prepared for IPCC by Working Group I. J. T. Houghton, G. J. Jenkins and J. J. Ephraums (eds.). University Press, Cambridge.

Jenkins, J. L., J. A. Powell, J. A. Logan, and B. J. Bentz. 2001. Low seasonal temperatures promote life cycle synchronization. *Bull. Math. Biol.* 63:573–595.

Logan, J. A., and B. J. Bentz. 1999. Model analysis of mountain pine beetle seasonality. *Environ. Entomol.* 28:924–934.

Logan, J. A., P. V. Bolstad, B. J. Bentz, and D. L. Perkins. 1995. Assessing the effects of changing climate on mountain pine beetle dynamics. Pp. 92–105 in R. W. Tinus (ed.), *Proc., Interior West Global Climate Workshop.* USDA Forest Service GTR-RM-262. Fort Collins, CO.

Logan, J. A., and J. A. Powell. 2001. Ghost forests, global warming, and the mountain pine beetle. *Am. Entomol.* 47:160–173.

———. 2004. Modelling mountain pine beetle phenological response to temperature. Pp. 210–222 in T. L. Shore, J. E. Brooks, and J. E. Stone (eds.), *Mountain Pine Beetle Symposium: Challenges and Solutions.* October 30–31, 2003, Kelowna, British Columbia, Canada. Natural Resources Canada, Canadian Forest Service, Pacific Forestry Centre, Victoria, British Columbia, Information Report BC-X-399.

Logan, J. A., J. Régnière, and J. A. Powell. 2003. Assessing the impacts of global climate change on forest pests. *Front. Ecol. Environ.* 1:130–137.

Logan, J. A., D. J. Wollkind, S. C. Hoyt, and L. K. Tanigoshi. 1976. An analytical model for description of temperature dependent rate phenomena in arthropods. *Environ. Entomol.* 5:1133–1140.

Malanson, G. P. 2001. Complex response to global change at alpine treeline. *Physiol. Geog.* 22:333–342.

Mattson, W. J. 1996. Escalating anthropogenic stresses on forest ecosystems: Forcing benign plant-insect interactions into new interaction trajectories. In H. Korpilahti, H. Mikkela, and T. Salonen (eds.), *Caring for the Forest: Research in a Changing World.* Congress Report vol. 2. IUFRO World Congress Organizing Committee, Finland.

Perkins, D. L., and T. W. Swetnam. 1996. A dendrochronological assessment of whitebark pine in the Sawtooth–Salmon River region, Idaho. *Can. J. For. Res.* 26:2123–2133.

Powell, J. A., J. Jenkins, J. A. Logan, and B. J. Bentz. 2000. Seasonal temperature alone can synchronize life cycles. *Bull. Math. Biol.* 62:977–998.

Powell, J. A., and J. A. Logan. 2005. Insect seasonality: Circle map analysis of temperature-driven life cycles. *Theor. Popul. Biol.* 67:161–179.

Raffa, K. F., and A. A. Berryman. 1987. Interacting selective pressures in conifer-bark beetle systems: A basis for reciprocal adaptation? *Am. Nat.* 129:234–262.

Ross, D. W., G. E. Daterman, J. L. Boughton, and T. M. Quigley. 2001. *Forest Health Restoration in South Central Alaska: A Problem Analysis.* USDA Forest Service Gen. Tech. Rep. PNW-GTR-523.

Safranyik, L. 1978. Effects of climate and weather on mountain pine beetle populations. Pp. 77–84 in A. A. Berryman, G. D. Amman, and R. W. Stark (eds.), *Proc. Symposium: Theory and Practice of Mountain Pine Beetle Management in Lodgepole Pine Forests.* April 25–27, 1978, Moscow, ID. University of Idaho Forest, Wildlife and Range Experiment Station, Moscow, ID.

Samman, S., and J. Logan (tech. eds.). 2000. *Assessment and Response to Bark Beetle Outbreaks in the Rocky Mountain Area: A Report to Congress from Forest Health Protection, Washington Office.* USDA Forest Service, RMRS-GTR-62.

9

A Synthesis of Recent Climate Warming Effects on Terrestrial Ecosystems of Alaska

Valerie A. Barber, Glenn Patrick Juday, Rosanne D'Arrigo, Edward Berg,
Brendan Buckley, Henry Huntington, Torre Jorgensen, David McGuire,
Tom Osterkamp, Brian Riordan, Alex Whiting, Greg Wiles, and Martin Wilmking

Abstract

The instrument-based climate record in Alaska displays a strong late-twentieth-century warming. Climate in Alaska also displays a record of sudden regime shifts. Precipitation there is highly variable and shows no strong trends. Effective moisture (P–PET), however, has decreased, resulting in widespread shrinkage and drying of lakes and ponds in regions of low or moderate precipitation. Overall glacial mass balance is negative, and most show ice margin retreat, although some glacial systems are in positive mass balance. Permafrost is warming across the state, and ground subsidence associated with thawing of ice-rich permafrost is commonly observed. Since buildings and infrastructure, as well as natural disturbances, can cause warming of the permafrost, it is difficult to distinguish from climatic warming in some cases. The annual period of snow and ice cover is decreasing, and growing season is increasing in length with greater normalized difference vegetation index (NDVI) greenness in the tundra region. North of the Brooks Range, tall shrubs have advanced into the tundra, and warming experiments show that low shrub cover would significantly increase with additional warming. White spruce populations at treeline include trees that grow more with warming as well as others that grow less with warming. Major species in the boreal forest region also include populations with similar responses, but growth on many of the most productive sites has declined. Recent high temperatures have caused widespread tree stress. Major outbreaks of tree-damaging insects have occurred due to both tree stress and direct temperature controls on insects. Millions of acres of beetle-killed trees on the Kenai Peninsula are a potential fire hazard. The extent of forest fires in Alaska is positively associated with specific temperature factors. These changes are confronting people with a variety of challenges, ranging from obtaining subsistence food and potable water to maintaining health and safety. Scenarios of future Alaska climate produced by general circulation models project significant future warming, which would exceed the apparent tolerance of some component species of current ecosystems.

Introduction

Climate warming caused by the increase of greenhouse gasses comes about from both anthropogenic additions to the atmosphere (fossil fuel combustion, novel industrially produced greenhouse gasses, cement production) and from reduction in the Earth's photosynthethic capacity, either naturally or through human land use changes that usurp primary production and thus carbon uptake from the atmosphere. The Earth's high latitudes, especially the Arctic and Subarctic regions, have long been projected as the areas likely to experience the greatest magnitude of climate warming from these processes.

In situ formation of warm and cool air can occur radiatively in this region, and even though changes in the distribution and movement of air mass patterns either import heat or export cold across specific portions of the region, general circulation models (GCMs) indicate that, as a whole, higher latitudes will experience the greatest warming from global climate change (Houghton et al. 1995, IPCC 2001). As a result, a warming trend in the northern high latitudes should first be detectable against the background noise of the Earth's variable climate system.

Climate change is naturally amplified in the Arctic and Subarctic through several processes. Changes in Arctic Ocean ice cover and boreal land cover affect albedo, or reflective power, with feedbacks that are significant at the planetary level (Bonan et al. 1992). Increased forest disturbances such as fire and insect-caused tree death can promote decomposition of fixed carbon in biomass and soils, and its release as greenhouse gasses. A particular vulnerability that amplifies climate-warming effects in the far north is the potential for abrupt change as physical systems rapidly convert from the frozen state, or biological systems become less limited by frozen conditions.

Although the records are spatially sparse and limited in duration, the instrument-based record of temperatures in Alaska clearly demonstrates recent warming. Observations of temperature-dependent phenomena are particularly valuable, providing additional time perspective of climate variability and offering insight into some of the effects of warming. Change in Alaska is occurring rapidly, and consistent evidence of warming is seen in changing hydrology, permafrost, forests, disturbances, and other features.

In this chapter we present some of the evidence of climatic change in Alaska and mention some of its potential impacts on northern ecosystems and their human inhabitants. The general consensus of the Arctic Climate Impact Assessment (ACIA) is that warming is occurring at an alarming rate, and although natural variability plays a role, most warming is from anthropogenic sources: the increase of greenhouse gases in the atmosphere from the combustion of fossil fuels and changes in land use (Symon et al. 2005).

Synoptic Climate and Instrument Record

A recent synthesis of climate evidence across the Arctic region in general confirms widespread (although not quite universal) warming over the last 30 years (Serreze et al. 2000, Symon et al. 2005). Mean annual temperatures in Alaska, in particular, have increased significantly over that period, as represented by mean annual temperature data from widely separated weather stations (Fig. 9.1). The data reveal that the climate in interior, south-central, and southeastern Alaska has switched from a predominantly cool and moist period to hot and dry after a Pacific-wide regime shift in 1977 (Barber et al. 2004, Ebbesmeyer et al. 1990). Although the greatest magnitude of warming occurred seasonally in winter, temperatures during the ecologically important spring and summer (May–August) increased as well (Fig. 9.2, top). Autumn temperatures, however, have decreased slightly overall.

In addition to seasonal unevenness in the distribution of recent temperature increases, there was a considerable diurnal difference. In the highly continental climate region of central Alaska, the mean of daily high temperatures during the warm season increased only very slightly during the twentieth century, but the mean of daily low temperatures increased more than 3°C (Fig. 9.2, bottom). Given this pattern of temperature change, the increase in daily minima during the growing season has been disproportionate, so that the increase in growing-season length in central Alaska, while still highly variable, has been particularly impressive (Fig. 9.3). The concentration of warming in the daily temperature minima is consistent with a process operating to dampen heat loss (greenhouse gases) rather than amplify energy input (e.g., increased solar luminosity).

Records of direct precipitation over the last 50 to 100 years do not show the same strong trend as temperature. Some regions of Alaska (along portions of the southern coast) had increased precipitation, but the North Slope, interior Alaska, and the Kenai Peninsula did not experience a noticeable increase. Across much of central and interior Alaska, annual total precipitation is quite low. Precipitation in the 100-year Fairbanks climate record, one of the longest in Alaska, is about 280 mm/year.

The pattern of variability over time in temperature and precipitation is distinctive. The climate of interior Alaska (and the south-central coast) comes packaged in alternate decadal-scale periods of growth-year precipitation (September–August) and May–August temperature (Barber et al. 2004) (Fig. 9.4, top). A strong maritime influence results in a cool, moist climate, whereas a strong continental influence results in a warm, dry climate. The trend of growth-year precipitation in Fairbanks has been slightly negative, but the summer temperature trend has been strongly positive (Fig. 9.4, top).

This decadal-scale variability is superimposed on the longer-term century-scale warming. Instrument records reveal that the latest climate regime, begun in the mid-1970s, has been the most sustained period of hot and dry weather in more than a century (Barber et al. 2004)—and much longer as determined by climate reconstruction proxies (Mann et al. 1999, Overpeck et al. 1997) (Fig. 9.4, bottom).

A summer temperature reconstruction based on stable isotopes (carbon-13) and maximum latewood density of low-elevation upland white spruce from interior Alaska (Barber et al. 2004) show continuity of similar temperature regimes back through 1800. Periods of warm summer

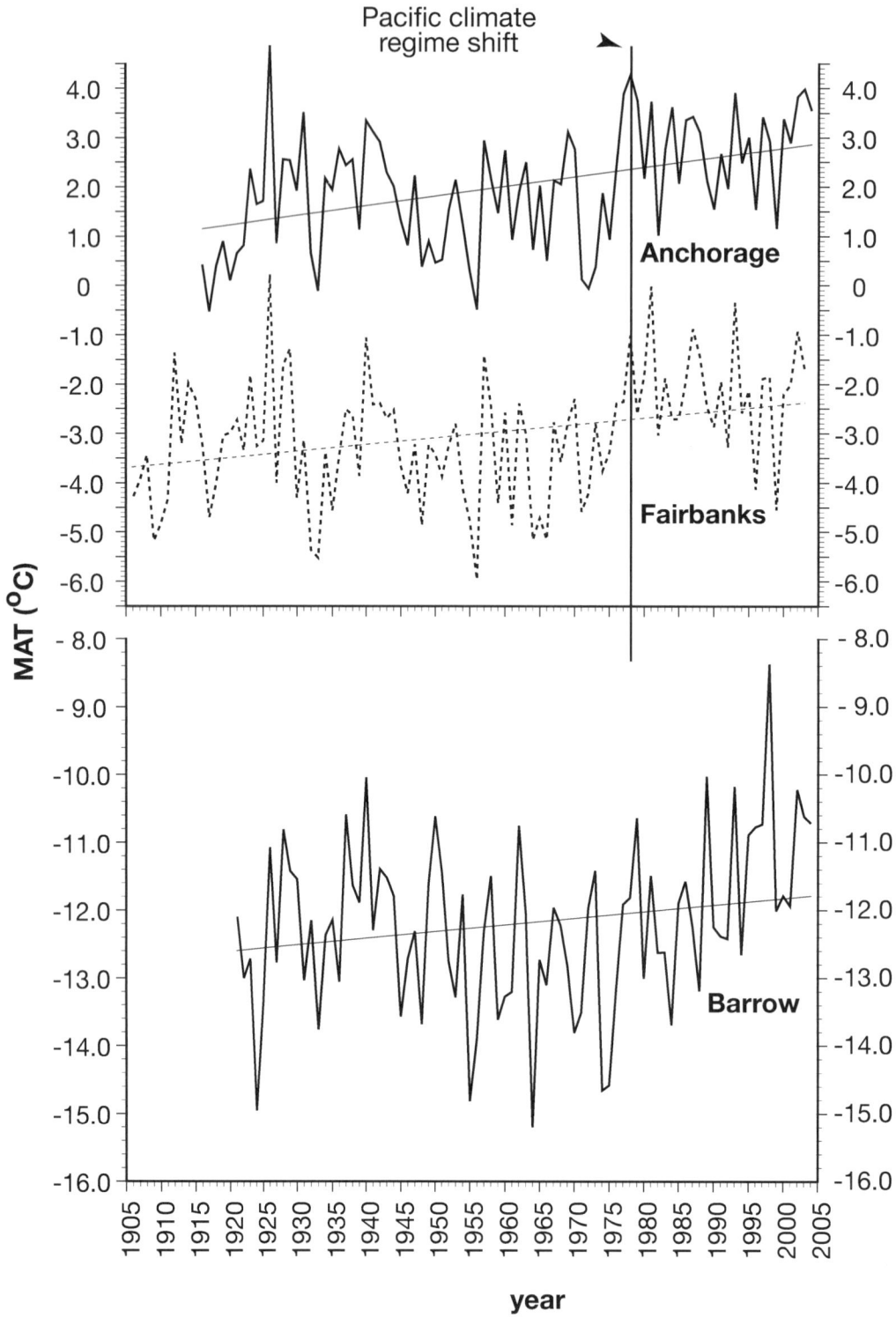

Figure 9.1. Mean annual temperatures at three locations in Alaska. The 1976–1977 Pacific climate regime shift is indicated in the upper panel.

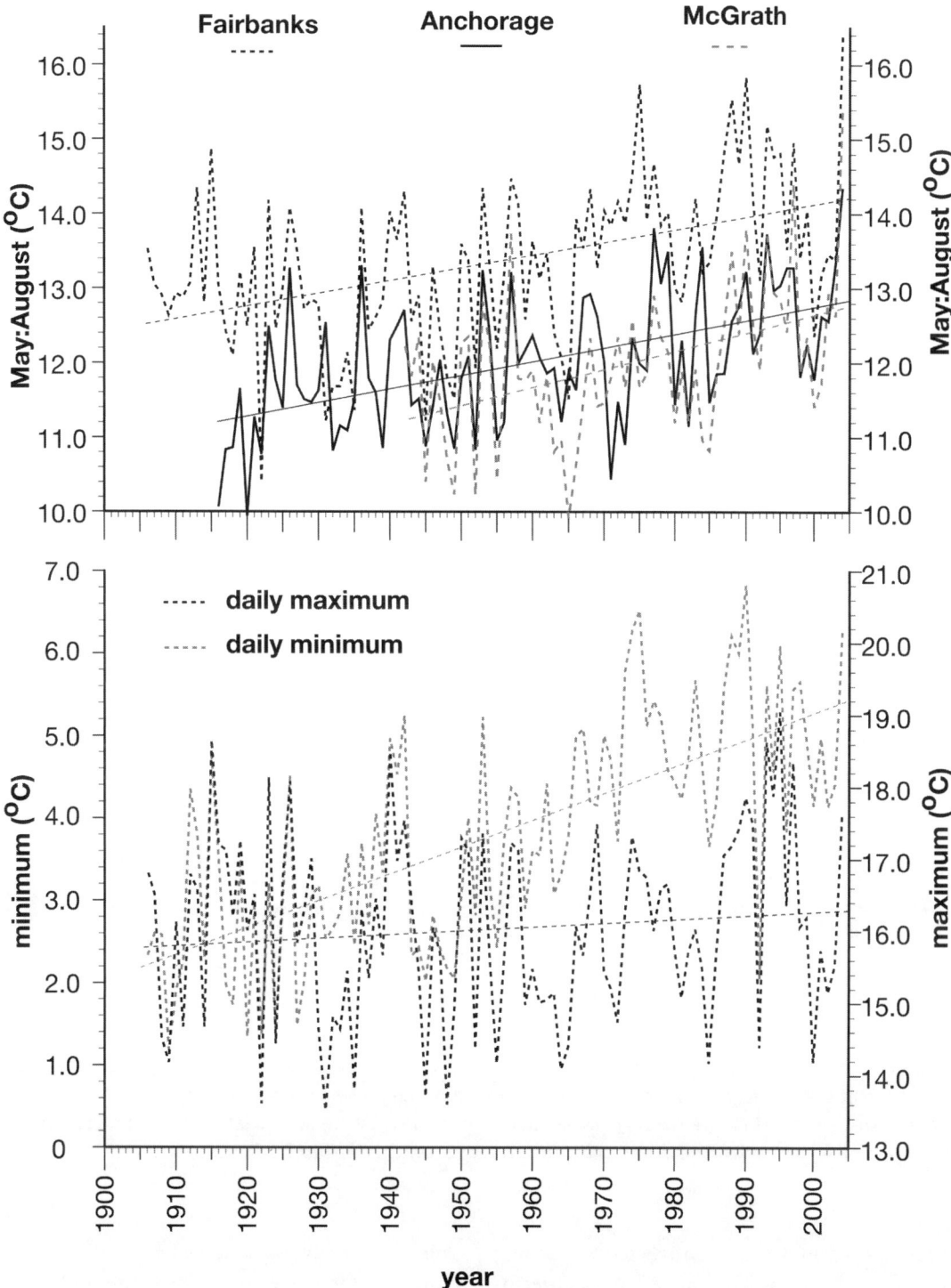

Figure 9.2. Warm season (May–August) temperatures at three localities in Alaska: Anchorage, 1916–2004 (south-central); Fairbanks, 1906–2004 (central interior); and McGrath, 1942–2004 (southwest interior). Anchorage and Fairbanks have experienced urban population buildup and heat island effects (not severe in the season depicted); McGrath has remained a small village with minimal anthropogenic heat input.

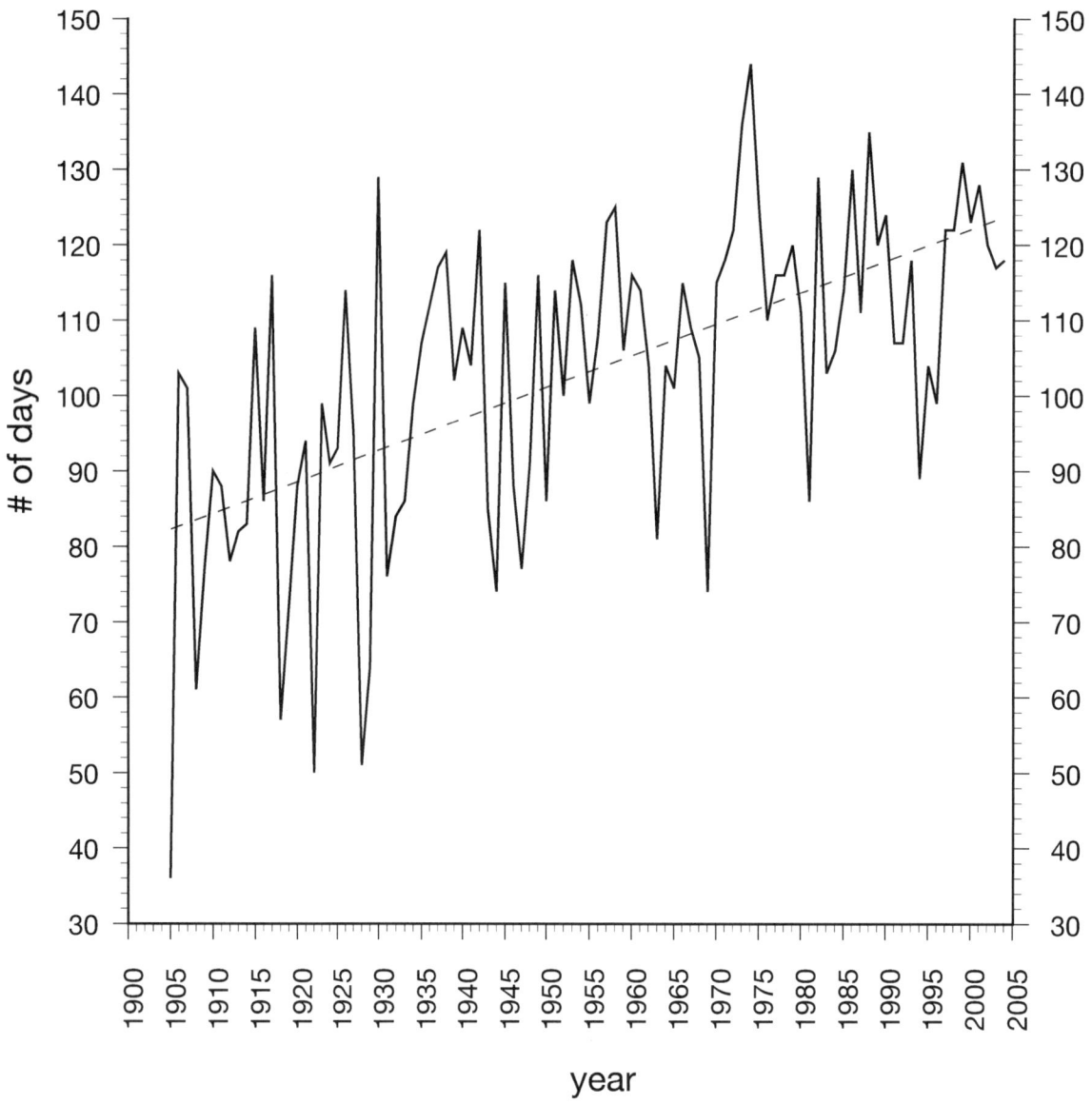

Figure 9.3. Growing-season length (greatest number of consecutive days between latest spring and earliest fall dates with temperatures equal to or below freezing) in Fairbanks. Regression line added to allow visualization of overall rate of change on long (century) time scale. Rate of increase in overall season length is 41 days per century, with an average 16-day earlier date of last spring frost and a 25-day delay in first fall frost.

temperatures versus cold can be reconstructed by tree-ring proxies as confirmed by recorded climate data, but precipitation is not sufficiently correlated to tree-ring properties to allow reconstruction.

The early part of the twentieth century was the coolest and wettest weather of the century in Alaska, which made the transition to the late-twentieth-century extremes of warmth particularly steep in the instrument-based records that span precisely that time period. Some of the

climatic trends are synergistic in their effects. For example, as temperature increases, the timing of snowmelt occurs earlier in the spring, which truncates the period of snowpack accumulation and extends the period of liquid-moisture evaporation. Fall initiation of snowpack has been delayed, although not to the same degree. As a result, even with steady or even slightly increasing annual precipitation, plants experience greater evapotranspiration demand under the warming climate regime.

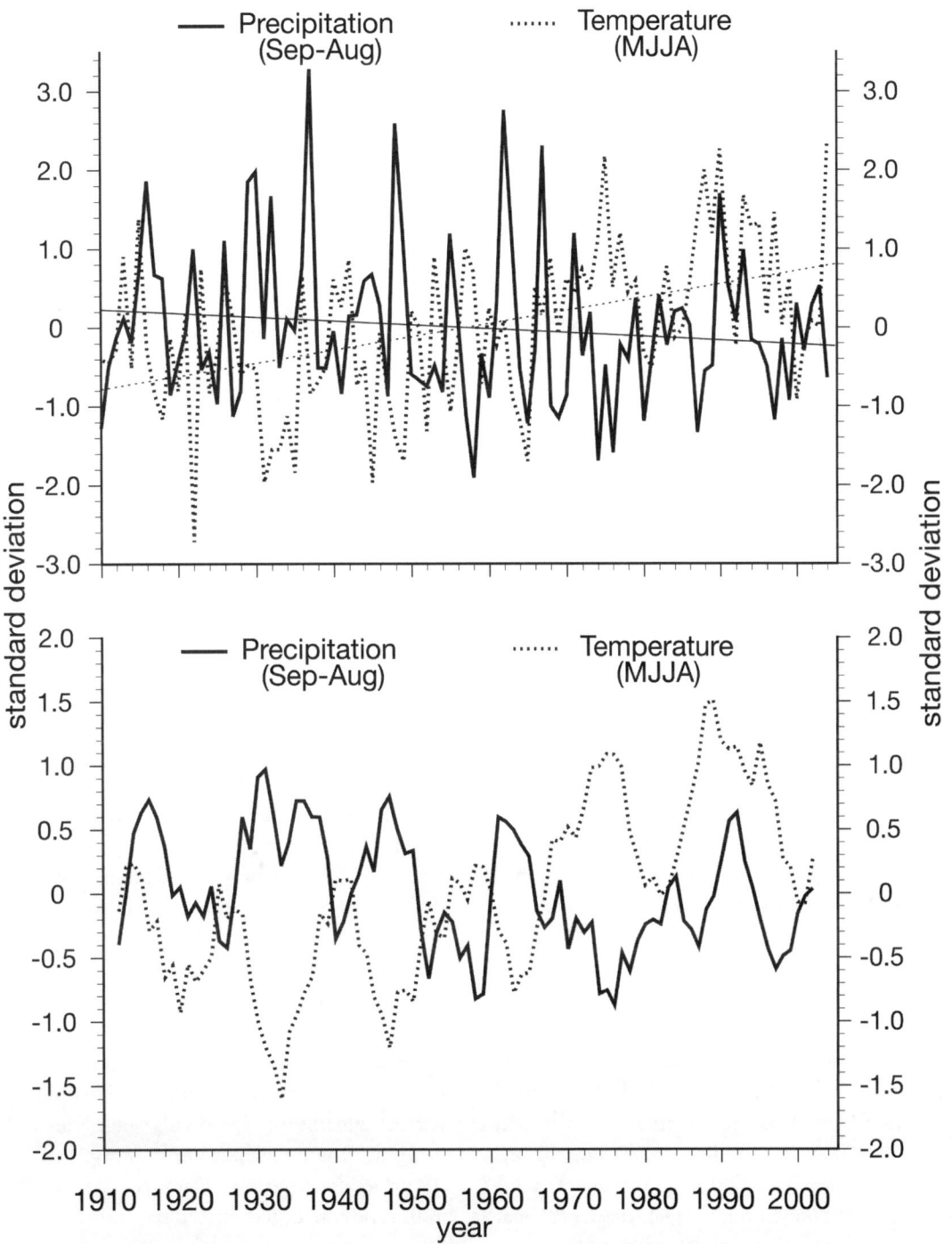

Figure 9.4. Normalized warm season (May–August) temperature versus normalized growth year (September–August) precipitation at Fairbanks. Top graph represents individual yearly values; bottom graph represents smoothed yearly values with five-year running mean.

Table 9.1 Area (ha) of water surface in closed-basin lakes and ponds by date and method of detection, with cumulative percentage change

Study Site	B&W Aerial Photography 1949–56	Color Infra-red Aerial Photography 1978–82	Landsat TM 1991	Landsat TM 1995	Landsat ETM+ 1999	Landsat ETM+ 2000	Landsat ETM+ 2001	Landsat ETM+ 2002	Total Percentage Change 1950–2000
Innoko Flats NWR	2,126	1,532				1,463			−31
Copper River Basin	130	95		85			93		−28
Minto Flats SGR	4,066	3,763	3,952			3,319		3,060	−25
Yukon Flats NWR	7,773	8,097				6,359			−18
Stevens Village	5,132	5,106	4,768			4,746		4,398	−14
Talkeetna	1,530	1,479				1,455			−5
Denali NP	1,758	1,964				1,681			−4
Tetlin NWR	4,811	4,566			4,597				−4
Prudhoe Bay ACP	3,162	3,127			3,208				1

Effects on Hydrology

Effective moisture, or surface-water balance, can be calculated as precipitation minus potential evapotranspiration (P–PET). Effective moisture has declined over central and northern Alaska as precipitation has failed to keep pace with warming, causing a water deficit in the summer months. For interior Alaska, the summer water deficit over the past 40 years has increased by almost 50 percent (−450 to −770 mm), while the coastal plain summer water deficit has increased by about 200 percent (−40 to −120 mm) (Hinzman et al. 2005). As a consequence of decreasing effective moisture, surface-water levels have been declining in closed-basin lakes (no inlet or outlet surface flow) in most of Alaska.

Aerial photos taken in the 1950s were compared against Landsat ETM images from 2000–2002 in a study of different regions in Alaska with surface water in the form of lakes and ponds (Fig. 9.5). The study determined the regional trend in surface-water area loss (Riordan 2005). The closed-basin water bodies were in nine regions across Alaska: (1) Copper River Basin, (2) Talkeetna, (3) Tetlin National Wildlife Refuge, (4) Denali National Park, (5) Innoko Flats National Wildlife Refuge, (6) Minto Flats State Game Refuge, (7) Stevens Village, (8) Yukon Flats National Wildlife Refuge, and (9) Prudhoe Bay/Arctic Coastal Plain. The study included more than 40,000 water bodies on approximately 1.8 million acres. Analysis was based on GIS and remote sensing techniques. Water body change was detected over a 50-year time period, with a minimum of three time periods used for each area.

All study regions in subarctic Alaska lost surface-water area from 1950 to 2002 (Fig. 9.6). The Prudhoe Bay/Arctic Coastal Plain region—which is underlain by deep, continuous permafrost (often in excess of 400 m)—had negligible change. Three areas experienced surface-water losses prior to 1977, three after 1977, and two before and after 1977 (Table 9.1). Copper River Basin and Innoko Flats National Wildlife Refuge had the most substantial reduction in surface-water area, with most of the loss occurring prior to 1977.

These observed surface-water losses are likely the result of three mechanisms: increased evapotranspiration, increased active-layer water-holding capacity, and enlarged taliks (unfrozen layers sandwiched between basal permafrost and overlying seasonally frozen soil layers). Temperature is the driving factor for all three mechanisms; both a growing active layer and enlarged taliks are responses to increased temperature and associated changes in permafrost dynamics (e.g., an enlarged volume of unfrozen material). These mechanisms may also lead to increased subsurface drainage in cases where water outflow from the basin is no longer obstructed by an impervious unfrozen layer. The consequences of these changes include changing plant communities and drying soils.

Clearly, from 1950 to 2002, there has been a substantial change in closed-basin water bodies in many areas of subarctic Alaska. While the exact causes of surface-water area loss are unknown, the change occurred during a period of increasing temperatures when precipitation levels did not change significantly. Because the landscape-level change has occurred in regions of discontinuous,

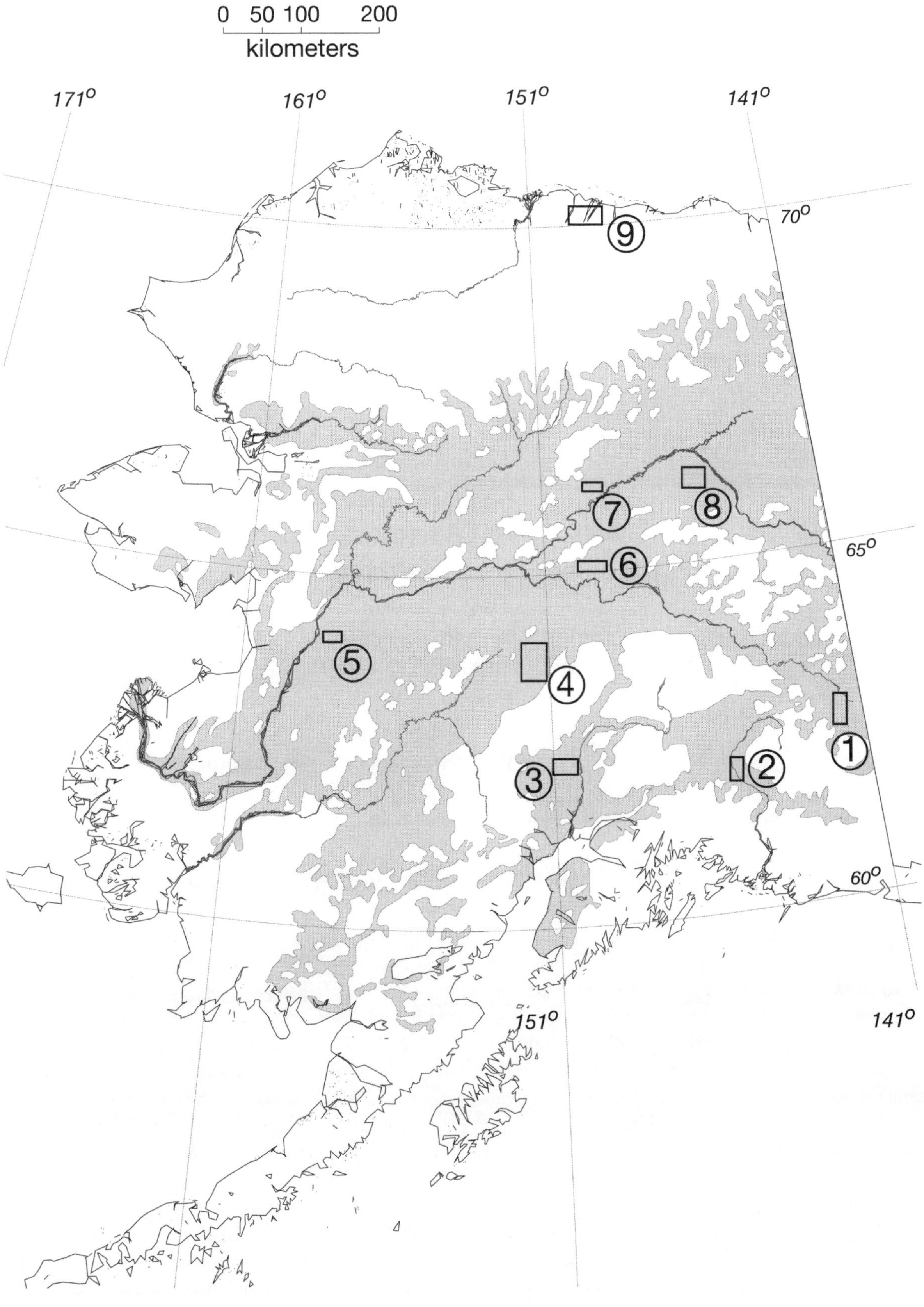

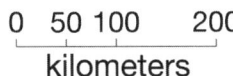

Figure 9.5. Nine study regions across Alaska, from south to north: (1) Tetlin National Wildlife Refuge, (2) Copper River basin, (3) Talkeetna, (4) Denali National Park, (5) Innoko Flats National Wildlife Refuge, (6) Minto Flats State Game Refuge, (7) Stevens Village, (8) Yukon Flats National Wildlife Refuge, and (9) Prudhoe Bay, Arctic Coastal Plain.

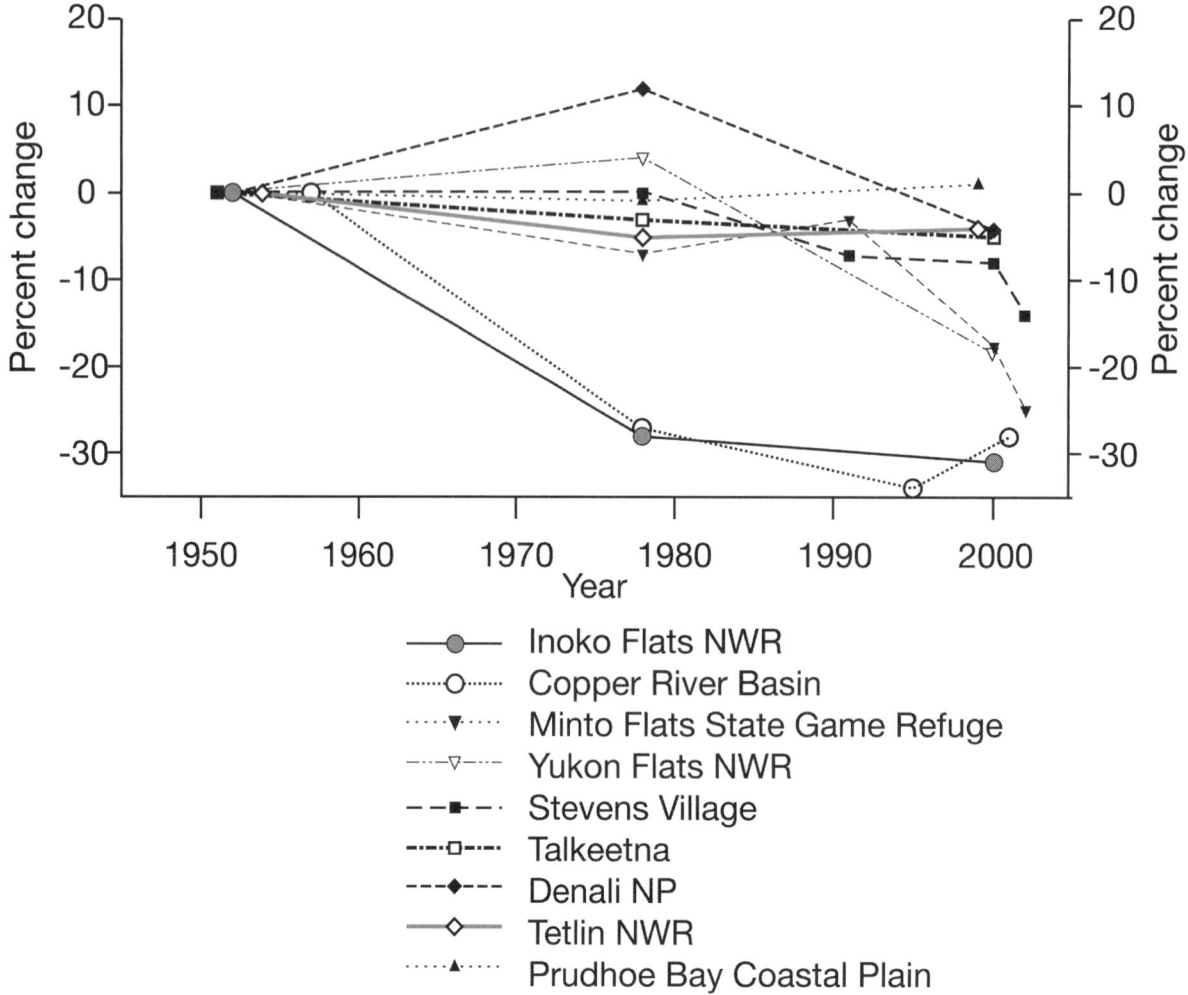

Figure 9.6. Percentage change in extent of open-water habitat in the nine study areas in Alaska.

warming permafrost, talik formation and growth may be an important mechanism for loss of surface water from closed-basin water bodies.

On the permafrost-dominated Seward Peninsula (western Alaska), 22 of 24 ponds shrank or dried up completely between 1950 and 2000 (Hinzman et al. 2005). Here, permafrost appears to be thawing, enabling taliks to merge below the ponds and providing pathways for the movement of surface pond water to groundwater or to nearby streams. There is also encroachment by vegetation as woody trees and shrubs advance onto former wetland surfaces, leading to increased shrubbiness and increased evapotranspiration.

Loss of open-water surface is also widespread in a non-permafrost area of Alaska, supporting the mechanism of a temperature-driven decrease in effective moisture. On the Kenai Peninsula south of Anchorage almost two-thirds of all water bodies studied showed some evidence of decrease in spatial extent (Klein et al. 2005). Some of these small lakes and ponds have disappeared.

Effects on Cryosphere

Sea Ice

The length of the cold season with snow and ice cover across the high-latitude north is decreasing, while at the same time the unfrozen season is increasing (Magnusom et al. 2000). In central and western North America and Eurasia during the period of reliable records (generally mid-1800s to present), there has been a strong trend toward shorter ice-covered seasons because of both later freeze-up and earlier breakup dates (Walsh et al. 2005). In the eastern portions of both continents, there has been little or no such trend. The trend toward earlier breakup has been particularly strong on the Tanana River at Ne-nana (central Alaska), where breakup has advanced earlier in the spring at the rate of 7.5 days per century since the record began in 1917 (Sagarin and Micheli 2001).

Other evidence of warming comes from changes in sea ice conditions in the Arctic. A 9 percent per decade decrease in perennial sea ice cover in the Arctic was documented between 1978 and 2000 using satellite data (Comiso 2002). In 2002, the floating sea ice cover of the Arctic Ocean shrank to record low levels (Serreze et al. 2003), but those were nearly matched the following two years, and a new record low was recorded in 2005.

Reduced sea ice cover represents a positive feedback to warming because open water has a lower albedo than ice, allowing more heat to be absorbed. Shrinking and dis-appearing sea ice has major ecological implications. Many fish, birds, and mammals live in, on, or under sea ice, or are otherwise constrained or aided by its presence. Changing sea ice conditions will impact these species as well as the people who rely on them for food and sustenance.

Sea ice affects coastal regions in other ways, and its presence protects against storm-driven shoreline ero-sion from fall and winter storms. Many coastal villages in northwest Alaska are experiencing increased erosion from later formation, earlier melt, and lower integrated annual extent of sea ice. Millions of dollars are needed to move these villages farther inland to protect life and property. Enhanced battering from these storms in ice-free seas erodes and exposes permafrost, promoting further erosion.

Glaciers

One of the most cited and striking examples of warming across Alaska in the twentieth century is the loss of glacier mass. The McCall Glacier in Alaska is an example of a gla-cier in negative mass balance. Its late-twentieth-century elevation is about 25 m lower than it was in 1969 (Rabus et al. 1995), and the glacier is continuing to lose mass at an increasing rate (Nolan et al. 2005). The retreat of glaciers is being seen all across Alaska. Retreats from the Little Ice Age maximum extended positions of the termini are measured in kilometers at many small land-terminating glaciers, and in excess of 10 km from southern coastal tide-water glaciers over the past several decades (Meier and Dyurgerov 2002).

The glacial geologic record (Wiles et al. 2004) and observation of recent changes in glacier length as well as mass-balance studies (Bitz and Battisti 1999, Trabant et al. 2003) reveal the sensitivity of Alaska's glaciers to multi-

decadal to century-scale climate variability. Although some modest gains in glacier mass from increased snow-fall have been documented (Bitz and Battisti 1999, Tra-bant et al. 2003), and some tidewater glaciers have surged and advanced, the overall trend has been retreat, with ac-celerated losses since 1988 (Arendt et al. 2002, Dyurgerov and Meier 1997 and 2000) (Fig. 9.7).

In the most comprehensive assessment, Arendt et al. (2002) used airborne laser altimetry to estimate thickness and volume change at 67 of Alaska's glaciers. This work showed that recent losses, logged at a few glaciers where mass-balance monitoring is underway (Trabant et al. 2003), are characteristic of the losses of this larger popula-tion of glaciers. Furthermore, their work has shown that previous estimates of world sea level significantly under-estimated the contribution of melting Alaskan glaciers to the global rise of sea level (Arendt et al. 2002, Meier and Dyurgerov 2002).

Results from this study, plus the accelerating loss of mass of the largest glaciers along the southern coast (Meier and Dyurgerov 2002), are strong reminders of the global effects of warming in Alaska and how it impacts the world by contributing to rising global sea levels. Extrapo-lating the thinning rate of the 67 measured glaciers to all of Alaska's glaciers indicates a contribution to world sea-level rise of 0.14 mm/yr (Walsh et al. 2005). More local impacts include the loss of fresh water supply, increased river temperatures, and potential effects on the North Pa-cific climate via fresh water influx.

Permafrost

Permafrost (soil or rock that remains frozen for two years or more) extends across 25 percent of the Northern Hemi-sphere (Brown et al. 1997) and is particularly sensitive to climate change and human disturbance. Of particular interest, however, is the 18 percent of the circumpolar area and the 38 percent of boreal forest region in Alaska (Nowacki et al. 2002) underlain by permafrost in lowland areas. These areas typically have high ice contents associ-ated with fine-grained surficial deposits and are at greater risk than mountainous areas for thermokarst (thawing and settling of the ground surface following evacuation of the frozen water). Thus, the potential consequences to hu-man infrastructure and ecological processes are also much greater.

Ground temperatures obtained from deep boreholes in Alaska provide a centuries-long record of rising perma-frost temperatures since the Little Ice Age (Lachenbruch

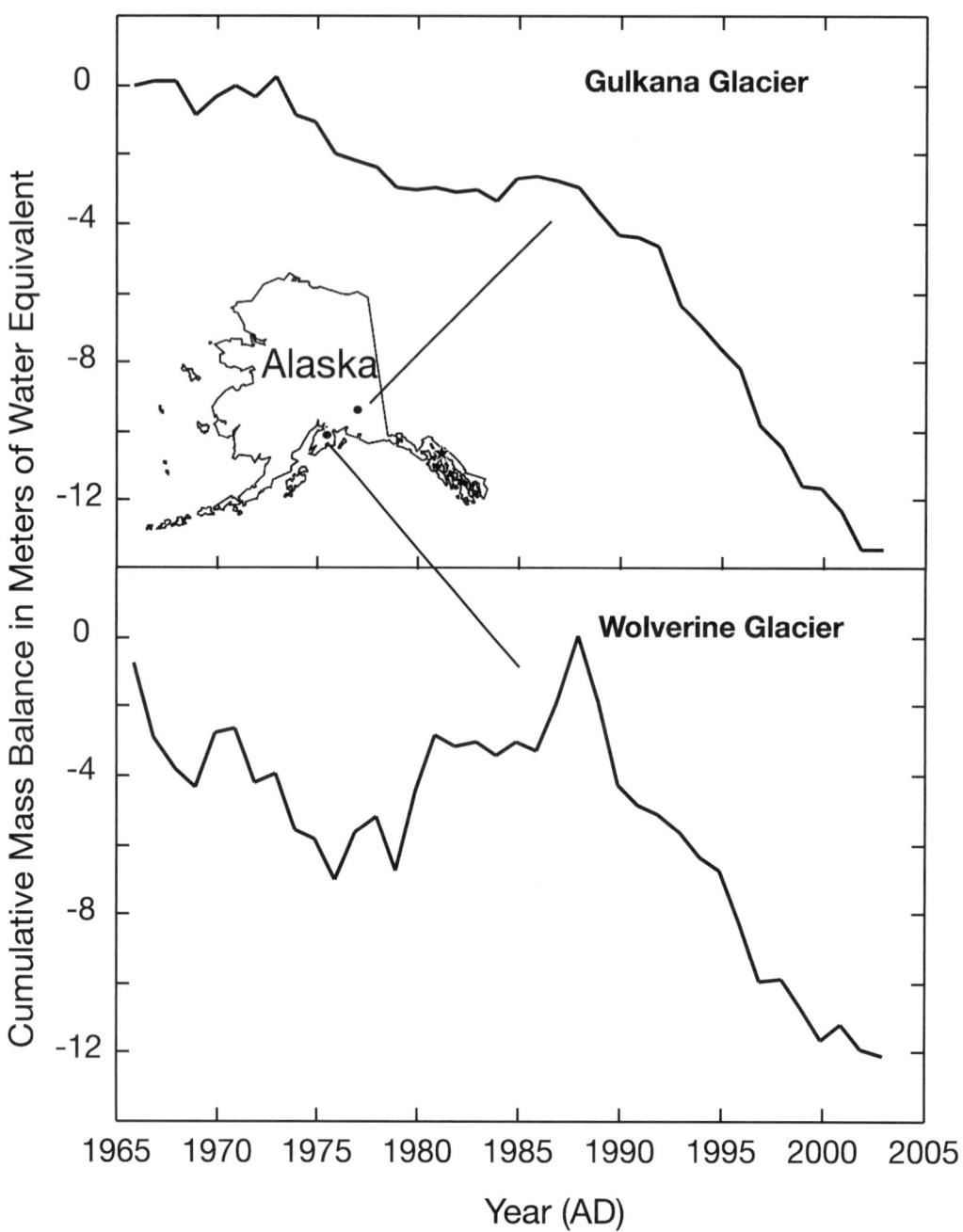

Figure 9.7. Change in mass balance at two Alaska glaciers (1966–2003). Water equivalent is the net mass of snow and ice gained or lost each year expressed as the depth of an equal amount of water averaged over the entire surface of the glacier.

and Marshall 1986). Permafrost temperatures in boreholes typically warmed 2–4°C during the early part of the nineteenth century. The data, which were obtained primarily along a north-south transect of Alaska from Prudhoe Bay to Gulkana (Osterkamp 2003), indicate only slight changes in permafrost temperatures into the mid-1980s. Since the late 1980s, however, permafrost temperatures along this transect and at other sites have generally warmed, initially in response to thicker snow covers.

Warming of the permafrost north of the Brooks Range up to 2°C (Fig. 9.8)—or even up to 3°C or more (Clow and Urban 2003, Osterkamp 2003)—over the past 30 years is comparable in magnitude to the century-long warming seen there (2–4°C) (Lachenbruch et al. 1988),

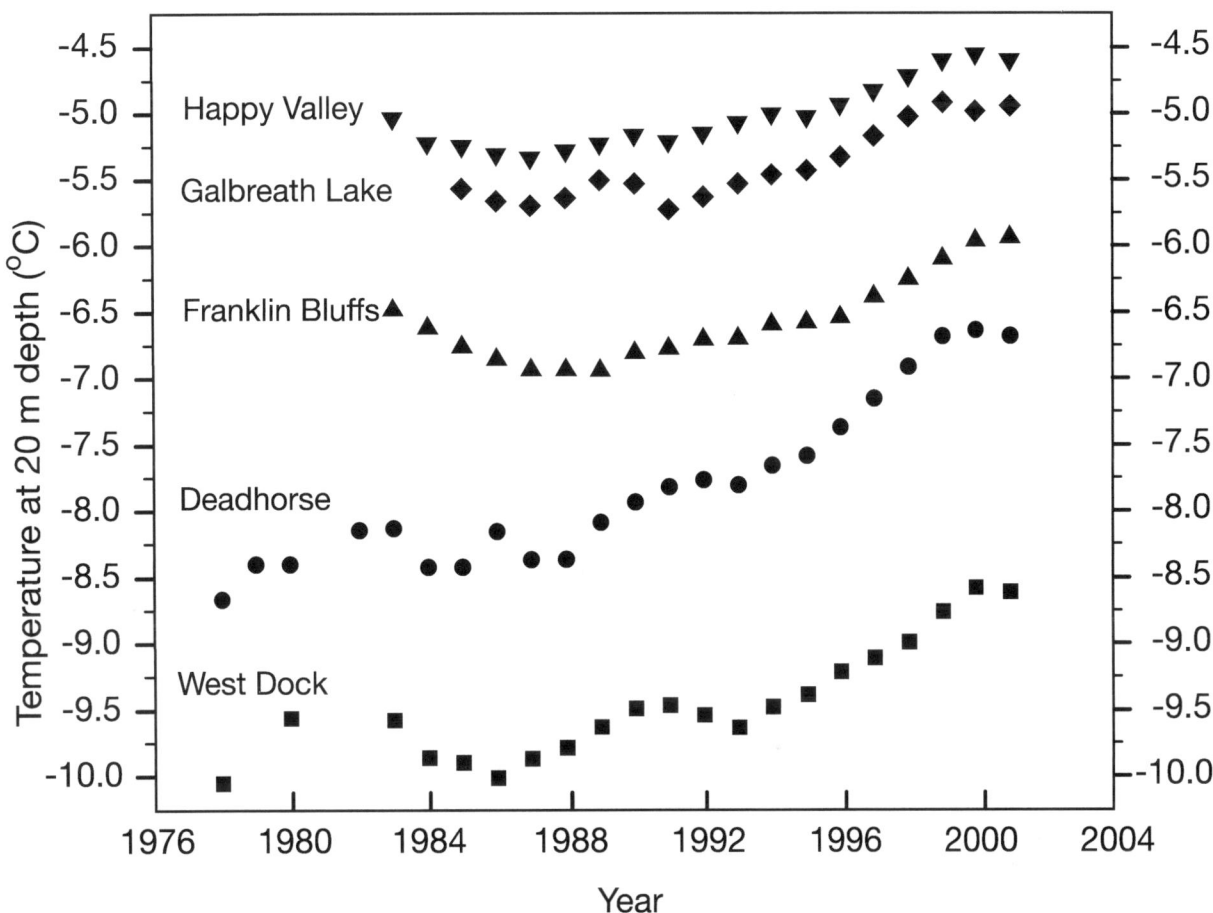

Figure 9.8. Temperature trend in continuous permafrost.

and about the same magnitude as the predicted warming of air temperatures for the next century. A calibrated permafrost temperature model (Osterkamp and Romanovsky 1999) calculates that soil temperatures in the active layer and in the permafrost have increased over the past four decades.

Although permafrost is generally considered stable in the continuous permafrost zone because mean annual ground temperatures (MAGT) are usually −7 to −11°C, permafrost in the discontinuous zone is already undergoing widespread degradation because MAGTs are near 0°C. In Alaska, the discontinuous permafrost zone falls between the Brooks Range to the north and the Alaska Range to the south. Warming of the discontinuous permafrost is typically 0.5°C (Fig. 9.9) to as much as 1.5°C (Osterkamp 2003, Osterkamp and Romanovsky 1999) over the past 30 years. Thermal offset has allowed mean annual temperatures at the permafrost table to remain below 0°C with ground surface temperatures up to +2.5°C. Recent modeling indicates that 10–17 percent of the

permafrost area of Alaska may thaw if climate warming continues as projected (Anisimov and Nelson 1996, Anisimov and Poliahov 2003).

The lateral boundaries of permafrost bodies in discontinuous permafrost are constrained to the phase equilibrium temperature, typically slightly less than 0°C. Any warming of the permafrost will cause thawing at these boundaries and contraction of the permafrost bodies (Osterkamp 2003). Consequently, in the discontinuous permafrost that has warmed, the boundaries of permafrost bodies must be thawing. Thawing permafrost has been observed at several sites with rates of about 0.1 m yr^{-1}, indicating time scales on the order of a century to thaw the top 10 m of ice-rich permafrost. Thin, discontinuous permafrost is thawing at the base of the permafrost layer at a rate of 0.04 m per year at one site in interior Alaska (Osterkamp 2003) (Fig. 9.10).

Thawing of ice-rich permafrost (in which ice exceeds the pore space of the soil) or massive underground ice can cause the surface to settle or liquefy. The amount of

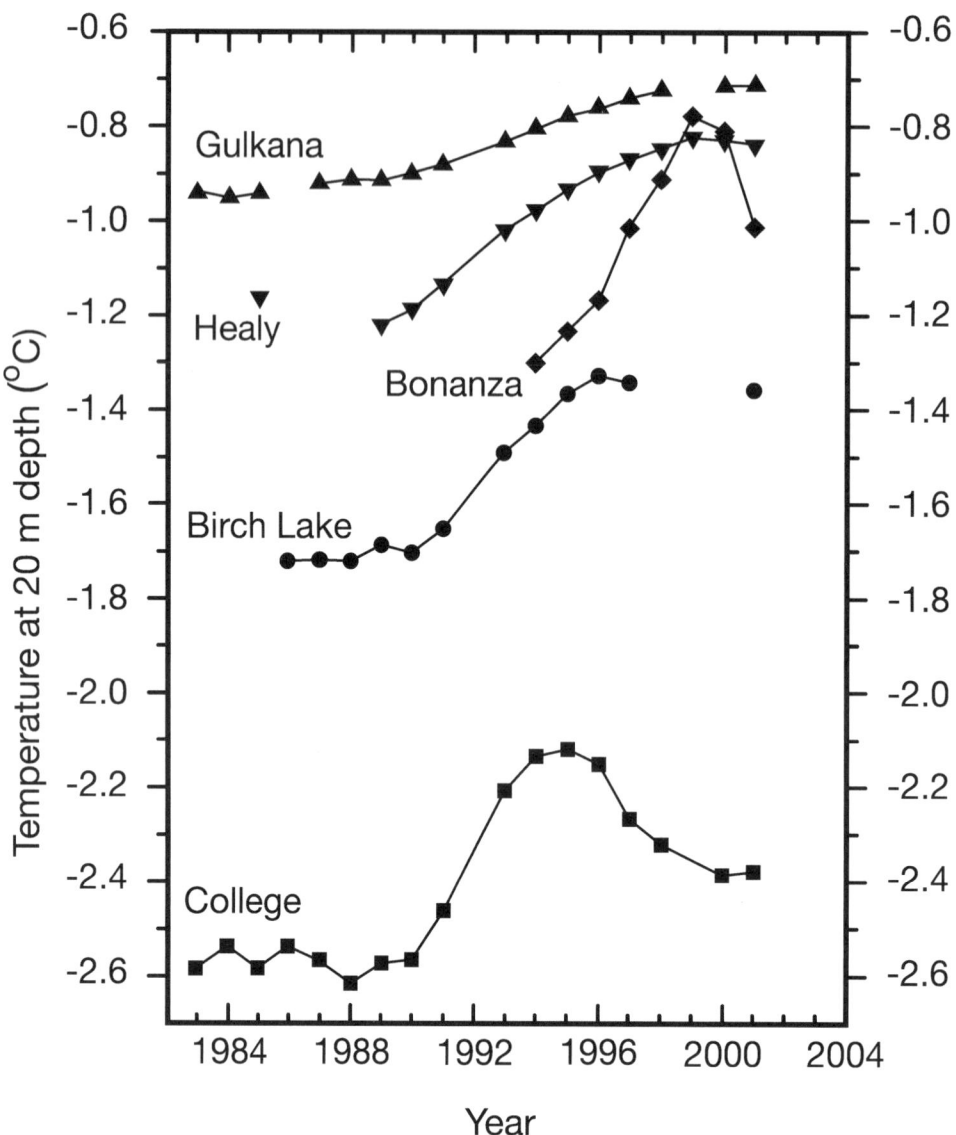

Figure 9.9. Temperature trend in discontinuous permafrost.

settlement is directly related to the amount of excess ice or massive ice; however, the mode and rate of permafrost degradation and its ecological consequences depend on complex interactions of slope position, soil texture, ice content, and hydrology (Jorgenson and Osterkamp 2004, Smith 1990).

The dominant modes of permafrost degradation and resulting thermokarst topography include: (1) rapid lateral degradation of very ice-rich soils by thermal and mechanical erosion, resulting in thermokarst lakes; (2) degradation of ice-rich soils connected to groundwater movement, creating linear collapse-scar fens; (3) degradation of ice-rich soils isolated from groundwater, leading to round collapse-scar bogs and pits; (4) differential set-

tlement from thawing of relic ice wedges, creating high-centered polygons; and (5) minor settlement of ice-poor silty soils, resulting in hummocks or irregular mounds.

Thermokarst is developing in the boreal forests of Alaska where ice-rich discontinuous permafrost is thawing (Jorgenson et al. 2001, Osterkamp et al. 2000). Thawing destroys the physical foundation (ice-rich soil) on which boreal forest ecosystems rest, causing discernable changes in the ecosystem, including tree toppling or drowning as the ground surface subsides. Impacts on the forest depend primarily on the type and amount of ice present in the permafrost and on drainage conditions. At sites generally underlain by ice-rich permafrost, forest ecosystems can be completely destroyed.

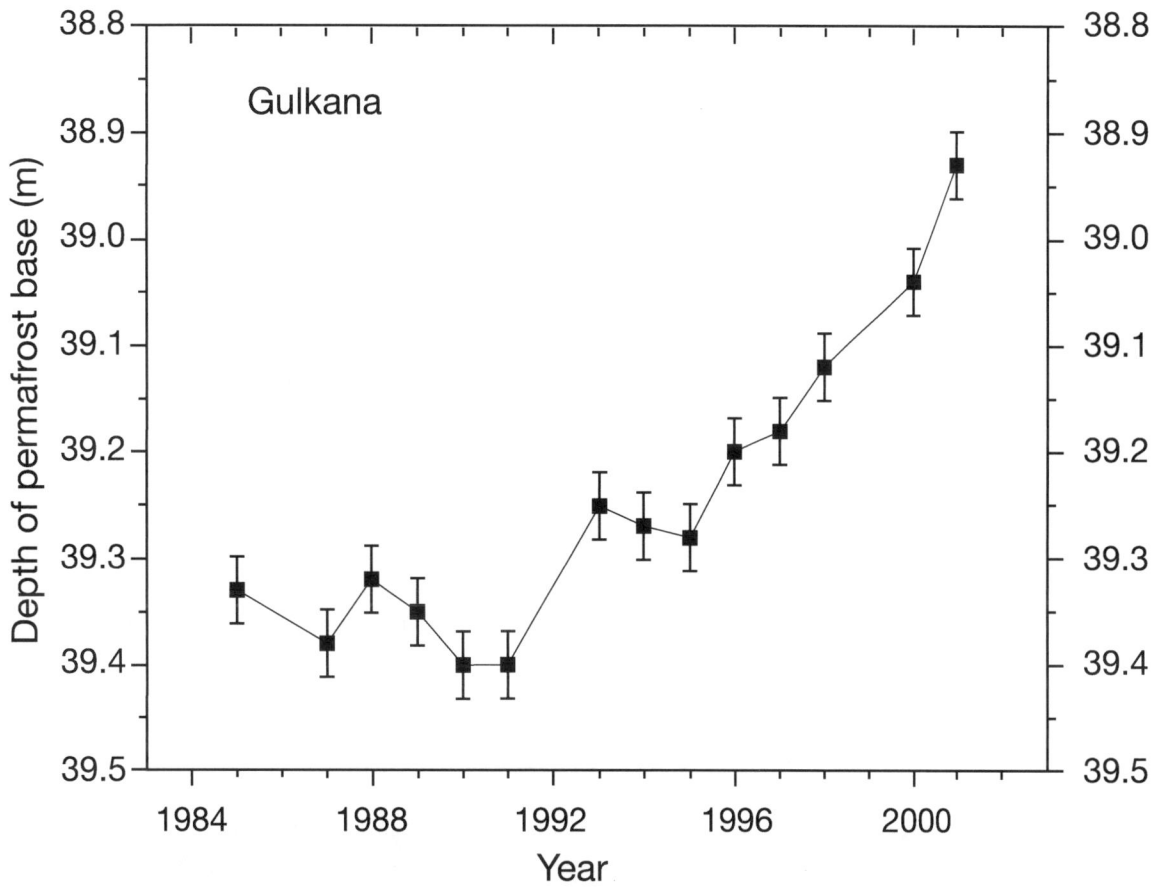

Figure 9.10. Change in depth to base of permafrost at Gulkana, Alaska. At a sufficient depth, the base or deepest front of permafrost yields to unfrozen soil or rock because of geothermal heat rising from the Earth. A rising permafrost base over time at a given location, as seen here, reflects cumulatively less cold infiltrating from the ground surface above.

In the Mentasta Pass area off the Tok Cutoff in eastern Alaska, wet sedge meadows, bogs, thermokarst ponds, and lakes are replacing forests. An upland thermokarst site on the University of Alaska campus consists of polygonal patterns of troughs and pits caused by thawing ice-wedge polygons. Trees are destroyed in corresponding patterns.

In the Tanana Flats, the ice-rich permafrost supporting birch forests is thawing rapidly, and the forests are being converted to minerotrophic floating mat fens. At this site, an estimated 83 percent of 2.6×10^5 ha was underlain by permafrost a century or more ago. About 42 percent of this permafrost has been influenced by thermokarst development within the last one to two centuries. Thaw subsidence at the above sites is typically 1 to 2 m, with some values up to 6 m. Much of the discontinuous permafrost in Alaska is extremely warm, usually within 1 or 2°C of thawing, and highly susceptible to thermal degradation.

Additional warming will result in the formation of new thermokarst, with its attendant impacts.

In contrast to the lakes on the Seward Peninsula, the lateral expansion of thermokarst lakes near Mentasta has been noticeable, with some islands decreasing in size by 24 percent from 1948 to 1981 (Osterkamp et al. 2000). On the Tanana Flats in central Alaska, where degradation is primarily through lateral expansion of collapse-scar bogs and fens in ice-rich silts, the area of totally degraded permafrost increased from 39 percent to 49 percent over 47 years (Jorgenson et al. 2001). In Canada, tree-ring data and peat landforms such as collapse scar bogs (Halsey et al. 1995) have shown that the distribution of continuous permafrost has retreated northwards during the last 100–150 years.

Thermokarst terrain with high thaw settlement (> 1 m) destroys forests in some areas, converting them to aquatic, fen, or bog ecosystems with no similarity in species composition to the original forests. Moderate thermokarst (high-centered polygons, water tracks, piping) with intermediate thaw settlement (0.5–1 m) produces a mosaic of well- and poorly drained soils, creating conditions for more diverse tree and understory species. Areas with negligible or hummocky thermokarst in thaw-stable soils become better drained and less anaerobic, and forest productivity can increase.

A quantitative determination of the impacts of thawing permafrost and thermokarst on infrastructure is difficult because the construction and habitation of a structure also have an impact on the permafrost, generally warming it. This makes it difficult to separate the effects of the structure itself from climatic warming on the underlying permafrost. Nevertheless, given the observed warming and thawing of undisturbed permafrost, it is clear that a portion of the observed damage to infrastructure must be associated with climatic warming. The economic and societal consequences of permafrost degradation are likely to be enormous because most human use of the land in boreal and arctic regions is in areas that have high potential for thermokarst (Nelson et al. 2001, Osterkamp et al. 1997).

The impacts of thawing permafrost on human activities and the physical environment will differ depending on the permafrost ice content (Osterkamp et al. 1997). In ice-poor permafrost, warming and thawing will be limited to thermal effects and the effects of converting ice to water. Impacts on the infrastructure in those areas will be minimal.

Thermokarst, however, is responsible for damage to houses, roads, airports, military installations, pipelines, and other facilities founded on ice-rich permafrost, creating severe maintenance and repair problems (Esch 1990, Osterkamp 1984, Osterkamp et al. 1997, Péwé 1982). Uneven topography from differential thaw settlement is a problem for agriculture and for recreational areas, such as ball fields and golf courses. In Russia, 0.4 percent of the forest zone has undergone permafrost degradation associated with human land use (Stolbovoi 1997).

Ecological Effects

Effects on Boreal Forest

The ecological effects of increasing temperatures, both observed and potential, on Alaskan forests are significant. In Alaska the boreal forest is a result of naturally occurring disturbances (especially fire and insects) that are triggered by warm temperatures, followed by development through successional stages influenced by characteristics of slope, aspect, elevation, drainage, and parent material. Boreal forest stands in Alaska are dominated by black spruce (*Picea mariana*, 55 percent of the forest cover), white spruce (*Picea glauca*, 26.2 percent of forest cover), and Alaska birch (*Betula neoalaskana*, 13.6 percent of forest cover). Only minor areas are dominated by cottonwood (*Populus balsamifera*), aspen (*Populus tremuloides*), or larch (*Larix laricina*) (Labau and Van Hees 1990).

Most white spruce trees in the boreal forest grow on low-elevation upland sites and show a negative radial growth response to summer temperatures (Barber et al. 2000, Barber et al. 2004, Juday et al. 2003). With warmer summer temperatures, there is less radial growth (Fig. 9.11) due to temperature-induced drought stress (Barber et al. 2000). Floodplain white spruce sites tend to be more productive, and growth is typically positively related to summer temperature (Adams and Juday 1997) since moisture is less limiting.

Black spruce trees occur mostly on permafrost-dominated sites. Those on well-drained, north-facing ridgetop sites show a negative radial growth response to summer temperature similar to white spruce, but with a different weighting of monthly temperature predictors of growth (Juday et al. 2003). Black spruce trees are also found on lower slopes and valley bottom sites, which are wet because of impeded drainage from the impervious permafrost layer. These trees show a positive response to winter temperature, with a strong negative response to April temperature, when above-ground temperatures can stimulate photosynthesis while below-ground water remains frozen (Juday et al. 2005).

In broad valley sites, black spruce trees also show a positive radial growth response to winter temperature (Fig. 9.12). The majority of black spruce sites studied to date in Alaska exhibit a negative growth response to increasing temperatures, which would lead to predictions of considerable reduction in the amount of this now-dominant species with a warming climate.

Aspen stands in Alaska have not been studied for their relationship to temperature; however, aspen across a large portion of the western Canadian interior experienced several cycles of reduced growth between 1951 and 2000, especially in 1976–1981, when basal-area growth decreased by nearly 50 percent (Hogg et al. 2005). This variation in aspen growth was explained by moisture status and insect defoliation, and another growth collapse likely occurred during a severe drought in the region during 2001–2003. There is no reason to believe that aspen on similar sites

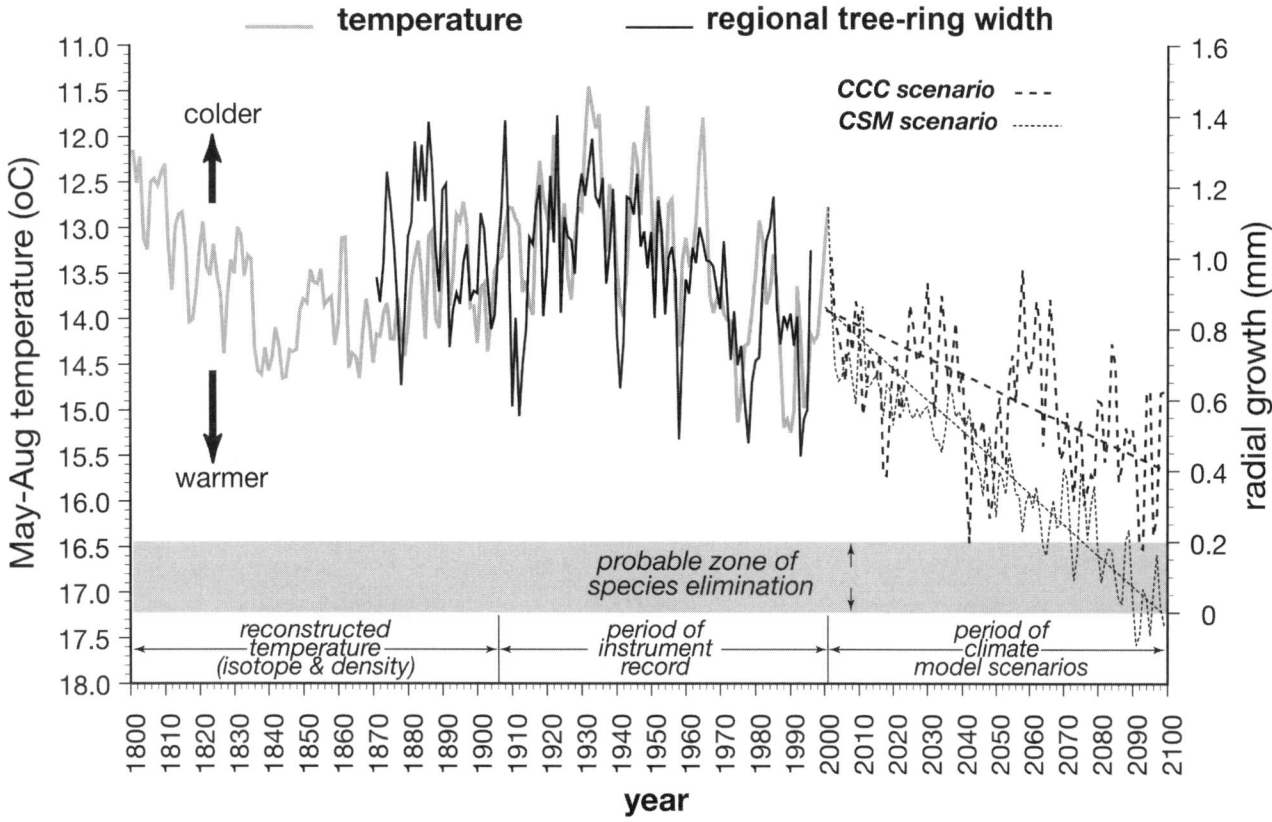

Figure 9.11. Relationship of warm season (May–August) temperature and radial growth of white spruce on productive upland sites in interior Alaska. Tree growth is mean of 169 trees in eight stands. Temperatures for the earliest period have been reconstructed (Barber et al. 2004), those for the middle period are from instrument records at Fairbanks, and twenty-first century temperatures are from general circulation model scenario output (Juday et al. 2005). Note the inverted temperature scale (because growth is negatively correlated to temperature).

in Alaska would respond differently to local droughts and defoliation episodes.

The integrated sum of photosynthesis over the period 1982–2003 in boreal North America confirms that the predominant growth anomaly during this time was reduced photosynthesis, although no systematic change in growing-season length occurred (Goetz et al. 2005).

The Arctic Climate Impact Assessment (Hassol 2004) used the output of five different general circulation models (GFDL—Geophysical Fluid Dynamics model, CSM—NCAR/Colorado State model, ECHAM—European Community Hamburg model, CCC—Canadian Climate Center model, and the HAD—UK Hadley Center model) to generate climate scenarios up to the year 2099. Future projections of radial growth of individual tree species have been modeled based on empirical relationships of growth to climate from previous studies and from projected temperature changes (Juday et al. 2005). Historical and reconstructed relationships between summer temperature

and white spruce growth on low-elevation upland sites in interior Alaska and projected growth based on climate-model data through 2099 are indicated (Fig. 9.11).

The scenarios indicate that by about 2070 or sooner, the climate will very likely no longer be able to sustain white spruce, based on past empirical relationships. Mortality would likely be due to temperature-induced drought stress or proximate agents, such as insects, that would be enhanced by the altered climate. White spruce will be able to persist in the landscape on floodplains and in regions where moisture is not limiting. The same relationships are indicated for black spruce trees, which display a negative sensitivity of radial growth to temperature (Fig. 9.12, bottom). The range of projected temperature increase to reach zero growth is 2–4°C.

Given these direct temperature controls of growth of the principal boreal tree species, a very different forest landscape may emerge in the next several decades in boreal Alaska.

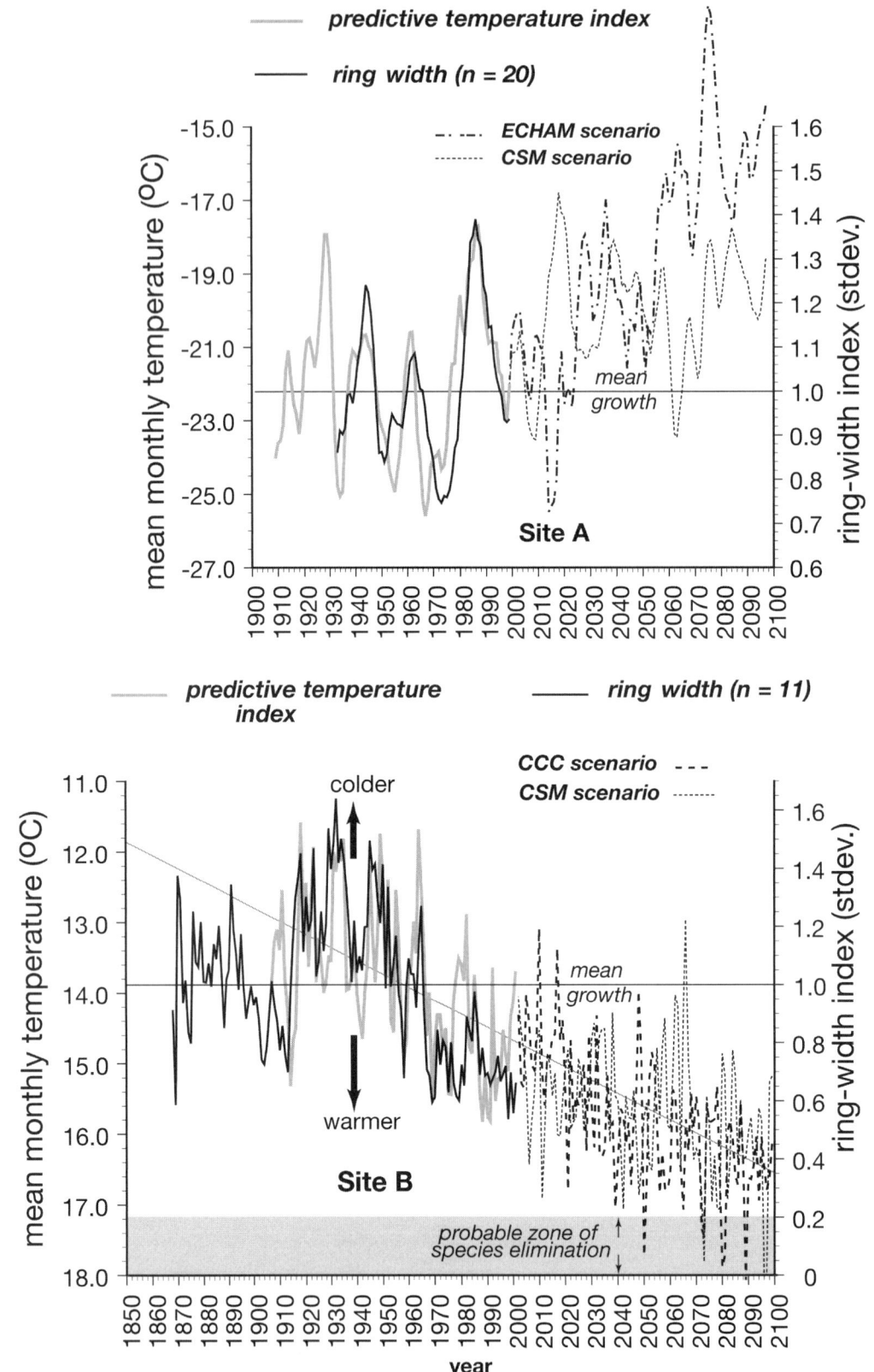

Figure 9.12. Relationship of winter temperature and radial growth of black spruce on a broad valley site (Site A, top) characterized by trees with a positive radial growth response to warming. The best predictive temperature index for this population is the mean of winter monthly temperatures (December and January in the year of ring growth, and the same months plus February in the previous year). Tree growth data at Site A have been smoothed with five-year running mean. A negative radial growth response of black spruce to summer monthly temperatures on a Tanana River terrace site (Site B) is shown in the bottom graph. The best predictive temperature index for this population is the mean of May, June, and previous June and July mean monthly temperatures. Tree-growth data at Site B are individual yearly values; the temperature scale is inverted because growth is negatively correlated. Twenty-first-century values are a scenario output from GCM climate models used in the Arctic Climate Impact Assessment (Juday et al. 2005).

Effects on Treeline

White spruce trees occurring at treeline in Alaska have mixed growth responses. Some trees have a positive radial growth response to spring temperature, some have a negative radial growth response to summer temperature, and others are not temperature sensitive (Wilmking 2003, Wilmking et al. 2004). The proportion of trees at each site belonging to the different responder types varies longitudinally across northern treeline in Alaska (Wilmking and Juday 2005). In the western region most trees respond with increased growth in response to recent warming; in the central region, populations of responder types are mixed; and in the eastern region, most trees have decreased radial growth in response to warming temperatures.

This pattern is consistent with a decrease in precipitation from west to east in Alaska, and points to drought stress as the factor for reduction of growth. Using correlation between tree growth and climate for the twentieth century, modeled carbon uptake of treeline white spruce populations for the twenty-first century predicts reduced carbon uptake in the eastern region in the twenty-first century compared to twentieth-century uptake under GCM scenarios (Wilmking and Juday 2005).

Spruce with negative radial growth responses to warming in the Brooks and Alaska Ranges still had greater total growth than the positive responders until the regime shift in 1977. After that shift, the positive responders grew more than the negative responders (Wilmking et al. 2004) (Fig. 9.13). These results were consistent across an east-west transect of both mountain ranges. The growth of negative responders was best correlated to prior July temperature, which explained most of the year-to-year variability in radial growth of this population of treeline white spruce.

More specifically, a threshold occurs for prior July temperature of around 16°C at Fairbanks (temperature at treeline in the Brooks or Alaska Ranges is about 4–5°C lower than at Fairbanks). This translates to an on-site threshold of about 11–12°C for July temperature in the Brooks and Alaska Ranges. In years with cooler temperatures, trees were not climatically sensitive, but in warmer years the trees became significantly sensitive and responded with decreasing radial growth (Wilmking et al. 2004).

Under future, warmer conditions this July temperature threshold will be crossed more often, and as a result, the diminished growth in these treeline populations is likely to reduce the potential for continuous treeline movement under a warming climate. Treeline movement is more likely in the western regions of Alaska (Lloyd and Fastie 2002), where wetter conditions reduce drought stress. Photographic evidence also indicates local shrub and treeline expansion during the last 50 years (Sturm et al. 2001) in western Alaska. A preliminary study (Wilmking et al., pers. comm.) of the northwestern-most white spruce in Alaska identified recent invasion of tundra by trees, whereas the central region of the Brooks Range seems to support stable treelines that have not changed during this time.

A tree-ring width chronology (FIRTH) of white spruce based on living and relict wood samples collected in the summer of 2003 from near latitudinal treeline in the Firth River area of northeastern Alaska spans the period from AD 1067 to 2002 and is based on 131 series (D'Arrigo et al. 2006). It shows very similar trends to those observed in an elevational treeline white spruce record (TTHH, AD 1099–2000, 89 series) from the nearby Yukon Territory (D'Arrigo et al. 2004, Jacoby and Cook 1981) (Fig. 9.14).

Both chronologies have been standardized to optimize retention of low-frequency trends, and they generally show common variability in radial growth, considered to be related to temperature changes over the past near millennium. Both series correlate significantly with Northern Hemisphere annual temperatures since the mid-nineteenth century. There is some indication, however, of decreased growth in recent decades in the TTHH chronology, which may relate to temperature-induced drought stress. Estimated temperature optima for tree growth at the Yukon site appear to have been exceeded in recent decades, coincident with the observed decline in growth (D'Arrigo et al. 2004).

Records of tree-ring widths obtained from old-growth white spruce trees at or near treeline on the Seward Peninsula in northwestern Alaska also show some evidence for decline and weakening of the relationship with growing-season temperature at Nome in recent decades (since ~1970). The meteorological trends at Nome, on the western Seward Peninsula, indicate that summer temperatures have risen in recent decades without any significant overall rise in moisture availability, which would be expected to increase evapotranspiration and the likelihood of drought stress. Maximum latewood density data from these same sites also show some weakening of climate response with Nome temperatures (D'Arrigo et al. 2004). Recent warming is also having an impact on white spruce at other sites on the Seward Peninsula (Lloyd and Fastie 2002).

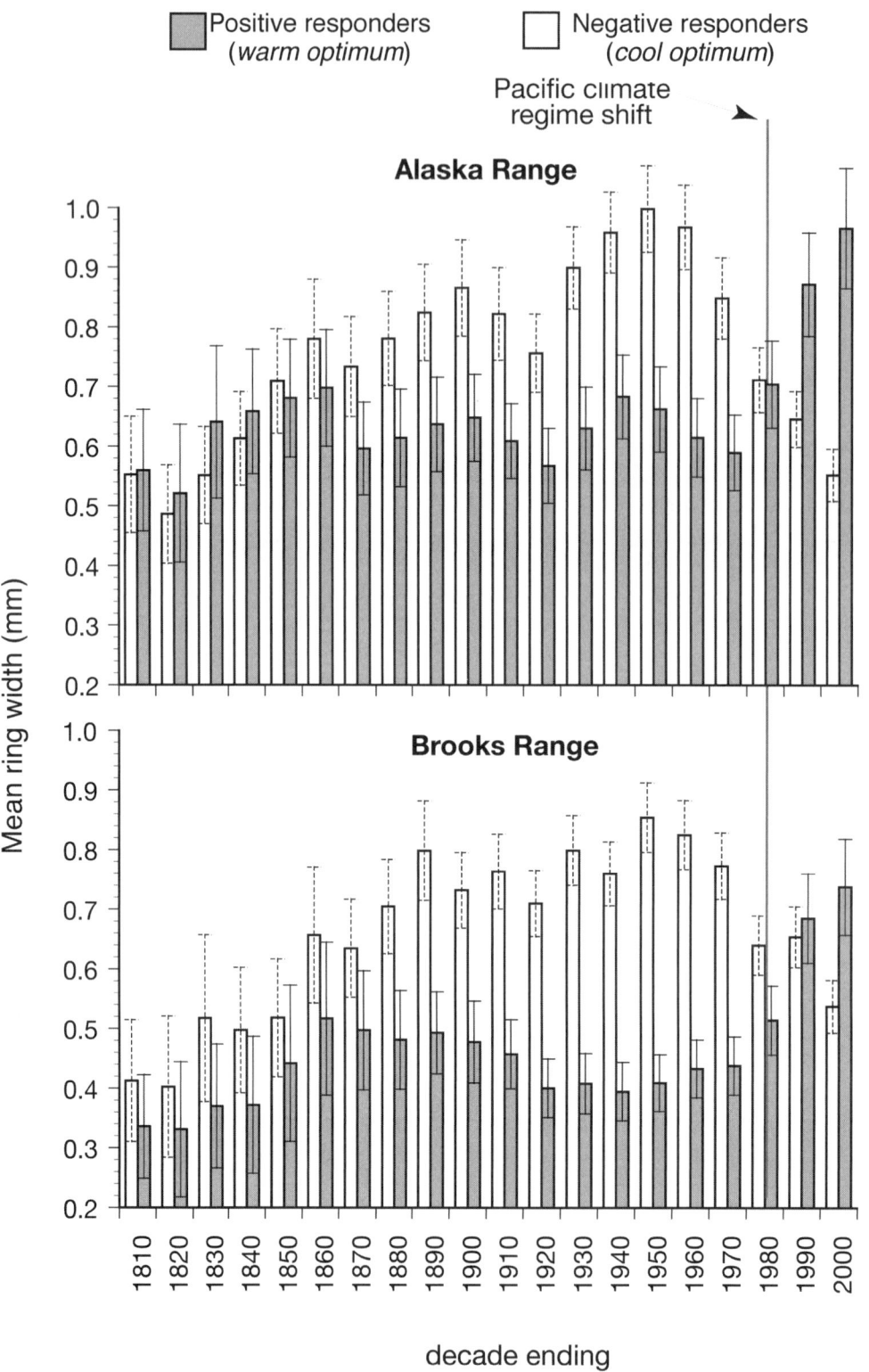

Figure 9.13. Mean decadal growth of white spruce at or near treeline ($n = 1{,}155$ trees; eight stands in the Brooks Range, seven stands in the Alaska Range). Note the crossover of growth relationship of the populations with a positive versus negative growth response to warmth at the time of the Pacific climate regime shift.

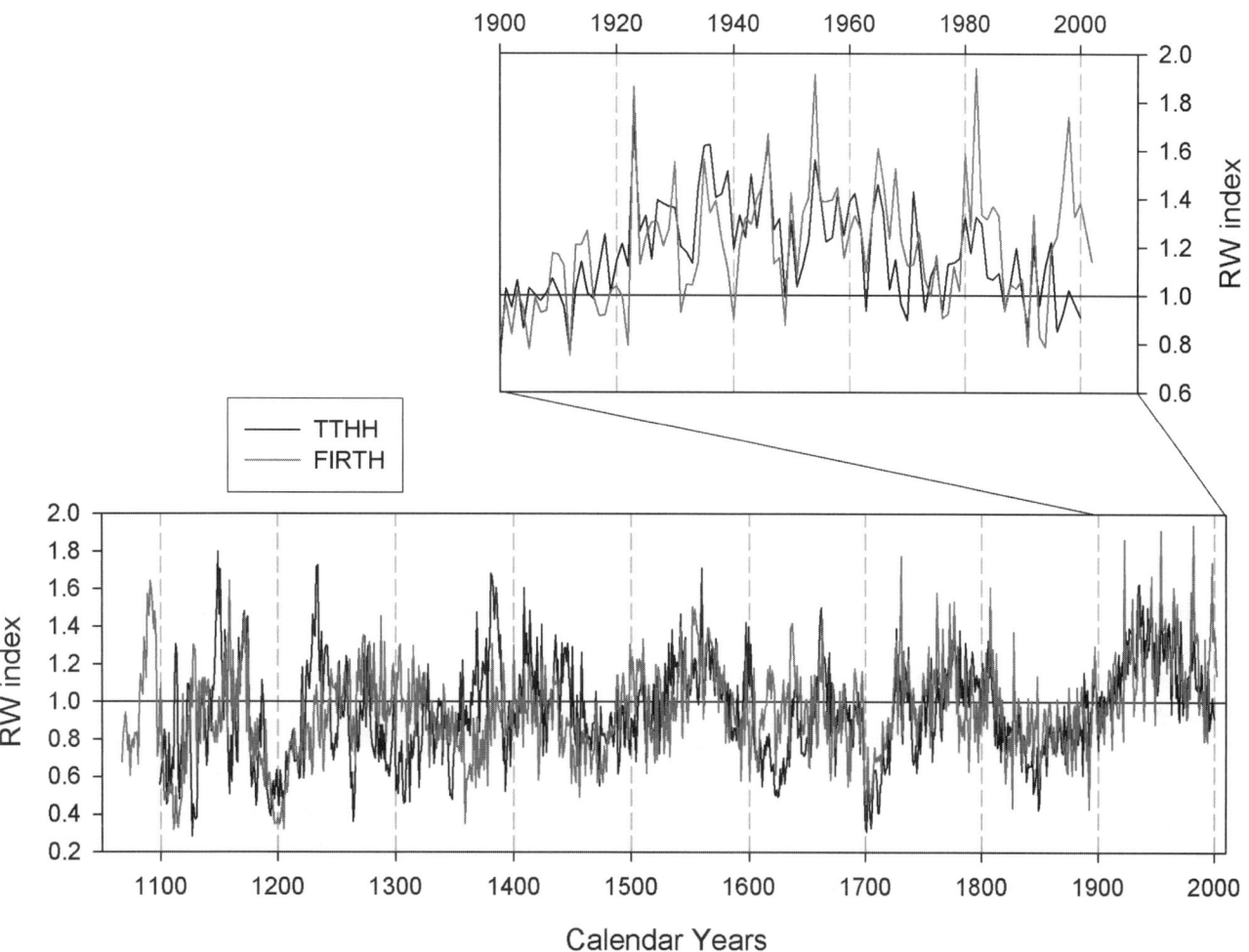

Figure 9.14. Tree-ring width (RW) chronologies for white spruce (*Picea glauca*) spanning the past millennium. The FIRTH record is from the Firth River area of northeastern Alaska (AD 1067–2002); the TTHH record is from the Yukon Territory, Canada (AD 1099–2000). Both series have been standardized to optimize retention of low-frequency trends using conservative methods (negative exponential function or regression line of negative or zero slopes). Note growth changes in the twentieth century, with some growth decline at TTHH site (inset).

Effects on Tundra

Shrub tundra is replacing tussock tundra in areas of warming climate (Bliss and Matveyeva 1992), and tundra is showing increased shrubbiness in Alaska on the North Slope and the Seward Peninsula. A study comparing and contrasting photos between 1949 and 2000 shows an increase in shrub density on the North Slope in the Chandler River drainage (Sturm et al. 2001). Most of the increased shrubbiness is a result of a "filling in" of shrubs where they were lacking earlier in 1949. A noticeable advance of shrubs was also documented on the Seward Peninsula along mountain valleys and riparian corridors between 1986 and 1992 using GIS and remote sensing (Silapaswan et al. 2001).

Increasing shrub cover suggests that directional changes are occurring on the Seward Peninsula consistent with experimental tundra warming (Chapin et al. 1995). Experimental warming of shrub tundra plots also suggests increased leaf area of existing shrubs (Arft et al. 1999, Bret-Harte et al. 2001, Chapin and Shaver 1996, Hobbie and Chapin 1998). Modeling studies suggest that shrub tundra will continue to expand if warming continues (Kaplan et al. 2003). Such expansion has implications for the exchange of water, energy, and trace-gas exchange with the atmosphere (McGuire et al. 2003, Thompson 2004 #1039).

Much of the landscape on the Seward Peninsula is in a transition zone between boreal forest and tundra, so it is expected that with warming, the treeline will advance onto the tundra, as it already has in some areas on the peninsula where disturbance has occurred (Lloyd et al. 2003a). Recent climate warming may be an important factor in influencing vegetation cover changes on the Seward Peninsula (Lloyd et al. 2003b).

Changing land cover also affects the surface energy budget by changing the albedo, both directly because of vegetation changes and indirectly (and of greater magnitude) due to changes in snow distribution. Shrubs tend to trap the snow and thus provide a positive feedback from albedo (less reflection of heat energy, and thus more absorption), which leads to warming. There has also been a documented increase in the normalized difference vegetation index (NDVI) measured since 1990 (Jia et al. 2003, Myneni et al. 2001, Myneni et al. 1997). This is evidenced by an increase in the growing season in interior Alaska (Barber et al. 2002) as well as north of the Brooks Range (Chapin et al. 1995).

The spring of 2004 saw the earliest leaf-out in Fairbanks (May 4) that has ever been recorded. With earlier and greater spring warming and later fall onset of freezing, the growing season has been extended (Barber et al. 2002). Consequently, the same amount of moisture is required over a longer growing season as was needed over a shorter growing season, presumably causing moisture stress. The recharge of water to the soil in spring, as contrasted with precipitation received during the growing season, is critical to vegetation growth through the summer months.

Modeling Results

A logistic regression model (Calef et al. 2005) predicts the potential equilibrium distribution of four major vegetation types in interior Alaska—tundra, deciduous forest, black spruce forest, and white spruce forest—based on elevation, aspect, slope, drainage type, fire interval, average growing-season temperature, and total growing-season precipitation. The hierarchical logistic regression model was used to evaluate how scenarios of changes in temperature, precipitation, and fire interval may influence the distribution of the four major vegetation types found in this region. The model was verified for interior Alaska (the region used to develop the model), where it predicted vegetation distribution among the steps with an accuracy of 60–83 percent. When the model was independently validated for northwest Canada, it predicted vegetation distribution among the steps with an accuracy of 53–85 percent.

Black spruce remains the dominant vegetation type under all scenarios, potentially expanding most under warming coupled with increasing fire interval. White spruce is clearly limited by moisture once average growing-season temperatures exceed a critical limit (+2°C). Deciduous forests expand their ranges the most when any two of the following scenarios are combined: decreasing fire interval, warming, and increasing precipitation. Tundra can be replaced by forest under warming, but expands under scenarios of precipitation increase.

Effects on Forest Disturbance

Insects

On average, insect infestations annually affect a similar amount of area as that affected by fire in the Alaskan forests (Werner et al. 2006). Records in Alaska indicate that insect infestation affects approximately 300,000 ha annually (Werner et al. 2006), but the effective area of insect-induced stand mortality is less than 300,000 ha because successive years of attack are required to cause stand mortality, and only a portion of the stands affected are host species for the attacking insects. In Alaska the insects responsible for major infestations include the spruce beetle, the spear-marked black moth, the large aspen tortrix, the eastern spruce budworm, and the larch sawfly.

Insect infestations in Alaska peaked during the late 1970s (impacting more than 600,000 ha annually) with outbreaks of the spear-marked black moth, the large aspen tortrix, and the larch beetle, and during the 1990s (approximately 500,000 ha annually) with outbreaks of the spruce beetle, the eastern spruce budworm, and the larch sawfly (Werner et al. 2006). An outbreak of the larch sawfly during the late 1990s is estimated to have killed most of the mature larch in interior Alaska (> 650,000 ha) (Werner et al. 2006).

Spruce bark beetles are found in background levels throughout Alaska, but a major outbreak has yet to occur in interior Alaska, and there is no evidence of prior infestations. Controls over the population dynamics of insects responsible for tree mortality are not well understood. Some insects seem to cycle on a regular basis, such as the spruce budworm in eastern Canada (~30 years, Kurz and Apps 1999), and the large aspen tortrix (~12 years) and the spear-marked black moth (~10 years) in Alaska (Werner et al. 2006). In contrast to infestations in eastern Canada, outbreaks of the eastern spruce budworm in Alaska have been observed only in the 1990s, when it infested around 600,000 ha (Werner et al. 2006). Werner et al. (2006) have hypothesized that this infestation may have been due to the temperature-induced drought stress experienced by white spruce in interior Alaska in association with warmer summers (Barber et al. 2000).

Factors that stress trees appear to be important in influencing insect outbreaks in western Canada as well, where decreasing effective moisture has caused multiple-year defoliation of aspen by the forest tent caterpillar as well as reduced tree growth (Hogg et al. 2005, Hogg et al. 2002). In Alaska (Werner et al. 2006), *Ips engraver* beetles tend to infest stressed and dying trees near the edges of fires in the interior (Werner and Post 1985).

Effects of warming temperatures on the life cycles of insects also play an important role in insect outbreaks. For example, a single generation of spruce beetle can require either one or two years to mature, and this generally depends on temperature (Werner and Holsten 1985). Warmer summers in south-central Alaska during the late 1980s and 1990s caused a shift from a two-year to a one-year life cycle for spruce beetles, which played a role in the large outbreak of the 1990s on the Kenai Peninsula (Werner et al. 2006). Warmer winters also allowed a larger population of bark beetles to survive the winter, leading to a higher rate of subsequent population increase. Finally, unfavorable climate conditions of warm temperatures and low precipitation have stressed regional populations of trees (Berg et al. 2006). Healthy trees can "pitch out" wood-boring beetles (unless overwhelmed by numbers), but a slow-growing, weak tree with low growth reserves is less able to defend itself.

On the Kenai Peninsula, the buildup of the population of beetles to overwhelming numbers was assisted by an already stressed population of trees (Berg et al. 2006). This combination of events led to a massive outbreak of the spruce bark beetle (*Dendroctonous rufipennis* Kirby) that killed most of the mature white and Sitka/Lutz spruce in the Kenai Peninsula forests, affecting two to three million acres over the last ten years (Berg et al. 2006, Wittwer 2004). Lack of host trees in this area is now the main limitation on future outbreaks. Under continued warming, we can expect to see more insect infestations spreading to previously unaffected forest regions in Alaska.

Fire

Fire is another disturbance in the boreal forest associated with warm temperature anomalies. In Alaska, the area burned in a particular year is positively correlated with mean June temperature, and negatively correlated with the depth of snowpack near the end of winter (Duffy et al. 2005). Occasional large fires account for most of the area burned in Alaska (Kasischke et al. 2006). Approximately

68 percent of the total area burned in Alaska from 1940 to 1998 occurred during 15 years characterized by a moderate-to-strong El Niño, which resulted in above-normal temperatures throughout the year, and normal precipitation from February to August in Alaska (Hess et al. 2001). Among the state's different ecoregions, fire cycle decreases with increasing growing-season temperature, decreasing growing-season precipitation, and increasing lightning frequency (Kasischke et al. 2006).

Humans affect fire regimes in the boreal forest through the ignition and suppression of fires, and land use that alters the spatial distribution of fuels. In Alaska, lightning ignitions are largely distributed throughout the interior between the Brooks and Alaska Ranges, while human-caused ignitions are largely confined to the road network (Gabriel and Tande 1983, Kasischke et al. 2006). Although humans are an important source of fire ignition in boreal forests, they are not responsible for the majority of area burned. In Alaska, humans cause more than 60 percent of all ignitions, but these ignitions result in only 10 percent of the area burned (Kasischke et al. 2006).

The low amount of area burned from human-caused fires is the result of poor burning conditions at the time of ignition and responsive fire-suppression efforts: the fires started by humans are generally quickly detected and accessed because they are located in proximity to transportation networks. In Alaska, recent analyses indicate that fire cycles are longer near population centers (Chapin et al. 2003).

Fire suppression is important in promoting longer fire cycles near population centers, but the alteration of fuels by human land use likely plays a role as well. For example, near Fairbanks, substantial harvesting of spruce forests during the settlement period in the early 1900s led to the replacement of many of these forests, through natural succession, by faster-growing aspen and birch forests, which are less flammable than spruce forests and inhibit the spread of fire. The combination of fire suppression and spatial alteration of fuels is presumably responsible for a fire cycle near Fairbanks that is estimated to be four times longer than the fire cycle of areas in interior Alaska located away from population centers. The fire regime may also be influenced by insect infestations; affected stands may be more vulnerable to fire where flammability of forest stands increases with tree mortality.

Alaska has a reliable 56-year record of area burned (1950–2005). Total area burned has increased markedly and has been concentrated in specific fire years. Burning conditions, in general, are characterized by sustained periods with high daily-maximum temperatures ($< 25°C$)

under the influence of high atmospheric pressure, which causes low humidity, plus a sufficient number of ignitions caused by lightning strikes. Major fire years, in particular, experience a strong initial burning season and then the persistence of warmth and drought into a later secondary burning season in which the season-ending precipitation event simply does not occur. Conditions causing major fire years have become more frequent in Alaska in the most recent decades.

During the record fire year of 2004, more than 2.7 million ha burned; the third-highest total was in 2005, with 1.6 million ha burned. Well over a quarter of forest land in the northeast quadrant of Alaska burned in those two years. Adjacent Yukon Territory experienced more than 1.8 million acres burned. The total area burned in Alaska and Yukon Territory in 2004 and 2005 was greater than the area of Ireland—a rate of burning that is unsustainable.

Area burned has increased in the remainder of northwest Canada as well, with an increase in wildland fire frequency in the latter half of the twentieth century associated with a warming climate (Kurz and Apps 1999, Podur et al. 2002, Stocks et al. 2000). Analyses of the response of fire weather to projections of future climate change suggest that climate change has the potential to substantially increase fire frequency throughout much of the boreal forest in North America (Flannigan and Wotton 2001, Flannigan et al. 2001, Stocks et al. 1998, Wotton and Flannigan 1993), assuming the fuels have not already been consumed.

More frequent and widespread fire and insect disturbances will reduce the average age of the forest at any given time. The increased frequency of disturbance can also affect the relative abundance of species by systematically reducing the abundance of insect- or fire-susceptible species. These indirect temperature controls of insect infestation and fire are causing a very different forest landscape to emerge in boreal Alaska.

Effects on Soils and Carbon

Boreal forests around the world contain approximately 27 percent of the world's vegetation-carbon inventory and 28 percent of the world's soil-carbon inventory (McGuire and Hobbie 1997). Much of the soil carbon has accumulated simply because of cold and/or anaerobic soil conditions, which slow or halt decomposition (Neff and Hooper 2002, Weintraub and Schimel 2003).

This carbon is highly labile, and warming over the past few decades in the high latitudes could trigger the release of carbon that has accumulated over thousands of years. Important trace gas feedbacks associated with the exchange of carbon dioxide (CO_2) and methane (CH_4) with the atmosphere operate in the higher latitudes (Roulet 2000). Changes in these fluxes could either enhance warming (positive feedbacks) or mitigate warming (negative feedbacks) (Chapin et al. 2000, Clein et al. 2002, McGuire et al. 2000, McGuire et al. 1997, Smith and Shugart 1993).

Although the warming of aerobic soils will tend to increase the release of CO_2 from soils of northern latitudes, the net effect of warming depends on the balance between production and decomposition. The changing length of the growing season in Alaska may have important effects on production and carbon storage (Frolking et al. 1996). In temperate forests, carbon storage is enhanced by about 6 g C m^{-2} for each day that the growing season is lengthened (Baldocchi et al. 2001), suggesting that terrestrial carbon storage should also be enhanced with longer growing seasons in boreal forests (see also Frolking et al. 1996). Photosynthetic activity of the canopy in the boreal forest designates the start of the growing season and is tightly coupled to soil thaw in conifer stands, as transpiration is greatly diminished in frozen soils (Frolking et al. 1999, Zhuang et al. 2003). Deciduous stands begin net carbon uptake following leaf-out, which is also keyed to the timing of snowmelt.

With warming, soil carbon will increase in boreal forests if the rate of carbon storage is faster than the rate of carbon lost through respiration from enhanced decomposition. If this does occur, then the long-term rate of soil carbon storage will depend on the rate of decomposition (Hobbie et al. 2000, Clein et al. 2002). Our lack of understanding of soil carbon and nitrogen transformations in response to warming is a key gap and limits our ability to make projections of the long-term response of soil carbon to warming (Clein et al. 2002).

Fire activity is highly variable from year to year in specific regions of the boreal forest (Amiro et al. 2001, Conard et al. 2002, Kasischke et al. 2006, Kurz and Apps 1999, Stocks et al. 2000, Stocks et al. 2002). It has only recently been demonstrated that fire in the boreal forest region as a whole may play a role in the interannual variability in the growth rate of atmospheric CO_2 (Langenfelds et al. 2002).

It is useful to examine the effects of fire on carbon storage at both the stand and regional levels in order to understand how variability in regional-scale fire activity influences carbon storage of the atmosphere. After a stand-level disturbance such as a fire, there is a period of carbon loss to the ecosystem, and production is less than decomposition. This is followed by a period of carbon gain to the ecosystem once there is enough growth and production exceeds decomposition (Kasischke et al. 1995, Kasischke et al. 2006, Zhuang et al. 2003). An increase or decrease in disturbance frequency will generally cause losses or gains, respectively, in carbon from terrestrial ecosystems to the atmosphere.

Together, both insect disturbance and fire have likely released substantial amounts of carbon into the atmosphere from Canada's forests in the latter part of the twentieth century (Amiro et al. 2001, Chen et al. 2000, Kurz and Apps 1999) and from Alaska during the twenty-first century. If stands are salvage logged after fire, an additional loss of carbon may occur. The level of fuel consumption is an important factor in carbon dynamics of the North American boreal forest (Harden et al. 2000). Insect disturbance and forest harvest also result in carbon loss to the atmosphere, but the consequences of these disturbances on regional carbon dynamics is less clear than for fire. The consequences of these disturbances on ecosystem processes and successional change through the cycle of disturbance and subsequent forest regrowth are not fully understood either.

Disturbance from fire and insects causes changes in vegetation, resulting in either carbon storage or loss depending on the type of transition (McGuire and Hobbie 1997). For example, in areas where tundra is replaced with boreal forest, the result is likely to be a net uptake of carbon from the atmosphere, whereas the transition of boreal forest to grassland is likely to result in a net release of carbon to the atmosphere.

Equilibrium modeling studies imply that the replacement of tundra with boreal forest in Canada and Alaska has the potential to substantially increase storage of carbon in vegetation (gains of 14 to 18 Pg C among different scenarios [McGuire and Hobbie] 1997), while storage of carbon in soil will likely be less affected (loss of 1 Pg C to a gain of 3 Pg C among different scenarios [McGuire and Hobbie 1997]). Soil carbon was relatively insensitive to the transition from tundra to boreal forest in these simulations because increases in production were offset by increases in rates of decomposition associated with increased soil temperature. This potential increase in ecosystem carbon storage by migration of boreal forest onto the tundra is likely to proceed at a very slow pace, likely taking many centuries (Chapin and Starfield 1997, Lloyd et al. 2003a, Starfield and Chapin 1996).

Although soil and lake drainage may be especially vulnerable to the melting of permafrost in response to climatic warming, the net effect on radiative forcing of the atmosphere is not clear because drainage can either be enhanced or retarded by permafrost degradation, and the response of drainage is likely to affect the release of CO_2 and CH_4—but in opposite directions (Roulet 2000, Roulet et al. 1992a, Roulet et al. 1992b).

For example, if permafrost warming results in a drop in the water table, the release of CO_2 from aerobic decomposition is likely to be enhanced (Oechel et al. 1995, Christensen et al. 1998), but CH_4 emissions will likely decrease because methanogenesis is an anaerobic process (Roulet 2000, Roulet et al. 1992a, Roulet et al. 1992b). In contrast, if permafrost melting results in the expansion of lakes and wetlands as the result of water collection in areas of surface subsidence, then CH_4 emissions are likely to be enhanced (Reeburgh and Whalen 1992, Zimov et al. 1997) while CO_2 emissions are reduced.

Observations of Changing Climate and Effects on Indigenous People

In recent decades, indigenous people in northern and western Alaska have observed notable changes in terrestrial ecosystems used by people (Huntington 2000, Huntington et al. 1998). Vegetation has shown accelerated rates of migration and succession, with trees moving into formerly treeless areas on the Seward and Baldwin Peninsulas (Pungowiyi 2000) and willows becoming larger and more common in the Kotzebue region (Whiting 2002). Warmer summers have led to a profusion of many kinds of insects, including ones not seen before in certain areas (Pungowiyi 2000).

Geese and songbirds have been arriving earlier than in living memory, making them more susceptible to sudden cold snaps (Pungowiyi 2000). Some new birds, which have no Iñupiaq names, have been seen in the Kotzebue region (Whiting 2002). Winters have been relatively mild, with little snow, allowing ptarmigan populations to rise (Pungowiyi 2000). Increases in lightning in northwestern Alaska have led to more frequent forest and tundra fires (Whiting 2002). Moose have expanded their range into the northwestern Arctic (into the area around Kotzebue) (Whiting 2002). Beaver and muskrat have extended their ranges in northwestern and western Alaska (Whiting 2002, Huntington et al. 1999).

These phenomena appear connected to changes in climate and are recognized by Alaska Native peoples as large-scale, directional changes of a kind and magnitude not seen before (Pungowiyi 2000). Many other changes have been reported—for example, in populations of caribou and furbearers—but these may be the result of population cycles, other environmental impacts, or other natural or human-caused phenomena.

The locations of settlements along the northern and western coasts of Alaska were chosen primarily to provide residents with the most convenient access to subsistence marine resources. In the past, indigenous people in these areas were nomadic and had multiple seasonal camps as needed. In the modern era, residents have become established at fixed locations that must be served by infrastructure such as airports, schools, roads, and permanent homes.

With recent climate warming, sea ice forms later in the cold season. This later onset of sea ice formation allows for accelerated coastal erosion because the shore is no longer protected from the battering of fall and winter storms. Even a thin layer of ice prevents the generation of high waves and storm surges produced by fall storms in particular. The recent extended ice-free season has removed this protection. Permafrost also exists in many of these coastal areas as well as at other interior sites. Warming, and in some cases melting permafrost, further accelerates coastal erosion, causing problems of infrastructure stability.

Rising sea level due to thermal expansion and melting of land-based glaciers is already contributing to coastal erosion. Because measures to stabilize the shoreline are not practical in rural Alaska, whole villages are now being destroyed by shoreline erosion and need to be relocated. The expenses to do so are prohibitive, and it is uncertain how these costs will be met.

While there are still many uncertainties in predicting the magnitude and direction of many of the changes from a warming climate, some of the effects are already impacting the inhabitants of Alaska in other ways. One of the more important effects of climate warming is the social impact on the native cultures for whom subsistence is a way of life. Most rural Alaskans depend on food resources harvested from the land for a major part of their diet. They generally lack the high levels of income needed to import food, and most of their communities have only limited, if any, handling and holding facilities.

The social structure in these communities is typically organized around harvesting, preparing, and sharing local food resources. Such traditional diets are now known to provide significant health benefits, but climate change is posing significant challenges for people who want to follow these traditional subsistence lifestyles.

As another result of climate warming, customary loca-

tions for the congregation of wildlife, such as sea ice openings that attract or support marine mammals, are shifting in unprecedented and unpredictable ways. New diseases triggered by climate warming are also appearing in traditionally harvested resources, such as the Ichthyophonus (wasting disease) in Yukon River salmon (Richard et al. 2004). An additional risk is the greater probability of the transmission of diseases among migratory birds and to humans under a warming climate (Webster 2002).

The warmer temperatures, in general, also lead to later onset of ice formation, and thinner and more dangerous ice on the ocean, lakes, and rivers, which are important travel routes for accessing subsistence resources. Snow and ice are the main travel routes for people in rural Alaska, where roads are generally lacking. Travel accidents have traditionally caused a significant number of deaths in these communities, and climate warming exposes them to greater risks (Berner et al. 2005).

Another problem created by climate warming is that traditional knowledge no longer correlates with reality. Consequently, traditional hunters are having difficulty predicting the changing timing of availability of potential food resources (Fox 2002). Missing the harvest of traditional resources (because of early bird migrations, late or non-returning salmon runs, the timing change of the ripening of berries, and so on) can leave a village without a significant part of its food requirements. Many rural Alaska communities only have a limited ability to pursue their key food resources and thus depend on a high success rate of capture and harvest.

Another problem is that the water and sanitation systems in many of these rural communities still do not meet national standards. A concerted effort is underway to address these deficiencies, but it is extremely expensive, and funds are becoming severely limited. Given the uncertainties of climate warming, it's hard to predict the future needs of rural communities and what it will take to provide them with adequate water to meet their current and impending needs.

Concluding Remarks

A number of the changes now detectable as a result of climate warming in Alaska are interacting, and some of the interactions are producing strong positive feedbacks; for example, warming temperatures cause increased melting of sea ice, warmer sea temperatures, and lower albedo, promoting further warming. Increased temperatures also lead to warming of cold soils and, as a result, the decomposition of stored carbon, which is then exported to the atmosphere as the greenhouse gases, carbon dioxide and methane. These positive feedbacks will accelerate climate warming. Other feedbacks are not so obvious, and it is yet unknown how cloud cover will be affected by warming in the Arctic and how this, in turn, may affect radiation. Finally, critical decisions about the management, harvest, and allocation of resources must be made as a changing environment makes previous knowledge obsolete.

There is still much to be learned.

References

Adams, C., and G. P. Juday. 1997. Growth rate responses of white spruce to river discharge patterns on a boreal forest floodplain. *Ecol. Soc. Amer.* (Suppl.-Ann. Meeting Abstracts) 78(4):43.

Amiro, B. D., J. B. Todd, B. M. Wotton, K. A. Logan, M. D. Flannigan, B. J. Stocks, J. A. Mason, D. L. Martell, and K. G. Hirsch. 2001. Direct carbon emissions from Canadian forest fires, 1959–1999. *Can. J. For. Res.* 31:512–525.

Anisimov, O. A., and F. E. Nelson. 1996. Permafrost distribution in the Northern Hemisphere under scenarios of climatic change. *Glob. and Planet. Change* 14:59–72.

Anisimov, O. A., and V. Y. Poliahov. 2003. GIS assessment of climate-change impacts in permafrost regions. Pp. 9–14 in M. Phillips, S. M. Springman, and L. U. Arenson (eds.), *Permafrost*. Balkema Publishers, Lisse, Netherlands.

Arendt, A. A., K. A. Echelmeyer, W. D. Harrison, C. S. Lingle, and V. B. Valentine. 2002. Rapid wastage of Alaska Glaciers and their contribution to rising sea level. *Science* 297:382–386.

Arft, A. M., M. D. Walker, J. Gurevitch, J. M. Alatalo, M. S. Bret-Harte, M. Dale, M. Diemer, F. Gugerli, G. H. Henry, M. H. Jones, R. D. Hollister, I. S. Jonsdottir, K. Laine, E. Levesque, G. M. Marion, U. Molau Molgaard, U. Nordenhall, V. Raszhivin, C. H. Robinson, G. Starr, A. Stenstrom, M. Stenstrom, O. Totland, L. Turner, L. J. Walker, J. Webber, J. M. Welker, and P. A. Wookey. 1999. Responses of tundra plants to experimental warming: Meta-analysis of the international tundra experiment. *Ecol. Monog.* 69:491–511.

Baldocchi, D., E. Falge, and K. Wilson. 2001. A spectral analysis of biosphere-atmosphere trace gas flux densities and meteorological variables across hour to multi-year time scales. *Agr. and For. Meteor.* 107:1–27.

Barber, V. A., G. P. Juday, and E. Berg. 2002. Assessment of recent and possible future forest responses to climate in boreal Alaska. In *Workshop on Northern Timberline Forests: Environmental and Socio-economic Issues and Concerns,* pp. 102–105, 288–289. The Finnish Forest Research Institute, Res. Paper 862. Kolani Research Station.

Barber, V. A., G. P. Juday, and B. P. Finney. 2000. Reduced growth of Alaskan white spruce in the twentieth century from temperature-induced drought stress. *Nature* 405:668–673.

Barber, V. A., G. P. Juday, B. P. Finney, and M. Wilmking. 2004. Reconstruction of summer temperatures in interior Alaska from tree ring proxies: Evidence for changing synoptic climate regimes. *Clim. Change* 63:91–120.

Berg, E. E., J. D. Henry, C. L. Fastie, A. D. D. Volder, and S. M. Matsuoka. 2006. Spruce beetle outbreaks on the Kenai Peninsula, Alaska, and Kluane National Park and Reserve, Yukon Territory: Relationship to summer temperatures and regional differences in disturbance regimes. *For. Ecol. Manage.* 227:219–232.

Berner, J., C. Furgal Bjerregaard, M. Bradley, T. Curtis, E. De Fabo, J. Hassi, W. Keatinge, S. Kvernmo, S. Nayha, H. Rintamaki, and J. Warren. 2005. Human Health. Chapter 15, pp. 863–906, in C. Symon, L. Arris, and B. Heal (eds.), *Arctic Climate Impact Assessment*. Cambridge University Press, Cambridge.

Bitz, C. M., and D. S. Battisti. 1999. Interannual to decadal variability in climate and the glacier mass balance in Washington, western Canada, and Alaska. *J. Clim.* 12:3181–3196.

Bliss, L. C., and N. V. Matveyeva. 1992. Circumpolar arctic vegetation. Pp. 59–89 in F. S. Chapin III, R. L. Jefferies, J. F. Reynolds, G. R. Shaver, and J. Svoboda (eds.), *Arctic Ecosystems in a Changing Climate: An Ecophysiological Perspective*. Academic Press, Orlando, FL.

Bonan, G. B., D. Pollard, and S. L. Thompson. 1992. Effects of boreal forest vegetation on global climate. *Nature* 359:716–718.

Bret-Harte, M. S., G. R. Shaver, J. P. Zoerner, J. F. Johnson, J. L. Wagner, A. S. Chavez, R. F. Gunkelman, S. C. Lippert, and J. A. Laundre. 2001. Developmental plasticity allows *Betula nana* to dominate tundra subjected to an altered environment. *Ecology* 82:18–32.

Brown, J., O. J. Ferrians, J. A. Heginbottom, and E. S. Melnikov. 1997. Circum-Arctic map of permafrost and ground ice conditions. U.S. Geol. Surv., Reston, VA.

Calef, M. P., A. D. McGuire, H. E. Epstein, T. S. Rupp, and H. H. Shugart. 2005. Analysis of vegetation distribution in interior Alaska and sensitivity to climate change using a logistic regression approach. *J. Biogeog.* 32:863–878.

Chapin, F. S., A. D. McGuire, J. Randerson, R. Pielke, D. Baldocchi, S. E. Hobbie, N. Roulet, W. Eugester, E. Kasischke, E. B. Rastetter, S. A. Zimov, and S. W. Running. 2000. Arctic and boreal ecosystems of western North America as components of the climate system. *Glob. Change Biol.* 6:211–223.

Chapin, F. S., T. S. Rupp, A. M. Starfield, L. DeWilde, E. S. Zavaleta, N. Fresco, J. Henkelman, and A. D. McGuire. 2003. Planning for resilience: Modeling change in human-fire interactions in the Alaskan boreal forest. *Front. Ecol. and Env.* 1:255–261.

Chapin, F. S., and G. R. Shaver. 1996. Physiological and growth responses of arctic plants to a field experiment simulating climate change. *Ecology* 77:822–840.

Chapin, F. S., III, G. R. Shaver, A. E. Giblin, K. G. Nadelhoffer, and J. A. Laundre. 1995. Response of arctic tundra to experimental and observed changes in climate. *Ecology* 76:694–711.

Chapin, F. S., III, and A. M. Starfield. 1997. Time lags and novel ecosystems in response to transient climatic change in arctic Alaska. *Clim. Change* 35:449–461.

Chen, J., W. Chen, J. Liu, and J. Cihlar. 2000. Annual carbon balance of Canada's forests during 1895–1996. *Glob. Biogeochem. Cycles* 14:839–849.

Christensen, T. R., S. Jonasson, T. V. Callaghan, and M. Havström. 1999. On the potential CO₂ releases from tundra soils in a changing climate. *Appl. Soil Ecol.* 11:127–134.

Clein, J. S., A. D. McGuire, X. Zhuang, D. W. Kicklighter, J. M. Melillo, S. C. Wofsy, G. Jarvis, and J. M. Massheder. 2002. Historical and projected carbon balances of mature black spruce ecosystems across North America: The role of carbon-nitrogen interactions. *Plant and Soil* 242:15–32.

Clow, G. D., and F. E. Urban. 2003. GTN-P monitoring network: Detection of a 3 K permafrost warming in northern Alaska during the 1990s. SEARCH Open Science Meeting, Fairbanks, AK. Arctic Research Consortium of the United States.

Comiso, J. C. 2002. A rapidly declining perennial sea ice cover in the Arctic. *Geophys. Res. Lett.* 29:17-1–17-4.

Conard, S. G., A. I. Sukhinin, B. J. Stocks, D. R. Cahoon, E. P. Davidenko, and G. A. Ivanova. 2002. Determining effects of area burned and fire severity on carbon cycling and emissions in Siberia. *Clim. Change* 55:197–211.

D'Arrigo, R., R. K. Kaufmann, N. Davi, G. Jacoby, C. Laskowski, R. Myneni, and P. Cherubini. 2004. Thresholds for warming-induced growth decline at elevational treeline in the Yukon Territory. *Glob. Biogeochem. Cycles* 18: 7 pp.

D'Arrigo, R., R. Wilson, and G. Jacoby. 2006. On the long-term context for late twentieth century warming. *J. Geophys. Res.* 111, 12 pp.

Duffy, A., J. E. Walsh, J. M. Graham, D. H. Mann, and T. S. Rupp. 2005. Impacts of large-scale atmospheric-ocean variability on Alaskan fire season severity. *Ecol. Appl.* 15:1317–1330.

Dyurgerov, M. B., and M. F. Meier. 1997. Year-to-year fluctuations of global mass balance of small glaciers and their contribution to sea-level changes. *Arc. and Alp. Res.* 29:392–402.

———. 2000. Twentieth century climate change: Evidence from small glaciers. Proceedings of the National Academy of Science 97:1406–1411.

Ebbesmeyer, C. C., D. R. Cayan, D. R. McLain, F. H. Nichols, D. H. Peterson, and K. T. Redmond. 1990. 1976 step in the Pacific climate: Forty environmental changes between 1968–1975 and 1977–1984. *Proceedings of the Seventh Annual Pacific Climate (PACLIM) Workshop.*

Esch, D. C., and T. E. Osterkamp. 1990. Cold regions engineering: Climatic warming concerns for Alaska. *J. Cold Reg. Eng.* 4:6–14.

Flannigan, M., I. Campbell, M. Wotton, C. Carcaillet, P. Richard, and Y. Bergeron. 2001. Future fire in Canada's boreal forest: Paleoecology results and general circulation model–regional climate model simulations. *Can. J. For. Res.* 31:854–864.

Flannigan, M., and B. M. Wotton. 2001. Climate, weather and area burned. Pp. 335–357 in E. A. Johnson and K. Miyanishi (eds.), *Forest Fires: Behavior and Ecological Effects.* Academic Press, San Diego.

Fox, S. 2002. These are things that are really happening. Pp. 12–53 in I. Krupniki and D. Jolly (eds.), *The Earth Is Faster Now: Indigenous Observations of Arctic Environmental Change.* Arctic Research Consortium of the United States, Fairbanks, AK.

Frolking, S., M. Goulden, S. Wofsy, S.-M. Fan, D. Sutton, J. Munger, A. Bazzaz, B. Daube, P. Crill, J. Aber, L. Band, X. Wang, K. Savage, T. Moore, and R. Harriss. 1996. Modelling temporal variability in the carbon balance of a spruce/moss boreal forest. *Glob. Change Biol.* 2:343–366.

Frolking, S., K. McDonald, J. Kimball, R. Zimmermann, J. B. Way, and S. W. Running. 1999. Using the space-borne NASA Scatterometer (NSCAT) to determine the frozen and thawed seasons of a boreal landscape. *J. Geophys. Res.* 104:27895–27907.

Gabriel, H. W., and G. F. Tande. 1983. *A Regional Approach to Fire History in Alaska.* Bureau of Land Management, Anchorage, AK.

Goetz, S. J., A. G. Bunn, G. J. Fiske, and R. A. Houghton. 2005. Satellite-observed photosynthetic trends across boreal North America associated with climate and fire disturbance. *Proc. Nat. Acad. Sci.* 102:13521–13525.

Halsey, L. A., D. H. Vitt, and S. C. Zoltai. 1995. Disequilibrium response of permafrost in boreal continental western Canada to climate change. *Clim. Change* 30:57–73.

Harden, J. W., S. E. Trumbore, B. J. Stocks, A. Hirsch, S. T. Gower, K. P. O'Neill, and E. S. Kasischke. 2000. The role of fire in the boreal carbon budget. *Glob. Change Biol.* 6:S174–S184.

Hassol, S. J. 2004. ACIA, Impacts of a Warming Arctic. *Arctic Climate Impact Assessment.* Cambridge University Press, Cambridge, UK.

Hess, J. C., C. A. Scott, G. L. Huford, and M. D. Fleming. 2001. El Niño and its impact on fire weather conditions in Alaska. *Int. J. Wild. Fire* 10:1–13.

Hinzman, L. D., N. D. Bettez, W. R. Bolton, F. S. Chapin, M. B. Dyurgerov, C. L. Fastie, B. Griffith, R. D. Hollister, A. Hope, H. P. Huntington, A. M. Jensen, G. J. Jia, T. Jorgenson, D. L. Kane, D. R. Klein, G. Kofinas, A. H. Lynch, A. H. Lloyd, A. D. Mcguire, F. E. Nelson, W. C. Oechel, T. E. Osterkamp, C. H. Racine, E. Romanovsky, R. S. Stone, D. A. Stow, M. Sturm, C. E. Tweedie, G. L. Vourlitis, M. D. Walker, D. A. Walker, J. Webber, J. M. Welker, K. S. Winker, and K. Yoshikawa. 2005. Evidence and implications of recent climate change in terrestrial regions of the Arctic. *Clim. Change* 72:251–298.

Hobbie, S. E., and F. S. Chapin, III. 1998. The response of tundra plant biomass, aboveground production, nitrogen, and CO₂ flux to experimental warming. *Ecology* 79:1526–1544.

Hobbie, S. E., J. P. Schimel, and S. E. Trumbore. 2000. Controls over carbon storage and turnover in high-latitude soils. *Glob. Change Biol.* 6:196–210.

Hogg, E. H., J. P. Brandt, and B. Kchtubajda. 2002. Growth and dieback of aspen forests in northwestern Alberta, Canada, in relation to climate and insects. *Can. J. For. Res.* 32:823–832.

———. 2005. Factors affecting interannual variation in growth of western Canadian aspen forests during 1951–2000. *Can. J. For. Res.* 35:610–622.

Houghton, J. J., L. G. Meiro Filho, B. A. Callander, N. Harris, A. Kattenberg, and K. Maskell. 1995. *The Science of Climate Change.* Contribution of Working Group I to the Second Assessment Report of the Intergovernmental Panel on Climate Change (IPCC), Climate Change 1995, Cambridge University Press, Cambridge, UK.

Huntington, H. P. 2000. Using traditional ecological knowledge in science: Methods and applications. *Ecol. Appl.* 10:1270–1274.

Huntington, H. P., and the Communities of Buckland, Elim, Koyuk,

Point Lay, and Skaktoolih. 1999. Traditional knowledge of beluga whales (*Delpinapterus leucas*) in the eastern Chukchi and northern Bering areas, Alaska. *Arctic* 52:49–61.

Huntington, H. P., J. H. Mosli, and V. P. Shustov. 1998. Peoples of the Arctic. In *The Assessment Report: Arctic Pollution Issues, Arctic Monitoring and Assessment Programme*, pp. 141–182. Oslo.

IPCC. 2001. *Climate Change 2001: The Scientific Basis*. Contribution of Working Group I to the Third Assessment Report of the Intergovernmental Panel on Climate Change. Cambridge University Press, Cambridge.

Jacoby, G. C., and E. R. Cook. 1981. Past temperature variations inferred from a 400-year tree-ring chronology from Yukon Territory, Canada. *Can. Arc. Alp. Res.* 13:409–418.

Jia, G. J., H. E. Epstein, and D. A. Walker. 2003. Greening of Arctic Alaska. *Geophys. Res. Lett.* 30:2067.

Jorgenson, M. T., and T. E. Osterkamp. 2004. Response of boreal ecosystems to varying modes of permafrost degradation. *Climate Disturbance Interactions in Boreal Forest Ecosystems*. International Boreal Forest Research Association, Univ. of Alaska, Fairbanks.

Jorgenson, M. T., C. H. Racine, J. C. Walters, and T. E. Osterkamp. 2001. Permafrost degradation and ecological changes associated with a warming climate in central Alaska. *Clim. Change* 48: 551–579.

Juday, G. P., V. Barber, P. Duffy, H. Linderholm, H. Rupp, S. Sparrow, E. Vaganov, and J. Yarie. 2005. Forests, land management, agriculture. Chapter 13, pp. 781–862, in C. Symon, L. Arris, and B. Heal (eds.), *Arctic Climate Impact Assessment*. Cambridge University Press, Cambridge, UK.

Juday, G. P., V. S. Barber, S. Rupp, J. Zasada, and M. Wilmking. 2003. A 200-year perspective of climate variability and the response of white spruce in interior Alaska. Pp. 226–250 in D. Greenland, D. Goodin, and R. Smith (eds.), *Climate Variability and Ecosystem Response at Long-Term Ecological Research (LTER) Sites*. Oxford University Press, New York.

Kaplan, J. O., N. H. Bigelow, I. C. Prentice, S. P. Harrison, J. Bartlein, T. R. Christensen, W. Cramer, N. V. Matveyeva, A. D. McGuire, D. F. Murray, Y. Razzhivin, B. Smith, D. A. Walker, M. Anderson, A. A. Andreev, L. B. Brubaker, M. E. Edwards, and A. V. Lozhkin. 2003. Climate change and arctic ecosystems II: Modeling, paleodata-model comparisons, and future projections. *J. Geophys. Res.* 108:12-1–12-17.

Kasischke, E. S., N. L. Christensen, and B. J. Stocks. 1995. Fire, global warming, and the carbon balance of boreal forests. *Ecol. Appl.* 5:437–451.

Kasischke, E. S., T. S. Rupp, and D. L. Verbyla. 2006. Fire trends in the Alaskan boreal forest region. In F. S. Chapin III et al. (eds.), *Alaska's Changing Boreal Forest*. Oxford University Press, New York.

Klein, E., E. E. Berg, and R. Dial. 2005. Wetland drying and succession across the Kenai Peninsula Lowlands, south-central Alaska. *Can. J. For. Res.* 35:1931–1941.

Kurz, W. A., and M. J. Apps. 1999. A 70-year retrospective analysis of carbon fluxes in the Canadian forest sector. *Ecol. Appl.* 9:526–547.

Labau, J., and W. Van Hees. 1990. An inventory of Alaska's boreal forests: Their extent, condition, and potential use. *Proc. Int. Symp. Boreal Forests: Climate, Dynamics, Anthropogenic Effects.* Archangelsk, Russia.

Lachenbruch, A. H., T. T. Cladouhos, and R. W. Saltus. 1988. Permafrost temperature and the changing climate. In *Fifth International Conference on Permafrost*, pp. 9–17.

Lachenbruch, A. H., and B. V. Marshall. 1986. Changing climate: Geothermal evidence from permafrost in the Alaskan Arctic. *Science* 234:689–696.

Langenfelds, R. L., R. J. Francey, B. C. Pak, L. P. Steele, J. Lloyd, C. M. Trudinger, and C. E. Allison. 2002. Interannual growth rate variations of atmospheric CO_2 and its del^{13}C, H_2, CH_4, and CO between 1992 and 1999 linked to biomass burning. *Glob. Biogeochemical Cycles* 16:1–22.

Lloyd, A. H., and C. L. Fastie. 2002. Spatial and temporal variability in the growth and climate response of treeline trees in Alaska. *Clim. Change* 52:481–509.

Lloyd, A. H., T. S. Rupp, C. L. Fastie, and A. M. Starfield. 2003a. Patterns and dynamics of treeline advance on the Seward Peninsula, Alaska. *J. Geophys. Res.* 108:2-1–2-15.

Lloyd, A. H., K. Yoshikawa, C. L. Fastie, L. Hinzman, and M. Fraver. 2003b. Effects of permafrost degradation on woody vegetation at arctic treeline on the Seward Peninsula, Alaska. *Permaf. and Perigla. Proc.* 14:93–102.

Magnuson, J. J., D. M. Robertson, B. J. Benson, R. H. Wynne, D. M. Livingstone, R. A. Assel, R. D. Barry, E. Card, E. Kuusisto, N. G. Granin, D. T. Prowse, K. M. Stewart, and S. Vuglinski. 2000. Ice cover phenologies of lakes and rivers in the Northern Hemisphere and climate warming. *Science* 289:1743–1746.

Mann, M. E., R. S. Bradley, and M. K. Hughes. 1999. Northern Hemisphere temperatures during the past millennium: Inferences, uncertainties and limitations. *Geophys. Res. Lett.* 26:759–762.

McGuire, A. D., and J. E. Hobbie. 1997. Global climate change and the equilibrium responses of carbon storage in arctic and subarctic regions. *Modeling the Arctic System: A Workshop Report of the Arctic System Science Program*, pp. 53–54.

McGuire, A. D., R. A. Meier, Q. Zhuang, M. Macander, T. S. Rupp, E. Kasischke, D. Verbyla, D. W. Kicklighter, and J. M. Melillo. 2000. The role of fire disturbance, climate, and atmospheric CO2 in the response of historical carbon dynamics in Alaska from 1950 to 1995: A process-based analysis with the Terrestrial Ecosystem Model. IBFRA International Science Conference: The Role of Boreal Forests and Forestry in the Global Carbon Budget.

McGuire, A. D., J. M. Melillo, D. W. Kicklighter, Y. Pan, X. Xiao, J. Helfrich, B. Moore III, C. J. Vorosmarty, and A. L. Schloss. 1997. Equilibrium responses of global net primary production and carbon storage to doubled atmospheric carbon dioxide: Sensitivity to changes in vegetation nitrogen concentration. *Glob. Biogeochem. Cycles* 11:173–189.

McGuire, A. D., M. Sturm, and F. S. Chapin III. 2003. Arctic transitions in the land-atmosphere system (ATLAS): Background, objectives, results, and future directions. *J. Geophys. Res. Atmos.* 108(02), 8166, doi:10.1029/2002JDOOR367.

Meier, M. F., and M. B. Dyurgerov. 2002. Sea level changes: How Alaska affects the world. *Science* 297:350–351.

Myneni, R. B., J. Dong, C. J. Tucker, R. K. Kaufmann, E. Kauppi, J. Liski, L. Zhou, V. Alexeyev, and M. K. Hughes. 2001. A large carbon sink in the woody biomass of northern forests. *Proc. Nat. Acad. Sci.* 98:14784–14789.

Myneni, R. B., C. D. Keeling, C. J. Tucker, G. Asrar, and R. R. Nemani. 1997. Increased plant growth in the northern high latitudes from 1981–1991. *Nature* 386:698–702.

Neff, J. C., and D. U. Hooper. 2002. Vegetation and climate controls on potential CO_2, DOC, and DOC production in northern latitude soils. *Glob. Change Biol.* 8:872–884.

Nelson, F. E., O. A. Anisimov, and N. I. Shiklomanov. 2001. Subsidence risk from thawing permafrost. *Nature* 410:889–890.

Nolan, M., A. Arendt, B. Rabus, and L. Hinzman. 2005. Volume change of McCall Glacier, Arctic Alaska, USA, 1956–2003. *Ann. Glaciol.* 42:409–416.

Nowacki, G., P. Spencer, T. Brock, M. Fleming, and T. Jorgenson. 2002. *Ecoregions of Alaska and Neighboring Territories.* Open File Report 02-297. USGS, Washington, D.C.

Oechel, W. C., G. L. Vourlitis, S. J. Hastings, and S. A. Bocharev. 1995. Change in Arctic CO_2 flux over two decades: Effects of climate change at Barrow, Alaska. *Ecol. Appl.* 5:846–855.

Osterkamp, T. E. 1984. Potential impact of a warmer climate on permafrost in Alaska. *The Potential Effects of Carbon Dioxide-induced Climatic Changes in Alaska,* pp. 106–113. Alaska Agr. Exp. Sta. Misc. Publ. 83-1.

———. 2003. Establishing long-term permafrost observatories for active-layer and permafrost investigations in Alaska, 1977–2002. *Permaf. Perigla. Proc.* 14:331–342.

Osterkamp, T. E., D. C. Esch, and E. Romanovsky. 1997. Permafrost. In *Proc. Implica. Global Change in Alaska and the Bering Sea Region,* pp. 115–127. University of Alaska, Fairbanks.

Osterkamp, T. E., and E. Romanovsky. 1999. Evidence for warming and thawing of discontinuous permafrost in Alaska. *Permaf. and Perigla. Proc.* 10:17–37.

Osterkamp, T. E., L. Viereck, Y. Shur, M. T. Jorgenson, C. H. Racine, A. P. Doyle, and R. D. Boone. 2000. Observations of thermokarst in boreal forests in Alaska. *Arc., Antarc., and Alp. Res.* 32:303–315.

Overpeck, J., K. Hughen, D. Hardy, R. Bradley, R. Case, M. Douglas, B. Finney, K. Gajewski, G. Jacoby, A. Jennings, S. Lamoureux, A. Lasca, G. MacDonald, J. Moore, M. Retelle, S. Smith, A. Wolfe, and G. Zielinski. 1997. Arctic environmental change of the last four centuries. *Science* 278:1251–1256.

Péwé, T. L. 1982. Geologic hazards of the Fairbanks area. College, AK, Alaska Div. Geol. and Geophys. Surv. Spec. Rept. 15: 109 pp.

Podur, J., D. L. Martell, and K. Knight. 2002. Statistical quality control analysis of forest fire activity in Canada. *Can. J. For. Res.* 32:195–205.

Pungowiyi, C. 2000. Native observations of change in the marine environment of the Bering Strait region. In H. P. Huntington (ed.), *Proc. Conf. Marine Mammal Comm.* Girdwood, AK.

Rabus, B., K. Echelmeyer, D. Trabant, and C. Benson. 1995. Recent changes of McCall Glacier, Alaska. *Ann. of Glaciol.* 21:231–239.

Reesburgh, W. S., and S. C. Whalen. 1992. High latitude ecosystems as CH_4 sources. *Ecol. Bull.* 42:67–70.

Richard, K., P. Hershberger, and J. Winton. 2004. Ichthyophoniasis: An emerging disease of Chinook salmon in the Yukon River. *J. Aquat. Anim. Health* 16:58–72.

Riordan, B. A. 2005. Using remote sensing to examine changes of closed-basin surface water area in interior Alaska from 1950–2002. M.S. thesis. University of Alaska, Fairbanks.

Roulet, N. T. 2000. Peatlands, carbon storage, greenhouse gases and the Kyoto Protocol: Prospects and significance for Canada. *Wetlands* 20:605–615.

Roulet, N. T., R. Ash, and T. R. Moore. 1992a. Low boreal wetlands as a source of atmospheric methane. *J. Geophys. Res.* 97:3739–3749.

Roulet, N. T., T. R. Moore, J. Bubier, and P. Lefleur. Northern fens: Methane flux and climate change. *Tellus* 44B:100–105.

Sagarin, R., and F. Micheli. 2001. Climate change in nontraditional data sets. *Science* 294:811.

Serreze, M. C., J. A. Maslanik, T. A. Scambos, F. Fetterer, J. Stroeve, K. Knowles, C. Fowler, S. Drobot, R. G. Barry, and T. M. Haran. 2003. A record minimum in Arctic sea ice extent and area in 2002. *Geophys. Res. Lett.* 30:1110.

Serreze, M. C., J. Walsh, F. S. Chapin, T. Osterkamp, M. Dyurgerov, V. Romanosky, W. Oechel, J. Morison, T. Zhang, and R. Barry. 2000. Observational evidence of recent change in the northern high-latitude environment. *Clim. Change* 46:159–207.

Silapaswan, C. S., D. Verbyla, and A. D. McGuire. 2001. Land cover change on the Seward Peninsula: The use of remote sensing to evaluate potential influences of climate change on historical vegetation dynamics. *Canad. J. Remote Sens.* 5:542–554.

Smith, M. W. 1990. Potential responses of permafrost to climatic change. *Cold Regions Eng.* 4:29–37.

Smith, T. M., and H. H. Shugart. 1993. The transient response of terrestrial carbon storage in a perturbed climate. *Nature* 361:523–526.

Starfield, A. M., and F. S. Chapin. 1996. Model of transient changes in arctic and boreal vegetation in response to climate and land use change. *Ecol. Appl.* 6:842–864.

Stocks, B. J., M. A. Fosberg, T. J. Lynham, L. Mearns, B. M. Wotton, Q. Yang, J.-Z. Jin, K. Lawrence, G. R. Hartley, J. A. Mason, and D. W. McKenney. 1998. Climate change and forest fire potential in Russian and Canadian boreal forests. *Clim. Change* 38:1–13.

Stocks, B. J., M. A. Fosberg, M. B. Wotten, T. J. Lynham, and K. C. Ryan. 2000. Climate change and forest fire activity in North American boreal forests. Pp. 368–376 in E. S. Kasischke and B. J. Stocks (eds.), *Fire, Climate Change, and Carbon Cycling in North American Boreal Forest.* Springer-Verlag, Berlin.

Stocks, B. J., J. A. Mason, J. B. Todd, E. M. Bosch, B. M. Wotton, B. D. Amiro, M. D. Flannigan, K. G. Hirsch, K. A. Logan, D. L. Martell, and W. R. Skinner. 2002. Large forest fires in Canada, 1959–1997. *J. Geophys. Res.* 107:8149.

Stolbovoi, V. 1997. *Degradation of Forest Land in Land-use/Cover Patterns of Russia.* Rept. No. IR-97-070: 20 pp. Int. Inst. for Appl. Systems Anal., Laxenburg, Austria.

Sturm, M., C. Racine, and K. Tape. 2001. Increasing shrub abundance in the Arctic. *Nature* 411:546–547.

Symon, C., L. Arris, and B. Heal. 2005. *Arctic Climate Impact Assessment.* 1042 pp. Cambridge University Press, Cambridge, UK.

Thompson, C., J. Beringer, F. S. Chapin III, A. D. McGuire. 2004. Structural complexity and land-surface energy exchange along a gradient from arctic tundra to boreal forest. *J. Veg. Sci.* 15:397–406.

Trabant, D. C., R. S. March, L. H. Cox, W. D. Harrison, and E. G. Josberger. 2003. *Measured Climate Induced Volume Changes of Three Glaciers and Current Glacier-Climate Response Prediction.*

SEARCH Open Science Meeting, Arctic Research Consortium of the United States (ARCUS), Seattle, Washington.

Walsh, J. E., O. Anisimov, J. O. M. Hagen, T. Jakobsson, J. Oerlemans, T. D. Prowse, E. Romanovsky, N. Savelieva, M. Serreze, A. Shiklomanov, I. Shiklomanov, S. Solomon, et al. 2005. Cryosphere and Hydrology. Chapter 6, pp. 183–242, in C. Symon, L. Arris, and B. Heal (eds.), *Arctic Climate Impact Assessment*. Cambridge University Press, Cambridge, UK.

Webster, R. 2002. The importance of animal influenza for human disease. *Vaccine* 20:S16–S20.

Weintraub, M. N., and J. P. Schimel. 2003. Interactions between carbon and nitrogen mineralization and soil organic matter chemistry in Arctic tundra soils. *Ecosystems* 6:129–143.

Werner, R. A., and E. H. Holsten. 1985. Effect of phloem temperature on development of spruce beetles in Alaska: The role of the host in the population dynamics of forest insects. *Proceedings of the IUFRO Conference*, pp. 155–163. Canadian Forestry Service, Banff, AB.

Werner, R. A., and K. E. Post. 1985. Effects of wood-boring insects and bark beetles on survival and growth of burned white spruce. Pp. 14–16 in G. P. Juday and C. T. Dyrness (eds.), *Early Results of the Rosie Creek Fire Research Project*. University of Alaska, Agricultural and Forestry Experiment Station, Fairbanks.

Werner, R. A., K. F. Raffa, and B. L. Illman. 2006. Dynamics of phytophagous insects and their pathogens in Alaskan boreal forests. Pp. 133–146 in F. S. Chapin, M. W. Oswood, K. V. Cleve, L. A. Viereck, and D. L. Verbyla (eds.), *Alaska's Changing Boreal Forest*. Oxford University Press, New York.

Whiting, Alex. 2002. Documenting Zikiktagnumiut knowledge of environmental change. Native Village of Kotzebue, AK.

Wiles, G. C., R. D. D'Arrigo, R. Villalba, P. E. Calkin, and D. J. Barclay. 2004. Solar forcing of century-scale mountain glaciations over the past millennium in Alaska. *Geophys. Res. Lett.* 31: 4 pp.

Wilmking, M. 2003. The treeline ecotone in interior Alaska: From theoretical concept to planning application and the science in between. M.S. thesis. University of Alaska, Fairbanks.

Wilmking, M., and G. P. Juday. 2005. Longitudinal variation of radial growth at Alaska's northern treeline: Recent changes and possible scenarios for the twenty-first century. *Glob. Planet. Change* 47: 282–300.

Wilmking, M., G. P. Juday, A. Barber, and H. S. J. Zald. 2004. Recent climate warming forces contrasting growth responses of white spruce at treeline in Alaska through temperature thresholds. *Glob. Change Biol.* 10:1724–1736.

Wittwer, D. 2004. Forest Health Conditions in Alaska—2003, Forest Health Protection. USDA, DNR GTR R10-TP-123.

Wotton, B. M., and M. D. Flannigan. 1993. Length of the fire season in a changing climate. *Forest. Chron.* 69:187–192.

Zhuang, Q., A. D. McGuire, J. M. Melillo, J. S. Clein, R. J. Dargaville, D. W. Kicklighter, R. B. Myneni, J. Dong, E. Romanovsky, J. Harden, and J. E. Hobbie. 2003. Carbon cycling in extratropical terrestrial ecosystems of the Northern Hemisphere during the twentieth century: A modeling analysis of the influences of soil thermal dynamics. *Tellus* B 55:751–776.

Zimov, S. A., Y. V. Vonopaev, I. P. Semiletov, S. P. Davidov, S. F. Prosiannikov, F. S. Chapin III, M. C. Chapin, S. Trumbone, and S. Tyler. 1997. North Siberian lakes: A methane source fueled by Pleistocene carbon. *Science* 277(5327):800–802.

PART III

Overview

10

Climate Warming and
Environmental Effects in the West

Evidence for the Twentieth Century
and Implications for the Twenty-First

Frederic H. Wagner

Abstract

Analyses of twentieth-century weather records for west-ern North America show air temperature increases over the entire region, from the southwestern United States to Alaska. Two air-temperature proxies—temperature pro-files in boreholes in the earth and ice cores from the Upper Fremont Glacier in Montana—also reflect the increases, which range mostly between 1 and 2°C (1.8–3.6°F). Evi-dence from several investigators shows greater increase in night-time minimum temperatures than in daytime max-ima, declining interannual variability, a south-to-north gradient in temperature increase, and greater increases at high elevations than at low.

General circulation models (GCMs)—developed for simulating dynamics of the earth's climate system and modified for regional projections of climate change—project mean annual temperature increases for western North America of 2.0–6.5°C (3.6–11.7°F) by 2100 with an assumed twofold increase in atmospheric CO_2.

Confidence that the model projections are valid, at least in direction if not in magnitude, is based on (1) sim-ilar directions projected by numerous models in differ-ent laboratories around the world; (2) the known physical effects of CO_2 on the earth's climate system; (3) the fact that the models project similar directional changes for the twenty-first century as the *measured* changes of the twen-tieth when atmospheric CO_2 rose by ~32 percent; and (4) retrospective modeling experiments in which the model simulations follow the *measured* twentieth-century tem-perature trends almost exactly when programmed with *measured* twentieth-century climatic variables and rise in atmospheric CO_2. These experiments also provide evidence that the CO_2 increase was implicated in the twentieth-century temperature rise.

Analyses of twentieth-century weather records show no precipitation increases in some parts of the southwest-ern United States, but increases in others. Consistent increases have been recorded in the northern half to two-thirds of the remaining western states and Alaska. The most recent models project northward expansion of the subtropical arid zone in the Southwest, reduced precipita-tion, and more extreme aridity, but they project precipita-tion increase in the mid-latitudes.

Three-fourths of water used by humans in the West comes from streams fed by meltwater from mountain snowpacks. These snowpacks are the water sources not only for western streams, but also for the Missouri, Rio Grande, Platte, and Arkansas Rivers.

The amount of water available for use in the West is a function of both precipitation, which is projected to decline in the Southwest and increase in the northern portion of western Lower 48, and of evapotranspiration loss, projected to increase as temperatures rise during the twenty-first century. There is a distinct possibility of re-duced water resources in the Southwest in this century.

Seasonal availability of water resources is a function of streamflow timing, itself a function of precipitation and timing of snowpack melt. This timing is changing as tem-peratures warm, creating problems of predicting stream-flows and, potentially, of managing water resources.

Effects of climate change on cultivated agriculture in the West will be mixed. Higher temperatures and in-creased evapotranspiration may foreclose cultivation of some crops and require substitution of new ones. As one example, the optimum climate for producing quality wine grapes may shift northward from California to Oregon, Washington, and British Columbia. A longer growing season, however, may make some double cropping possible and increase the number of hay cuttings per season. With much of western agriculture dependent on irrigation, the future of the industry will be strongly influenced by the effects of future trends in precipitation and evapotranspiration, and ultimately the availability of water.

Livestock ranching, primarily of cattle, operates on a thin profit margin in the West and is declining. Forage production of native vegetation is certain to decline at lower elevations where precipitation declines and evapotranspiration increases, but it could increase at higher elevations with prolonged growing seasons. The cost of supplementing with hay, and hence the effects on ranchers' operating costs, will depend on the availability of irrigation water.

A burgeoning literature is reporting climate-change effects on the natural biota. Single-species effects include northward and upslope movement of species' distribution; advancing phenology (timing) of migration, reproduction, foliage growth, and flowering; and changing population densities. The inference of causation in these correlational observations can be strengthened by (1) examination of the physiological mechanisms involved in the changes; (2) parallel observations in numerous cases; and (3) meta-analysis.

The single-species responses elicit changes in their interactions with other species or establish new interactions such as predation, herbivory, competition, and provision of food and habitat. These are interactions that structure whole communities and ecosystems; thus climate change is likely to alter the latter and, in some cases, they are being observed to do so. Changes are occurring in terrestrial, fresh-water, and marine systems. Some observers are predicting extensive alterations of the world's entire biota and widespread extinctions.

Introduction

The subject of climate change is one of extreme breadth, ranging from the complex scientific basis of the phenomenon to the human role in the change; the environmental, social, and economic consequences of change; technological and policy solutions to address the problem; mitigation and adaptation; and the economic and political considerations surrounding remedial action.

These issues span two time scales. One is the reality of climate change, its causation, and associated environmental effects during the twentieth century. This aspect has been subjected to extensive scientific investigation and measurement, and the resulting evidence sheds light on its reality.

Projected changes for the twenty-first century are based on century-long scientific knowledge of the physical effects of greenhouse gases, especially CO_2, on the earth's climate; analogy with changes during the twentieth-century, when CO_2 rose by approximately a third; and climate simulation models that make climate-change projections based on hypothesized levels of greenhouse gas emissions during the twenty-first century. Projections have also examined potential environmental effects and their costs, including possible mitigation, and the social and political aspects of climate change.

The chapters in this book focus largely on climate change during the twentieth century and associated environmental effects intrinsic to the unique climatic, topographic, ecological, and land use characteristics of western North America. But they also point to a number of likely effects during the twenty-first century if the climate continues to change in directions set in the twentieth—and as projected by the models.

The contributors to this volume do not engage in policy advocacy, but rather provide information useful for resource managers and policy makers in planning future actions. In particular, the western governors have entered into the Western Regional Climate Action Initiative, and it is our hope that the information in this volume will be useful in their deliberations.

Climate Change in the West

Evidence of Twentieth-Century Warming

Air Temperature Measurements

There is now overwhelming evidence of temperature increases in western North America during the twentieth century, primarily air-temperature measurements from hundreds of weather stations in the region. In Chapter 1, Wagner shows increases over the nine-state Rocky Mountain/Great Basin region during the 1900s:

1. Increase in all eight subregions in mean annual minimum temperatures, five of them statistically significant at $p \leq 0.05$.
2. Increase in six of eight subregions in mean annual maximum temperatures, two significant at $p \leq 0.20$.
3. Increase in all subregions in mean annual temperatures, six significant at $p \leq 0.10$ by Sen-slope analysis. Increases in the six range from 0.1 to 0.65°C (0.18–1.17°F) per 100 years.

As discussed in Chapter 1, Brian McInerny (2006, pers. comm.), of the Salt Lake City office of the National Weather Service, calculated mean annual temperature increase between 1895 and 2004, averaged over all Utah weather stations, at 1.67°C (3.0°F).

Sprigg, Hinkley et al. (2000) have generalized temperature increases for the Southwest (in this case including New Mexico, Arizona, southeastern California, Nevada, Utah, and Colorado) at 1.1–1.7°C (2.0–3.1°F) during the twentieth century. Field et al. (1999) have calculated twentieth-century temperature increase, averaged over all of California, at 1.1°C (2.0°F).

Kittel et al. (2002), using U.S. Historical Climatology Network station data for the Rocky Mountains, subdivided their data for the northern, central, and southern Rockies to calculate average maximum and minimum temperatures. All showed increase, with the minima rising from 0.22 to 0.86°C (0.40–1.55°F) per 100 years, and the maxima increasing from 0.10 to 0.27°C (0.18–0.40°F). Only the minima for the northern and central Rockies were statistically significant at $p < 0.05$.

Mote (1999) calculated mean annual temperature trends for the three-state Pacific Northwest (Washington, Oregon, Idaho) and part of southwestern Canada. His results showed a 0.86°C (1.48°F) increase per 100 years over the region. He also estimated increases for the southern and northern Canadian Rockies between 1950 and 1998 at 1.5°C (2.7°F) and 2.0°C (3.6°F), respectively. In Chapter 9, Barber et al. show twentieth-century temperature increases in several areas of Alaska, ranging mostly from 1 to 2°C (1.8–3.8°F).

More comprehensively, Peterson and Vose (1997) show a southeast-northwest gradient in temperature increase for the whole of North America between 1950 and 1997, with no increase in the southeastern and eastern United States, but grading to 2°C (3.6°F) in the entire continental west (the United States, Canada, and Alaska).

Thus, numerous sources of air temperature data for western North America leave no doubt that air temperatures rose during the twentieth century. Several subsidiary trends emerge from these analyses:

1. Subdivision of data into annual averages of daytime maxima and night-time minima show the latter to have increased more than the former (Chapters 1 and 9; see also Kittel et al. 2002). Thus day-night temperature differentials have narrowed, with night-time increases more pronounced than those of the daytime.
2. Although the trends are crude, there is some suggestion from Chapter 1 and the Kittel et al. (2002) results that there is a south-north increase in the magnitude of temperature rise. This is also indicated by the Peterson and Vose (1997) data.
3. The declining standard deviations around 30-year moving averages of mean minimum temperatures (Chapter 1) indicate that while temperatures rose during the twentieth century, year-to-year temperature variability declined.

Temperature Proxies

The extensive evidence of rising temperatures during the twentieth century in the West, based on direct measurements, is paralleled by measurements of air-temperature proxies, two of which are reported for western North America in Chapters 2 and 3. Chapman, Harris, and Bartlett (Chapter 2) show that temperature profiles in boreholes in the earth indicate air-temperature increases during the past century not only in the West, but more broadly over North America.

Naftz, Oswald, Schuster, and Miller (Chapter 3) show evidence of a 2.1°C (3.8°F) temperature rise between 1850 and 2000 in ice cores in the Upper Fremont Glacier, located in the Wind River Mountains of Wyoming. The authors infer that the rise up to 1900 represents recovery from the Little Ice Age (LIA) of the 1700s and 1800s, but they also note that half of the 150-year increase occurred in the last 30 years of the 1900s, well after the end of the LIA.

Thus, this temperature-proxy evidence adds to the great mass of air temperature evidence, and collectively they leave no doubt that environmental temperatures rose throughout most of the West during the twentieth century.

Temperature Trends Projected for the Twenty-First Century

Causation and Validity of the Models

Temperature predictions for the twenty-first century are based on the output of complex computer models (general circulation models, or GCMs) structured according to the physical factors that drive the earth's climate system, and on projections of twenty-first-century greenhouse gas emissions anticipated from predicted increases in human populations and their economies. Since skeptics on global warming often voice their doubts about these models, and about inferences that temperature changes are being forced by increasing levels of greenhouse gases, it seems useful to digress into a brief discussion of the validity of the models and the evidence used by scientists to infer the role of these gases. This information might be useful for policy makers in formulating action, and also in responding to critics of their efforts.

The earth's surface temperatures are largely determined by energy from the sun. The earth has a molten core, and its heat works up through its mantle, but as it reaches the surface, it has largely dissipated and only contributes a minor amount to the earth's surface temperature (see Chapter 2).

Most of the energy arriving from the sun is in the short, light wavelengths. Since the atmosphere is transparent to light, most of the incoming energy penetrates it and strikes land and water surfaces.

By the laws of physics, any light absorbed by an object is converted to heat. And by those same laws, any object containing heat radiates it. Thus the earth radiates the energy it has received from the sun out toward space. If it did not, the earth's temperature would continue to rise to the point of making the globe uninhabitable.

But the atmosphere is partially opaque to the long wavelengths of heat because several gases present in the atmosphere in small amounts, the greenhouse gases, absorb a portion of the radiated energy and radiate it back to earth. Thus the earth's temperature is a function of the energy received from the sun, plus the amount radiated back by these greenhouse gases, minus the amount that gets through the atmosphere and flows out into space. This greenhouse effect has been known to science since the 1800s, and physicists estimate that without it, the earth would be 54°F colder and uninhabitable.

The greenhouse gases include carbon dioxide (CO_2), methane, nitrous oxide, and water vapor. Methane and nitrous oxide have a much more powerful forcing effect on a unit volume basis, but because CO_2 is far more abundant than the others, it has a much larger greenhouse influence. Even so, it is only three to four one-hundredths of 1 percent of the atmosphere, making its influence far out of proportion to its minute presence. CO_2 is emitted into the atmosphere by the combustion of any organic material (especially fossil fuels), by volcanic eruptions, by rot and decay of organic materials, and by respiration of both plants and animals.

Three lines of evidence support the inference that CO_2 emissions have been the major determinant of the twentieth-century temperature rise:

1. The known physics of the CO_2 greenhouse effect and the emissions measured during the 1900s, during which atmospheric concentrations increased by approximately one-third, a level not reached in the preceding 650,000 years (Siegenthaler et al. 2005).

2. A 1,000-year correlation between atmospheric CO_2 concentrations and temperatures in the Northern Hemisphere (Mann et al. 1999, Crowley and Lowery 2000, Wahl and Amman 2005, Albritton et al. 2007). Although a correlation between two variables does not automatically imply cause and effect, the reality of a 1,000-year correlation plus the known physics of the CO_2 greenhouse effect leave little doubt that this is a causal relationship.

3. Retrospective model experiments with *measured* twentieth-century climate variables programmed into the models, and model simulations with and without *measured* greenhouse gas values that calculate twentieth-century temperatures. These experiments are illustrated in Figure 10.1.

In order to project twenty-first-century temperatures with the models, it is necessary to program into them *likely* or *anticipated* twenty-first-century values for the physical factors that shape the climate, based on historical norms for those values. It is also necessary to program *anticipated* twenty-first-century levels of atmospheric CO_2, based on *projected* world population increase and growth of world economies.

In Figure 10.1, however, the models are programmed with *known* values, measured during the *twentieth* century. They can be run with and without *measured* (observed) twentieth-century values for CO_2, and the output can be compared with *measured* (observed) values for twentieth-century temperatures.

This protocol is tantamount to the standard scientific procedure of experimentally holding constant all variables affecting a phenomenon except one, varying its values, and observing the effect on the dependent variable. That the "natural" (i.e., without CO_2) model simulation far underestimates the actual ("observed") temperature trend of the twentieth century, but follows it closely with the inclusion of CO_2 ("anthropogenic") is de facto experimental evidence that CO_2 has been the major factor driving the observed temperature increases.

These experiments have been repeated in several research centers with similar results (cf. Anthes 2003, Albritton et al. 2007). The new IPCC report (Albritton et al. 2007) concludes that there is a 66 to 90 percent probability that the observed warming of the last 50 years has been due to the increase in greenhouse gas concentrations.

Moreover, as commented above, the models are structured on predicted twenty-first-century greenhouse gas emissions based on projected population increases and

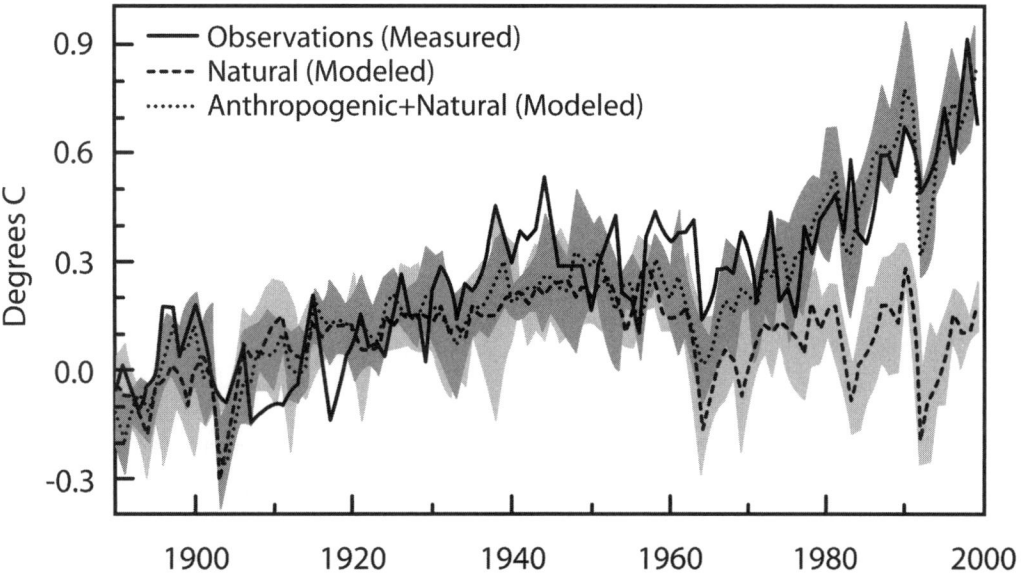

Figure 10.1. Retrospective modeling exercise depicting global temperature change based on *measured* twentieth-century climate variables. The dashed line is model output incorporating only measured "natural" climate forcings (volcanoes, variations in solar energy), but with no inclusion of values for greenhouse gases and sulphates ("anthropogenic" variables). The dotted line is model output with both natural and anthropogenic values included. The solid line is the mean global temperature trend based on world-wide weather station measurements. The simulations were conducted four times, and the shaded areas are the ranges of the four. The dotted and dashed lines are the averages of the four.

This exercise contributes two important points to analysis of the climate-change issue. (1) The close correspondence between the dotted and solid lines attests to the ability of the models to validly depict temperature change. (2) The exercise is tantamount to standard experimental procedure in which a single factor, the anthropogenic, is varied (included and not) while the others are held constant. The close correspondence between the solid and dotted lines is persuasive evidence that anthropogenic factors influenced the twentieth-century temperature rise. This figure is redrawn from a 2003 report by Anthes, who presented it to the U.S. Congress.

economic growth supported by combustion of fossil fuel. Climatologists readily acknowledge that the model outputs can be taken only as approximations of future values, but several lines of evidence indicate that they approximate actual climate behavior within limited margins of error:

1. The model projections are based on the assumption of a twofold increase in atmospheric CO_2 by the year 2100. Data showing that the twentieth-century climate changed when CO_2 rose approximately 32 percent, in the same directions projected by the models for the twenty-first century, provide evidence that the model structures validly approximate the directions of change in the climate system.

2. The retrospective modeling experiments discussed above (Fig. 10.1) provide experimental evidence not only of the role of greenhouse gases in forcing temperature changes, but of the ability of the models to simulate these changes. As commented above,

the models are simplifications of the actual climate system, and their simulations are taken as approximations of actual climate values. The shaded bands around the lines in Figure 10.1 represent the margins of error around the simulations, and indicate the probabilities that the simulations correctly estimate the actual values. Numerous research centers that have conducted similar experiments with their models have obtained similar results (cf. Albritton et al. 2007).

3. Beyond projected long-term trends in mean temperatures and precipitation, the models project several subsidiary climate behaviors that are borne out by actual twentieth-century measurements, as discussed above:

a. A south-north gradient in temperature increase.

b. More pronounced increases in night-time minima than in daytime maxima.

c. Temperature increases that are more pronounced

at high elevations than at low ones. The records presented by Naftz et al. in Chapter 3 show this to have been the case in the Wind River Mountains of Wyoming.

Temperature Projections for the West in the Twenty-First Century

As discussed above, the original general circulation models (GCMs) used to project twenty-first-century temperature trends were designed to simulate trends for the entire globe. Projecting trends at regional scales involves a degree of precision that includes provision for topographic variation, local wind patterns, proximity to large bodies of water, vegetation and land use characteristics, and other variables for which the models are not sufficiently detailed. Also, as discussed above, the models depend on anticipated twenty-first-century values for natural climate forcings and on predictions of CO_2 emissions for that period.

California is a prime example (Field et al. 1999). Its long latitudinal extent occasions a precipitation gradient from aridity in the south to rainforest conditions in the north. West of the Sierra Nevada, cool maritime conditions are produced by westerly winds flowing onshore over the cold California Current. East of the Sierra, in its rainshadow, the warm climate of the Great Basin produces the extreme summer-winter and day-night temperature differentials of continental climates. Climate also varies with elevation and slope aspect over the state's mountain ranges. Thus a single average temperature for the state has uncertain meaning.

Given these uncertainties, regional projections of temperature increases for the twenty-first century must be taken as very general indications. Based on earlier models that assume a twofold increase in CO_2, these projections, by region, are:

Southwest: 3.0–5.8°C (5.4–10.4°F) by 2090 (Sprigg, Hinkley et al. 1999).
Rocky Mountain/Great Basin: 3.6–6.5°C (6.5–11.7°F) by 2100 (Chapter 1).
California: 2.2°C (4.0°F) by 2050 (Field et al. 1999).
Pacific Northwest: 2.9°C (5.3°F) by 2050 (Mote 1999).
Alaska: 2.0°C (3.6°F) mean monthly by 2100 (Chapter 9, Fig. 9.12).

Thus the projections by several models of mean annual temperature increases in the western portion of the Lower 48 are quite consistent, ranging from 2.2 to 2.9°C

(4.0–5.3°F) by 2050, and 3.0 to 6.5°C (5.4–11.7°F) by the end of the century. These projections also accord well with the latest projections for the entire globe.

Winter temperature increases are consistently projected to exceed summer increases, as are night-time temperatures over daytime, and high elevations over low (Bradley et al. 2004). The projected increases exceed those of the twentieth century several fold, with the probability of much more marked environmental changes in the coming century than those of the twentieth described in the preceding chapters and in the discussions below.

Evidence of Twentieth-Century Precipitation Change

The investigators of the climate-change assessments analyzed twentieth-century weather records for their regions to determine whether precipitation changed during the century in directions projected for the twenty-first. The purpose, as with the temperature analyses, was to obtain one source of evidence for the validity of the models' twenty-first-century projections. If precipitation changed during the 1900s when CO_2 rose by a third, and it increased in the same directions as projected by models for the twenty-first century, when CO_2 is projected to increase more markedly, it could be taken as one tentative indication of model validity.

All of the regional assessments analyzed short-term "natural," interannual, and interdecadal variability in annual precipitation. Most influential in the West is the El Niño/Southern Oscillation (ENSO) and the Pacific Decadal Oscillation (PDO). ENSO is most influential in the Southwest, imparting strong year-to-year variation in annual precipitation (Sprigg, Hinkley et al. 2000). Its effect fades to the north, but is still measurable in the Pacific Northwest (Mote 1999). The PDO influence is most pronounced in the north, especially the Pacific Northwest (Mote 1999), and fades to the south.

As their names imply, these are quasi-cyclic changes, with ENSO varying between two to seven years, and averaging a periodicity of three to four years, and PDO averaging about a 24-year cycle. While inducing year-to-year variation in a long-term (e.g., 100 years) precipitation trajectory, there is no evidence of a secular trend in these short-term oscillations. Thus a longer-term trend embracing these fluctuations can reasonably be associated with other causation (Baldwin 2003).

While the assessment of the Southwest emphasized this year-to-year variability, especially that associated with ENSO, it found a patchwork pattern of century-long precipitation increase and decrease during the 1900s in its six-state region. Precipitation decreased in southern California, two-thirds of Arizona, and most of Colorado and eastern Utah (Sprigg, Hinkley et al. 2000), but it increased in much of New Mexico, two-thirds of Utah, and nearly all of Nevada.

The Rocky Mountain/Great Basin (RMGB) regional assessment analyzed precipitation trends in its eight subregions for the 1900s (Chapter 1). In five subregions, annual precipitation increased ($p \leq$ 0.10) during the century, particularly in June. Increases in annual precipitation ranged from 6 to 16 percent. Analysis of weather records just for Utah showed a 13 percent increase. Kittel et al. (2002) found statistically significant twentieth-century increases in summer precipitation in the northern and central Rockies, but not in the southern. RMGB investigators also analyzed interannual variability in precipitation by calculating standard deviations around 30-year moving averages in the annual precipitation time series. The standard deviations increased in all subregions with rising twentieth-century precipitation, indicating increasing interannual variability.

The Pacific Northwest assessment (Mote 1999) showed evidence of short-term influence on annual precipitation, both by ENSO and PDO, but the long-term trend in the region over the twentieth century was a net rise of 13.7 percent. Most of the increase came between October and March, the region's main precipitation season. There was almost no change between April and September, the traditional dry season (Mote 1999).

Barber et al. report in Chapter 9 that precipitation trends over the large state of Alaska have been variable, but that rising temperatures are creating water-stress problems in vegetation.

The RMGB assessment also examined twentieth-century annual flows of five western streams to determine whether the precipitation increases of the 1900s produced equivalent increases in streamflows. The results (Baldwin et al. 2003) showed statistically significant rise ($p <$ 0.05) in three streams—the Boise River (Idaho), the Humboldt River (Nevada), and Blacksmith Fork (Utah)—and increase just short of significance in the Yellowstone River (Montana). Flow in the San Juan (southern Colorado, New Mexico, southern Utah) did not rise significantly. Moreover, analyses of standard deviations around 30-year moving averages of annual streamflow showed increasing interannual variability of about the same magnitude as that in the annual precipitation.

The RMGB investigators also analyzed annual flows of the Colorado River, measured at Lee's Ferry, Arizona, with data provided by the U.S. Bureau of Reclamation. The flows since 1926, when actual flow measurements began (previous flows were estimated from tree-ring proxies), had not increased by 1996 on average. They also analyzed twentieth-century flows in five headwater streams of the Upper Colorado Basin: the Green (Montana); the Yampa, Colorado, and Gunnison (all in Colorado); and the San Juan (New Mexico, Colorado, southern Utah); none of these increased statistically during the 1900s. Again, these streams arise in that portion of the RMGB where precipitation did not increase during the 1900s.

In sum, weather records and streamflow analyses point to precipitation increases during the twentieth century in the northerly half to two-thirds of the Intermountain West and Pacific Northwest, but extensive areas without precipitation or streamflow increases in the southern and southeastern portions of the West.

Projected Twenty-First-Century Precipitation Changes

The general circulation models have for some time been projecting an increase in global precipitation driven by increasing temperatures that will evaporate more moisture off the surface of the oceans and project it into the global hydrologic cycle (Tebaldi et al. 2006). As climatologists have pointed out, however, it is difficult to predict how this precipitation will be distributed over the globe.

Despite the uncertainties of regional projections, the U.S. Global Change Research Program regional assessments of the 1990s used the modified GCMs to predict twenty-first-century changes in precipitation in the same way they did for temperatures. Based on a projected twofold increase in CO_2, five out of six regional simulations for the West projected precipitation increases. Using the British Hadley and Canadian models, Doherty and Mearns (1999) projected sizeable increases by 2090 in the Southwest (Sprigg, Hinkley et al. 2000), but using a higher resolution model, Mearns et al. (1999) have projected essentially no change for the region.

The Canadian and British models projected mean annual precipitation increases of 54 to 184 percent by 2100 for the Rocky Mountain/Great Basin region (Chapter 1). The same models projected an increase of 25 percent or more for California along the Cascade and Sierra cordillera and westward to the coast, but a decline for southeastern California (Field et al. 1999).

For the Pacific Northwest, the Canadian model projected a substantial twenty-first-century increase from

October to March, the main precipitation season for this region, but almost no change from April to September, the traditional dry season (Mote 1999). For the entire West, the models typically projected more increase in winter than in the other seasons.

The newer modeling efforts are more geographically explicit in their projections. The most recent IPCC report (Albritton et al. 2007) predicts that precipitation increase is likely at tropical latitudes (0–10° north and south) and "over mid- to high latitudes," but is likely to *decline* in the 10–30° north and south latitude zones.

Most recently, Seager (2007) described the likely shifts in the earth's Hadley cells between the equator and 30° north and south that will extend these already arid zones northward and southward, while at the same time their already meager precipitation will decline. Hence the result will be increasing area and intensifying aridity of the arid U.S. Southwest. The respective latitudes of the region's major cities are: El Paso, 32°; Albuquerque, 35°; Tucson, 32°; Las Vegas, 36°; and Los Angeles, 34°.

Seager comments that this shift should already be underway. It is possible that the spotty failure of precipitation and streamflows in the Southwest to increase during the 1900s, when they were increasing substantially over the remainder of the West, may be an early harbinger of Seager's projections. Jenkins (2007) notes that 2007 was the driest year on record in southern California, and that the Colorado River, from which the region gets much of its water, "has been gripped by drought for eight years running."

Environmental Effects of Climate Change in the West

Water Resources

The Unique Situation with Western Water Resources
Because much of the West is semiarid to arid—mean annual precipitation for the state of Nevada is 229 mm (9.0 in.), perhaps a metaphor for much of the region—water is a particularly critical resource. At the same time, the West has the fastest growing human population in the United States.

Three-fourths of water used by humans in the West is derived from streams fed by meltwater from mountain snowpacks, and managed by an engineering infrastructure of dams and reservoirs that control their magnitudes and seasonality of flow. And the importance of the western snowpacks extends beyond their values as sources for major western rivers like the Colorado, Sacramento, and Columbia. They are in essence the headwaters of the Mis-

souri, Rio Grande, Platte, and Arkansas Rivers. Thus the effects of climate change on western precipitation have far-reaching socioeconomic and ecological implications that extend beyond the region.

Two climate-change effects on western water resources are of concern. One is the effect on the magnitude of the resource, and hence the amount available for human use. The second is the seasonality of streamflow influenced both by the timing of montane snowpack development and spring thaw. Urban use, agricultural irrigation, electric power generation, and the up- and downstream movements of anadromous fish are all keyed to this seasonality of flow.

Warming Effects on Magnitude of Water Resources
The climate models now converge on projections of twenty-first-century temperature increase and precipitation increase in the northern half to two-thirds of the West, but precipitation decrease in the southern third to half, all assuming a twofold increase in CO_2. However, the amount of available water is also a function of evapotranspiration, a variable that rises exponentially with temperature increase.

Jeanne Ruefer (Nevada Division of Water Planning, Carson City) reports that 90 percent of Nevada's precipitation is lost to evapotranspiration (Baldwin et al. 2003). Nash and Gleick (1993) conclude, on the basis of their hydrologic simulations of the Colorado River basin, that a 2°C (3.6°F) rise in temperature with no change in precipitation would reduce runoff in the basin by 4–12 percent. A 4°C (7.2°F) temperature increase would reduce runoff by 9–21 percent.

Whether or not the newer models are correct that precipitation will decline during the twenty-first century in the arid Southwest, the rising temperatures and associated increase in evapotranspiration may well reduce water resources in the region. In that case the resource will need to be reallocated among several uses, including municipal use by four major cities: Las Vegas, San Diego, Phoenix, and Los Angeles, the latter two among the ten largest in the nation. It would seem highly desirable that the areas involved engage in contingency planning against these eventualities.

Warming Effect on Water-Resources Timing
With the heavy human reliance (~75 percent) on streams for water resources in the West, the seasonal patterns of human use are dependent on the seasonality of streamflow. Mote (1999) comments that the major water demands are for generating electric power, for which peak

uses are in the winter for heating and summer for cooling; for agricultural irrigation (80–85 percent of total western water use) between spring and fall; for recreation in summer; and for year-round municipal use, 60–70 percent of which in the West is for watering lawns and gardens in summer (Baldwin et al. 2003). Mote (1999) also describes the importance of high spring and early summer streamflows for salmon smolts migrating to the ocean in the Pacific Northwest.

Except for the Southwest, with its two precipitation peaks each year (winter and summer), precipitation over most of the West peaks between fall and spring. Mote (1999) distinguishes rain-dominated streamflows west of the Cascade Mountains, where temperatures are generally above freezing, and where streamflows peak in winter, paralleling the precipitation pattern; and snow-dominated streams east of the Cascade crest and, in fact, over most of the West. In these, streamflow is low in late summer and fall before precipitation resumes, remains low during winter because precipitation falls as snow and does not run off, increases in spring to a peak in early summer as the montane snowpacks melt, then declines to the late-summer and fall low.

Obviously these seasonal patterns of flow do not coincide well with seasonal patterns of human demand, and the extensive western engineering infrastructure has been constructed to capture and store the flow year-round, allowing it to be released at times of human need, and to control flooding risks at times of above-average precipitation. While the system has thus far been able to accommodate the needs of the western population, there are still risks associated with the interannual and seasonal variations in precipitation, and these have created occasional shortage and flooding problems. For example, in spring 1983, following a highly above-average snowfall year and an abrupt, early warm-spring thaw, there was concern whether Glen Canyon Dam on the Colorado—which impounds Lake Powell, the second-largest impoundment on the Colorado River system—would remain intact.

The historical patterns of streamflow variation with which the system now copes are changing under the influence of climate warming. As Mote describes in Chapter 4, and as reported by Dettinger (2006), western snowpacks have shrunk during the 1900s under rising temperatures. Glaciers are a form of snowpack, and Naftz et al. describe the shrinkage of Upper Fremont Glacier in the Wind River Mountains of Wyoming in Chapter 3. Hall and Fagre (2003) report that two-thirds of 150 glaciers in Glacier National Park, Montana, disappeared between 1850 and 1980, and they predict that all of the park's remaining glaciers will be gone by sometime between 2030 and 2050. Barber et al. describe widespread shrinkage of glaciers in Alaska in Chapter 9. Fyfe and Flato (1999) project that snowpacks in the northern Rockies of western Canada will disappear by approximately 2070 if temperatures rise in the twenty-first century as their models predict.

The hydrologic forces that are shrinking snowpacks are also altering the seasonality of western streamflows. Precipitation in late fall and early winter, which historically has fallen as snow and contributed to snowpacks, now falls as rain and runs off. Even precipitation in late winter and early spring, which in the past has contributed to the snowpacks, now falls as rain and runs off. Hence snowpack size is shrinking, and spring peaks in stream runoff are both shrinking and occurring earlier as a result of the warming-induced earlier thaws.

Thus Lundquist et al. point out in Chapter 5 that the snowpack shrinkage is altering the seasonality of flow of the Sacramento River by reducing the proportion of annual flow running off during the spring peak, and advancing the peak one to three weeks. Dettinger (2006) has shown the same for the combined flow of eight California streams (reproduced here as Figure 10.2), and the Clark Fork of the Yellowstone River in Wyoming. Cayan et al. (2001) have reported the same changes for streams more widely over the western United States.

Implications for Water Management in the West
As described above, the engineering patterns of water storage and release are keyed to the seasonality of human water needs. Water is released from the dams according to those needs, but also based on models that make assumptions about ensuing precipitation and runoff patterns through the remainder of the year. The latter are based on statistical probabilities derived from weather patterns of prior years. As Rick Wells, hydrologist with the U.S. Bureau of Reclamation in Boise, Idaho, stated at a workshop on February 22–24, 1999 (Baldwin et al. 2003): "The reservoir operating curves are statistically based, and the more atypical the precipitation and run-off distribution, the more likely the risk of flooding or failure to fill the reservoir system." And Knowles et al. (2006) remark:

> If warming continues and raises the mean winter wet-day minimum temperatures in more of the West…snowfall declines (and rainfall increases), combined with earlier melting of the remaining accumulations of snowpack, [it] will diminish the West's natural freshwater storage capacity. The shift from snow-fall to rainfall also may be

April-July Flows as Fraction of Water-Year Total

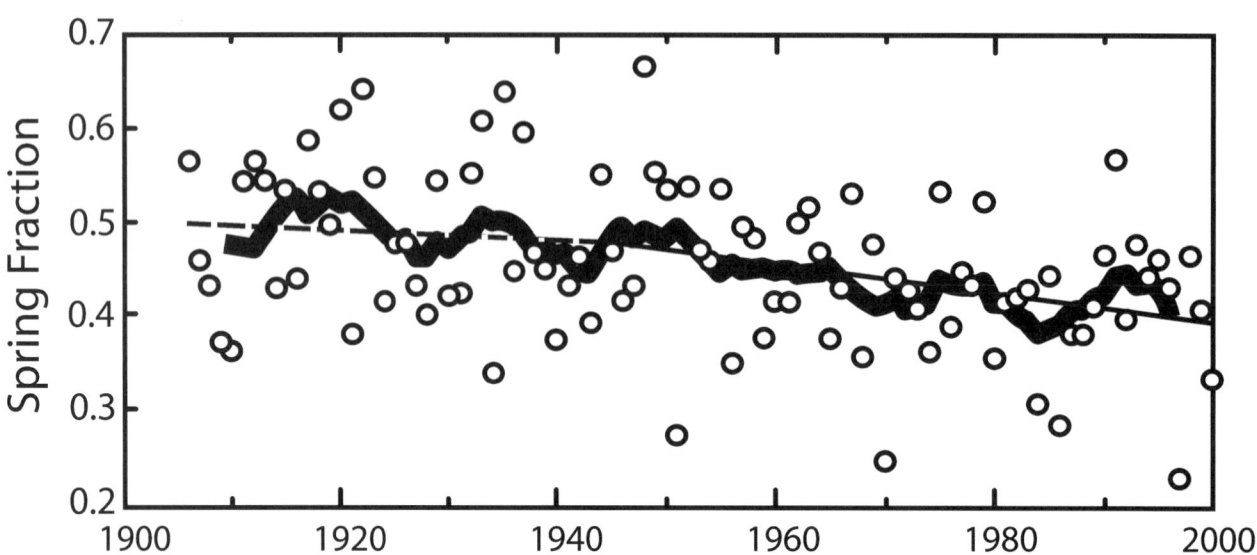

Figure 10.2. Fraction of total water year (October–September) streamflow during April–July in eight major rivers in the western Sierra Nevada, California. Circles are yearly flows; the heavy line graphs nine-year moving averages; and dashed and solid straight lines are linear trends before and after 1945. (Redrawn from Dettinger 2006.)

expected to increase risks of winter and spring flooding in many settings. The combination of greater flood risk and reduced natural storage threatens to exacerbate the tension between flood control and storage priorities that many western reservoir managers face. Better understanding of how flood risks will change, of the atmospheric conditions that control precipitation form, and of possible trends in those conditions are needed to project and accommodate future changes in the West's water supplies.

These comments resonate with the observation of Lundquist et al. in Chapter 5 that the changing seasonality of flow in the Sacramento River, driven by ongoing climate change, is not well predicted by existing, probability-based models. The authors urge expanded effort on the development of models structured on physical forces.

This discussion has focused on the implications for water management of changing streamflow seasonality, for which there is now extensive evidence in the West. It has not considered implications of major changes in precipitation *amount* on water availability for a region with a population likely to double in 30 to 40 years.

If precipitation declines in the Southwest, as the newer models are projecting, or does not increase but the resource declines in the face of increasing evapotranspira-

tion, there will clearly be a need for reallocation among the several uses. The resource is now fully allocated. If precipitation increases—as the earlier models cited in Chapter 1 projected, and the newer ones project for the northern half or two-thirds of the West—and the increases more than compensate for the heightened evapotranspiration, it may well pose storage and flood control problems. Thus there is a need for thorough analysis of the adequacy of the western infrastructure. Some water managers have been expressing a need for increased storage capacity (Jenkins 2007a).

Consequently, for several aspects of western water resources, there is a need for intensified research and analysis to provide a knowledge base for future policies and actions, depending on the nature of climate change. A comment by Miller (1997), made more than a decade ago, is even more relevant today:

Some would argue that the impacts of climate change are so far in the future that they pale in significance relative to more pressing concerns. However, the risk should not be ignored. In large part, climate change provides further reason to take actions that will improve resilience to the droughts and floods that arise from ongoing climate variability. However, climate change adds new twists to the uncertainties facing water users, water managers, and those who value preservation of environmental resources.… It is not too early to begin thinking about how to improve our capacity to manage these uncertainties and to respond efficiently and fairly to a range of possible streamflow changes, as well as to the effect of ongoing climatic variability.

Climate-Change Effects on Western Agriculture

Cultivated Agriculture

Except for the powerful farming industry of California's Central Valley, with its major contribution to both the state's and the nation's economies, the localized grain- and fruit-growing areas of Washington and Oregon, and the limited irrigated areas of Arizona and Idaho, cultivated agriculture is not a major economic venture in the West. Only 10 percent of the area of the nine-state Rocky Mountain/Great Basin region is in farms, and only 5 percent is cultivated.

Climate change is likely to affect crop production in two ways: directly through effects on crop physiology, and indirectly through effects on availability of irrigation water. Some 80–85 percent of western water use is agricultural irrigation. In California, it is 87 percent (Field et al. 1999).

The varieties of crops grown in an area are typically selected for the climatic conditions in that locale. Warming in some areas is likely to move temperatures out of a variety's optimum range for photosynthesis and transpiration (Berry and Björkman 1980, Amthor 1989). These effects could be countered to some degree by fertilization of rising atmospheric concentrations of CO_2 (Drake et al. 1997), but exact compensation would be pure chance.

The net result could be increase, no change, or decrease in crop production depending on the crop and the degree of change in these environmental variables. The per-acre yield of wheat, barley, and potatoes rose from 26 to 44 percent between the early 1980s and 1997 in the Rocky Mountain/Great Basin region (Wagner and Baldwin 2003). Some observers attribute some of this increase to the use of improved varieties, but at least the temperature increases did not prevent these production improvements.

In some regions temperature increases may prolong the growing season to the point of allowing double cropping. In the RMGB region, per-acre yield of alfalfa rose by 27 percent because growing-season length allowed a third or fourth cutting in areas that had traditionally experienced two or three during the twentieth century. In general, the magnitude of crop production across the United

States in the twenty-first century is not expected to decline under influence of the climate changes projected for the period (Reilley et al. 2001).

The influences on plant physiology, however, may be more subtle than those affecting the magnitude of the yield, and may affect the quality of the product. The effects on wine grapes described by White et al. in Chapter 6 are a case in point. Grapes are one of the two most valuable crops in California and, along with dairy products, account for one-fourth of the state's agricultural income (Field et al. 1999). Historically the Central Valley climate has been optimum for production of high-quality grapes. But White et al. now suspect that the warming projected for the years ahead will alter these conditions and move the optimum climatic conditions northward into Oregon, Washington, and British Columbia.

Similarly, investigators on responses of forage grasses to elevated atmospheric CO_2 in Colorado have shown declining protein levels, and hence lower quality, of livestock forage in range vegetation (Ojima and Lackett 2002). Similar effects have been observed in California grasslands (Field et al. 1997).

Availability of irrigation water is likely to be the greater concern for climate-change effects on western agriculture. A major fraction of western agriculture is irrigated, and western water resources are fully appropriated. With the western U.S. population expected to double in the next 30 to 40 years, demand for municipal water will increase.

Agricultural water is now, and will continue to be, the source filling that demand. Sprigg, Hinkley et al. (2000) report that the percentage of water use in Arizona allocated to agriculture declined from 93 percent in 1963 to 81 percent in 1995. Municipal use rose from 5 to 14 percent in the same period. In southern California, farmers in recent years have received subsidies to fallow portions of their land and avoid irrigation (Jenkins 2007b).

If southwestern water resources are reduced further by declining precipitation and increasing evapotranspiration, the situation will become more acute, and the region's agriculture may well hang in the balance. The problem is not only the amount of the resource, but the fact that, as demand rises, prices increase and raise operating costs for an industry that now operates on a thin profit margin.

If the precipitation increases projected for the northern portion of the West occur, and they are sufficient to offset rising evapotranspiration, water resources may be adequate to accommodate the growing population and sustain the region's agriculture. But if it is not, the industry is likely to decline.

Livestock Ranching

Although it arose in the West, cattle ranching today is a minor economic venture in the region, the industry having largely shifted to the southeastern United States, where 81 percent of the nation's beef cattle are produced (Donahue 1999). (Sheep ranching has largely disappeared following its high point in the early 1900s.) Nevertheless, much of that half of the 11 western states in federal ownership (primarily the USDA Forest Service in the montane elevations, and the U.S. Bureau of Land Management in the lower altitudes) is grazed by cattle owned by private ranchers who pay fees for the privilege.

The potential effects of climate change on western cattle ranching, which is already financially strapped and declining, in many ways parallel the effects on cultivated agriculture. Temperature increases will impair native forage production unless accompanied by enough precipitation increase to offset elevated evapotranspiration, something that seems improbable for the Southwest and the lower elevations generally. But the temperature increase can potentially extend the spring–fall period of vegetation growth, and hence the grazing seasons, in the higher elevations to which livestock are moved at that time of year.

The industry is also subject to the availability and cost of irrigation water used for hay crops that are fed to cattle between fall and spring. Precipitation increases that would add to water resources would lower water costs, reduce hay prices, and ultimately benefit the industry. But declining availability of water—whether from reduced precipitation, increasing evapotranspiration, and/or increasing competition from municipal demand—would ultimately increase operating costs for an industry that now operates on a razor-thin profit margin and is declining (Wagner and Baldwin 2003).

Ecological Effects of Climate Change on the Natural Biota

Context

The West is the most ecologically complex portion of North America, made so in part by the complexity of the climates in which the biota functions. The latitudinal span from the U.S.-Mexico border to Point Barrow, Alaska, is ~38°. There are continuous temperature, day-length, and growing-season gradients along this span. And precipitation ranges from aridity in the south, to rain-forest conditions in the middle, to low levels in the far north. Along this latitudinal extent, temperatures and growing-season lengths grade from high to low, and precipitation from low to high, along the many montane elevational spans

ranging from below sea level to mountaintops as high as 6,280 m (20,600 ft). Nevada alone has 120 mountain ranges.

Local climates differ at all points along these gradients. At each locale, an assemblage has developed of plant and animal species that are adapted to the microclimate of the area and to each other. For some species, the climate may be barely within their tolerance ranges. Climate change may shift conditions away from what they can tolerate, may improve conditions for others, and shift into a range now tolerable for some species not previously present. The net result is likely to be widespread restructuring of natural communities and, in the view of some investigators, numerous extinctions (cf. Kolbert 2006).

Most observations on responses of the biota to climate change, and predictions of likely changes during the twenty-first century, have only been made during the past 20 to 25 years. Hence insights into the effects on the biota are only beginning. Most observations have focused on the responses of individual species. No detailed, empirical analyses of changes to date or predicted for the next century in entire ecosystems have yet been published, although a number of abstract modeling efforts are underway. Nevertheless, the documented and predicted changes in individual species provide some basis for observing and predicting effects on the interspecific interactions that structure ecosystems, and thus provide a small beginning in the ultimate need of understanding community and ecosystem effects.

Reported Single-Species Changes

Three types of single-species response have pervaded much of the literature to date. One involves shifts in spatial distributions of both plant and animal species. The distribution of a number of plant and animal species is limited at the northern, or upper-elevation, edges of their ranges by low temperatures. A number have moved northward and/or upslope during the twentieth century.

Butterflies have been especially fruitful subjects for observing these changes. Probably the best chronicled is Edith's checkerspot butterfly (*Euphydryas editha*), discussed by Matthews and Parmesan in Chapter 7. The species' range extends from northern Baja California to southern British Columbia. Between the 1930s and 1990s, the northern edge of the species' range moved northward 92 km, while the upper elevational limit moved upslope 105 m (Parmesan 1996). Crozier (2002) documents the northward range extension of the sachem skipper (*Atalopedes campestris*) from its "historical" (pre-1967) range in California to southern Washington by 1998.

Matthews and Parmesan (Chap. 7) also describe the upslope movement of pika (*Ochotona princeps*) colonies in Nevada mountain ranges, as reported by Beever et al. (2003). This small mammal occupies intermediate and upper elevations of western mountains. Of 25 colonies mapped by E. Raymond Hall in 1946, 7 (those at lower elevations) had disappeared by the 1990s. The lower edges of the surviving 18 had moved upslope an average of 130 m, according to Beever et al.

In 2003, James Patton conducted a faunal survey at 3,100 m in the Lyell Canyon region of Yosemite National Park, where Joseph Grinnell had made observations in 1915 (Nijhuis 2005). Patton observed five species of small mammals at 3,100 m which Grinnell had only seen at 610 m. Patton also recorded 17 avian species at 3,100 m not observed by Grinnell.

There are also records of vegetation range extensions. In 1992, Smith et al. (1994) established ten, permanent 10 × 100 m plots in four watersheds at elevations ranging from 1,948 m (6,398 ft) to 3,060 m (10,037 ft) in Great Basin National Park on the Nevada-Utah border to measure vegetation change over time. Beever et al. (2005) resurveyed these plots in 2001 and found a reduction in the number of Engelmann spruce (*Picea engelmannii*) stems in the lower-elevation plot.

DiSilvestro (2005) describes 46–55 m upslope extension of spruces in Banff National Park in recent years. Barber et al. (Chap. 9) report the contemporary northward movement of shrubby species out onto the tussock tundra of northern Alaska.

A second form of single-species response to changing climate is advancing phenologies. Phenology is the study of the seasonal timing of biological events—for example, migration, maturation, and reproduction in animals; leaf emergence, flowering, fruiting, and so on in plants. Phenological advances are being reported for hundreds of plant and animal species over the world.

In the western United States, David Inouye has observed spring arrival dates of American robins (*Turdus migratorius*) in the vicinity of the Rocky Mountain Biological Laboratory (Colorado) at 2,946 m elevation since 1971 (Inouye et al. 2000). In recent years, robins have arrived in spring in the vicinity of the laboratory two weeks earlier than in his early years of observation. He has also observed a one-month-earlier emergence from hibernation of yellow-bellied marmots (*Marmota flaviventer*) in the lab's vicinity.

Cayan et al. (2001) report advances in spring blooming dates of two species of honeysuckles (*Lonicera tatarica* and *L. korolkowii*) and lilac (*Syringa vulgaris*) in the west-

ern states between 1970 and 1998. The blooming dates of the three species advance about 3 days per degree latitude. When these differences are corrected for, average blooming dates over the West for lilac advanced 7.5 days, and those for the honeysuckles 10 days, over the ~38-year observation period.

A third form of single-species response to climate change is changing population densities. As described above, the plant and animal species occupying a given area are adapted to its specific physical environment and biotic components. Range extensions and phenological changes such as those described above are either direct responses to that environment, and/or to other species in the assemblages which themselves have changed. The net result may be changes in the reproductive and/or mortality rates of a species, and ultimately its population density. A number of population changes have now been reported.

Matthews and Parmesan (Chap. 7) discuss the population changes in 62 species of intertidal invertebrates in Monterey Bay, California, between 1931 and the early 1990s, originally reported by Sagarin et al. (1999). During this period, the mean annual water temperature along the coast rose 0.79°C (1.4°F). As Sagarin et al. describe, 24 species censused in 1931 increased, while 22 decreased. Those that increased were species primarily occurring to the south along the Pacific coast, while those that decreased were more northerly species. The 16 that did not change significantly in numbers were species the authors considered to be "cosmopolitan" in their distribution.

Parmesan's studies of Edith's checkerspot butterfly (Parmesan 2006) show the species to be distributed in metapopulations (small subpopulations in habitat islands scattered over the landscape). During the 60-plus years through which the species has been observed to move northward and upslope, survival of the subpopulations has varied along the latitudinal and altitudinal gradients. Approximately 70 percent of the subpopulations in the southern portions of the species' range disappeared. Between 35 percent and 50 percent of the subpopulations in central and low-elevation portions of the range were eliminated. Less than 20 percent of those in the northern and high-elevation portions of the range vanished.

In Chapter 8, Logan and Powell show the recent northward and eastward extension of mountain pine beetle outbreaks from their historic range in British Columbia west of the Continental Divide. The outbreaks occur in mature lodgepole pine stands when the beetle populations reach sufficient densities to overpower the protective mechanisms of the trees. Achieving those densities requires synchrony of the beetle life cycles, and completion of those cycles in a single year (univoltinism). As the authors' models show, univoltinism can occur only above some minimum temperature like that attained in this northern extension of the species' range.

Inferring Causation

These examples are mainly correlational, and it is well recognized that correlation alone does not necessarily imply cause and effect. An entire volume discusses the problem in evaluating climate-change effects (Schneider and Root 2002). Matthews and Parmesan explore this point in Chapter 7, and discuss three procedures for strengthening the probability of an inference of causation.

Mechanistic studies are one procedure. These are investigations of the physiological or ecological mechanisms underlying a species' behavioral, population, or ecological response to temperature change. An excellent example is the Logan and Powell study (Chap. 8) of the northward extension of mountain pine beetle into southwestern Canada, which they infer is induced by climate warming. The authors hypothesize that, with further warming, the beetle outbreaks will move upslope in western mountains to elevations and pine species not previously affected because of lower temperatures, and will spread to other species of pines across central Canada and the northeastern, eastern, and southeastern United States.

The causal mechanism linking population outbreaks that have destroyed pine stands is, as described above, the life-history phenomenon of completing a life cycle in a single year (univoltinism). The beetles must undergo univoltinism in order to increase to outbreak numbers, and this is only possible above a critical temperature. Temperatures have not risen to levels permitting univoltinism—and allowing the species to undergo outbreaks farther north, and at higher elevations in western mountains—until recent years.

The authors have modeled the relationship and calculated that the beetles cannot achieve univoltinism and population outbreaks at 3,049 m (10,000 ft) until mean annual temperatures have risen 3°C (5.4°F).

In addition, a natural experiment, described by Logan and Powell in Chapter 8, has further strengthened the hypothesis of an outbreak-temperature link. The Stanley Basin in central Idaho is a cold air sink that historically has not had beetle outbreaks, but in recent years, temperatures have risen, and outbreaks have appeared in lodgepole pine in the basin.

As another example, Barber et al. (Chap. 9) have developed a model that predicts elimination of white spruce (*Picea glauca*) in Alaska if May–August temperatures rise another 2°C (3.6°F) during the twenty-first century (temperatures rose ~1.5°C [2.7°F] during the twentieth). The mechanism underlying the predicted mortality and programmed into the model is an inverse correlation between temperature and radial growth in the trees.

Matthews and Parmesan's second protocol for inferring causation is simple correlational studies, but based on numerous populations or species. The probability of inference for any one species is strengthened when several of its populations behave similarly. Thus 18 of 25 pika populations observed by Beever (2003) in Nevada mountain ranges, as discussed above, moved to higher elevations between 1946 and the end of the twentieth century.

An analogy is the situation when large numbers of species behave similarly. Price (1995) reports that the ranges of "most" Great Plains avian species have moved northward and upslope since 1960. An entire volume (Moller et al. 2006) discusses changing migratory and reproductive phenology in European birds.

The authors' third procedure, meta-analysis, is similar to the last in combining numerous individual cases, but involves a more formal quantitative procedure.

I suggest one other means for strengthening the inference of temperature causation for species in the West: concurrent northward and upslope movement of a species. Parmesan (1996) reported both a 92 km northward and 105 m upslope range extension of Edith's checkerspot butterfly within its range from Baja California to British Columbia.

Effects on Interspecific Interactions

All of the species discussed above are involved in innumerable interactions with other species in the communities/ecosystems they occupy: as predators, herbivores, food for other species, competitors, and users or providers of habitat. These interactions collectively structure their natural communities and ecosystems. The ultimate ecological goal is to measure and model the effects of climate change on whole communities and ecosystems. However, while a number of abstract modeling efforts are underway, the current knowledge is not sufficient to develop such system generalizations empirically. But as a step toward that understanding, it is useful to begin focusing on the responses of species within these interactions, and how the responses change those interactions and ultimately alter the structure of whole systems.

One example is the effects of changing phenologies. Each species has its own unique, genetically determined physical-environmental requirements. Hence each will respond differently to a given temperature change. In cases where two or more species interact by virtue of similar phenologies, a given temperature change may alter the phenology of one and weaken or eliminate the interaction. Several of these asynchronies have been reported.

I described above David Inouye's observations of a two-week advance in spring arrival phenology of robins at the Rocky Mountain Biological Laboratory. However, when the birds arrive, the insect species on which they have fed historically have not yet emerged, creating a food shortage. Similarly, the new one-month-earlier emergence of yellow-bellied marmots now precedes development of the plant species on which they have traditionally fed.

Parmesan (1996) attributes the higher disappearance rate of the checkerspot butterfly subpopulations at the southern limits of the species' range to a phenological asynchrony between the butterfly larval emergence and their food supply. The larvae feed on the annual plant *Plantago*. In recent years, the increasing temperatures have advanced the growth phenology of *Plantago*, and the plants go dormant before the larvae have fed sufficiently to advance in their life cycles.

Mote (1999) describes changing predation pressures on coho salmon smolts associated with phenological changes in predation pressure. The smolts move downstream into coastal waters in late spring and early summer. They are preyed upon by Pacific hake and Pacific mackerel, which move into coastal waters as spring water temperatures rise. In recent years, as coastal temperatures have warmed earlier, the predators have moved into the areas earlier and prolonged the predation period. There is some evidence of suppression in coho populations.

The extensions of species ranges discussed above are another example of climate-change responses evoking new interspecific interactions, and changing system structure in the process. As a species extends its range into a new area, it comes into contact with new ones and sets up predatory, herbivorous, competitive, commensal, food- and habitat-provision interactions that may be positive or negative for the species involved. Hersteinsson and MacDonald (1992) report the northward range extension of the red fox (*Vulpes vulpes*) in northern Canada into the range of the Arctic fox (*Alopex lagopus*), which has led to aggressive interactions. Their observations indicate that the red fox is the superior competitor, and that the Arctic fox has withdrawn its range northward.

In Chapter 7, Logan and Powell point out that if the northward extension of pine beetle outbreaks extend into the range of jack pine in Alberta, the latter may be severely affected because the two species have not historically interacted, and the jack pine has not evolved the protective mechanisms of lodgepole pine. Similarly, whitebark pine may have the same risks if temperatures at the higher elevations warm sufficiently to permit pine beetle outbreaks in this species, which also has not coevolved with the beetle.

Martin (2007) describes the sharp reduction of woody plant species between 1987 and 1995 at the higher elevations of the Mogollon Rim in central Arizona by elk (*Cervus elaphus*). Historically, the winter snowpacks forced the animals to winter at lower elevations where food plants were not snow covered. But with shrinking snowpacks (cf. Chap. 4), the elk have in recent years been able to winter at the higher elevations and have browsed out much of the woody vegetation. Five avian species that use the vegetation for nesting declined between 1985 and 2005, and a sixth disappeared from the research area.

Predicted System Changes in Response to Climate Change
A number of authors are predicting changes in entire ecosystems in response to changing temperatures and precipitation. Several (Romme and Turner 1991, Neilson 1998, Reiners 2003) are projecting elevational shifts in the western montane vegetation zones, depending on trends in both temperature and precipitation. If only temperatures rise with no change in precipitation, all zones are predicted to move upslope. With temperature increase in the alpine zones of the higher mountains, forests are predicted to occupy these elevations, obliterating the alpine vegetation and associated fauna (Romme and Turner 1991). Without precipitation increase, the lower timberline is predicted to move upslope and be followed by the shrub-steppe of the foothills due to the increased aridity produced by the temperature rise.

If precipitation increases significantly along with temperature rise, the lower timberline is projected to move downslope, conceivably extending woody vegetation out into the valleys of the Intermountain West. The result would be replacement of the valley shrub-steppe, and probably such faunal elements as sage grouse (*Centrocercus urophasianus*) and pronghorn antelope (*Antilocapra americana*).

In Chapter 9, Barber et al. discuss the declining radial growth of white spruce in Alaska and predict that a sufficient further temperature rise would produce widespread mortality of the species. At the same time, the drought-stressed trees are more vulnerable to insect attack. The accelerated tree mortality induced by these agents provides fuel for wildfires, which have increased in recent years and have destroyed entire white spruce forests. Inevitably, this loss eliminates the associated fauna of the vegetation type.

Western streams possess temperature gradients from warmer levels in the lower reaches near their mouths to lower temperatures toward their headwaters at higher elevations. The aquatic plant and animal species sort themselves out along these gradients according to their temperature preferenda. In Utah streams, the introduced brown trout (*Salmo trutta*) occupies the lower reaches, the rainbow trout (*Oncorhynchus mykiss*) the middle reaches, and the native cutthroat trout (*Oncorhynchus clarki*) the higher elevations.

Aquatic ecologists predict that stream temperature increases will move the entire faunal gradient upslope, with a resulting shift in stream habitats and reduction in the upper ones. Keleher and Rahel (1996) project that a 3°C (5.4°F) mean July air temperature increase would reduce stream habitat for salmonid species in the Rocky Mountains of Colorado by 21 percent. A 5°C (9.0°F) summer air temperature rise would reduce habitable stream lengths by 43 percent. The most vulnerable species are the headwater inhabitants, which can move no higher than the stream origins. Conceivably, stream reaches habitable for cutthroat trout in some intermediate-elevation Utah mountains would be foreclosed, and the species eliminated.

Perspective

Plant and animal species worldwide, living in natural environments, have adapted to climates in their geographic areas, and to coexistence with other species that have also accommodated to the local physical environments. Changing climates will affect them and their coexisting species directly and/or through changes in their interactions with others in innumerable ways, and with the result of universally altered natural communities and ecosystems. Changes are already being observed in terrestrial, freshwater, and marine systems. Ultimately, it is likely that the entire global biota will change in ways too diverse, and in many cases too profound, to predict at this time.

Perhaps the systems most vulnerable to change are those that have adapted to extreme physical-environmental conditions barely tolerable for living organisms. Barber et al., in Chapter 9, describe a plethora of profound changes already during the twentieth century in virtually every aspect of the Alaskan landscape: receding glaciers, thawing permafrost, land subsidence, disappearing ponds and lakes, shrinking streams, aerobic and anaerobic breakdown of ages-old soil organic matter, eroding coastlines. The consequences for the biota are no less extreme: declining photosynthetic rates in trees that make them more vulnerable to insect attack, both of which kill the trees and provide fuel for fires (over one-fourth of the forest land in the northeast quadrant of the state burned in 2004 and 2005); invasion of the tundra by woody vegetation; changing distributions and phenologies of the fauna; impacts on the lifestyles of indigenous people.

Obviously climates have changed in past eras of earth history. Many species have evolved to adjust to those changes. But the current rate of climate change is higher than most of the changes of the past, and a number of authors conclude that this rate is too high to enable many species to evolve sufficiently to accommodate the altered conditions (Barnosky and Kraatz 2007). Barber et al. point out that the Alaskan Native people, who have been on the continent for 12,000–20,000 years, have no terms in their lexicons for the changes underway. Widespread extinctions are being predicted.

References

Albritton, D. L., et al. 2007. *Summary for Policymakers: A Report of Working Group I of the Intergovernmental Panel on Climate Change.* I.P.C.C., Shanghai, PRC.

Amthor, J. S. 1989. *Respiration and Crop Production.* Springer-Verlag, New York.

Anthes, R. 2003. Congress looks at climate change. *UCAR Quarterly* (fall) 2003:5.

Baldwin, C. K. 2003. Historical climate analysis. Pp. 58–72 in F. H. Wagner (ed.), *Preparing for a Changing Climate: The Potential Consequences of Climate Variability and Change/Rocky Mountain/Great Basin Regional Climate-change Assessment.* Rept. for the U.S. Global Change Res. Prog. Utah State Univ. Press, Logan.

Baldwin, C. K., F. H. Wagner, and U. Lall. 2003. Water resources. Pp. 79–112 in F. H. Wagner (ed.), *Preparing for a Changing Climate/Rocky Mountain/Great Basin Regional Climate-change Assessment.* Rept. for the U.S. Global Change Res. Prog. Utah State Univ. Press, Logan, UT.

Barnosky, A. D., and B. P. Kraatz. 2007. The role of climatic change in the evolution of mammals. *BioScience* 57:523–532.

Beever, E. A., P. F. Brussard, and J. Berger. 2003. Patterns of apparent extirpation among isolated populations of pikas (*Ochotona princeps*) in the Great Basin. *J. Mammal.* 84:37–54.

Beever, E. A., D. A. Pyke, J.C. Chambers, F. Landau, and S. D. Smith. 2005. Monitoring temporal change in riparian vegetation of Great Basin National Park. *West. No. Amer. Nat.* 65:382–402.

Berry, J., and O. Björkman. 1980. Photosynthetic response and adaptation to temperature in higher plants. *Ann. Rev. Plant Physiol.* 31:491–543.

Bradley, R. S., F. T. Keimig, and H. F. Diaz. 2004. Projected temperature changes along the American cordillera and the planned GCOS network. *Geophys. Res. Lett.* 31, L16210:4pp.

Cayan, D. R., S. A. Kamerdiener, M. D. Dettinger, J. M. Caprio, and D. H. Peterson. 2001. Changes in the onset of spring in the western United States. *Bull. Amer. Meteor. Soc.* 82:399–415.

Chapman, D. S. 1998. *Global warming: Just Hot Air?* Gould 7th Ann. Distinguished Lecture in Technology and the Quality of Life. Univ. of Utah, Salt Lake City.

Crowley, T. J., and J. S. Lowery. 2000. How warm was the Medieval warm period? *Ambio* 29:51–54.

Crozier, L. 2002. Climate change and its effect on species range boundaries: A case study of the sachem skipper butterfly, *Atalopedes campestris*. Pp. 57–91 in S. H. Schneider and T. L. Root (eds.), *Wildlife Responses to Climate Change: North American Case Studies*. Island Press, Washington, D.C.

Dettinger, M. D. 2006. Changes in streamflow timing in the western United States in recent decades. *USGS Fact Sheet 2005–3018.* 6 pp.

DiSilvestro, R. 2005. The proof is in the science. *Nat'l. Wildlife* (April/May):22.

Doherty, R., and L. O. Mearns. 1999. *A Comparison of Simulations of Current Climate from Two Coupled Atmosphere-ocean Global Climate Models Against Observations and Evaluation of Their Future Climates*. Rept. Nat. Inst. Global Env. Change (NIGEC). Boulder, CO.

Donahue, D. L. 1999. *The Western Range Revisited: Removing Livestock from Public Lands to Conserve Biodiversity*. Univ. of Oklahoma Press, Norman.

Drake, D. B., M. A. Gonzalez-Meler, and S. P. Long. 1997. More efficient plants: A consequence of rising atmospheric CO_2? *Ann. Rev. Plant Physiol., and Plant Molec. Biol.* 48:607–640.

Field, C. B., G. C. Daily, F. W. Davis, S. Gaines, P. A. Matson, J. Melack, and N. L. Miller. 1999. *Confronting Climate Change in California: Ecological Impacts on the Golden State*. Union Con. Scient., and Ecol. Soc. Amer., Cambridge, MA, and Washington, D.C.

Field, C. B., C. P. Lund, N. R. Chiariello, and B. E. Mortimer. 1997. CO_2 effects on the water budget of grassland microcosm communities. *Global Change Biol.* 3:197–206.

Fyfe, J. C., and G. M. Flato. 1999. Enhanced climate change and its detection over the Rocky Mountains. *J. Clim.* 12:230–243.

Hall, M. H. P., and D. B. Fagre. 2003. Modeled climate-induced glacier change in Glacier National Park, 1850–2100. *BioScience* 53:131–140.

Hersteinsson, P., and D. W. MacDonald. 1992. Interspecific competition and the geographical distribution of red and Arctic foxes *Vulpes vulpes* and *Alopex lagopus*. *Oikos* 64:505–515.

Inouye, D. W., B. Barr, K. B. Armitage, and B. D. Inouye. 2000. Climate change is affecting altitudinal migrants and hibernating species. *Proc. Nat. Acad. Sci.* 97:1630–1633.

Jenkins, M. 2007a. Into thin air? *High Country News* 39 (8):6–7.

———. 2007b. L.A. bets on the farm. *High Country News* 39(21): 12–17.

Karl, T. R., R. W. Knight, D. R. Easterling, and R. G. Quale. 1996. Indices of climate change of the United States. *Bull. Amer. Meteor. Soc.* 77:279–292.

Keleher, C. J., and F. J. Rahel. 1996. Thermal limits to salmonid distributions in the Rocky Mountain region and the potential habitat loss due to global warming: A geographic information system (GIS) approach. *Trans. Amer. Fish. Soc.* 125:1–13.

Kittel, T. G. F., P. E. Thornton, J. A. Royle, and T. N. Chase. 2002. Climates of the Rocky Mountains: Historical and future patterns. Pp. 59–82 in J. Baron (ed.), *Rocky Mountain Futures: An Ecological Perspective*. Island Press, Washington, D.C.

Knowles, N., M. D. Dettinger, and D. R. Cayan. 2006. Trends in snowfall versus rainfall in the western United States. *J. Clim.* 19:4545–4559.

Kolbert, E. 2006. Butterfly lessons: Insects and toads respond to global warming. *The New Yorker,* Jan. 9, 2006:32–39.

Mann, M. E., R. S. Bradley, and M. K. Hughes. 1999. Northern Hemisphere temperatures during the past millennium: Inferences, uncertainties, and limitations. *Geophys. Res. Lett.* 26:759–762.

Martin, T. E. 2007. Climate correlates of 20 years of trophic changes in a high-elevation riparian system. *Ecol.* 88:367–380.

Mearns, L. O., I. Bogardi, F. Giorgi, L. Matyasovsky, and M. Paleki. 1999. Comparison of climate change scenarios generated from regional climate model experiments and statistical downscaling. *J. Geophys. Res.* 104:6603–6621.

Miller, K. A. 1997. *Climate Variability, Climate Change, and Western Water*. Rept. Western Water Policy Rev. Adv. Comm., Nat Tech. Info. Serv. Springfield, VA.

Moller, A. P., W. Fiedler, and P. Berthold. 2004. *Advances in Ecological Research: Birds and Climate Change*. Elsevier Academic Press, Oxford.

Mote, P. 1999. *Impacts of Climate Variability and Change in the Pacific Northwest*. A Rept. of the Pacific Northwest Regional Assessment Group for the U.S. Global Change Res. Prog., Univ. of Washington, Seattle.

Nash, L. L., and P. H. Gleick. 1993. *The Colorado River Basin and Climatic Change: The Sensitivity of Streamflow and Water Supply to Variations in Temperature and Precipitation*. A Report Prepared for the U.S. Env. Prot. Agency, EPA 230-R-93-009. Pacific Inst. for Studies in Dev., Env., and Security, Oakland, CA.

Neilson, R. P. 1998. Potential effects of global warming on natural vegetation at global, national and regional levels. Pp. 55–63 in F. H. Wagner and J. Baron (eds.), *Proc. Rocky Mountain/Great Basin Regional Climate-change Workshop, Feb. 16–18, 1998, Salt Lake City, UT*. Utah State Univ. Press, Logan.

Nijhuis, M. 2005. The ghosts of Yosemite. *High Country News* 37(19): 8–14, 19.

Ojima, D. S., and J. M. Lackett. 2002. *Preparing for a Changing Climate: The Potential Consequences of Climate Variability and Change/Central Great Plains*. A Rept. Of the Central Great Plains Regional Assessment Group for the U.S. Global Change Res. Prog., Colorado State Univ., Ft. Collins.

Parmesan, C. 1996. Climate and species range. *Nature* 382:765–766.

Peterson, T. C., and R. S. Vose. 1997. An overview of the Global Historical Climatology Network temperature data base. *Bull. Amer. Meteor. Soc.* 78:2837–2849.

Price, J. 1995. Potential impacts of global climate change on the summer distribution of some North American grassland birds. Ph.D. diss. Wayne State Univ., Detroit, MI.

Reilly, J., F. Tubiello, B. McCarl, and J. Melillo. 2001. Climate change and agriculture in the United States. Pp. 379–403 in National Assessment Synthesis Team, *Climate Change Impacts on the United States.* Rept. for the Global Change Res. Prog. Cambridge Univ. Press, Cambridge, UK.

Reiners, R. P. 2003. Natural ecosystems: 1. The Rocky Mountains. Pp. 145–184 in F. H. Wagner (ed.), *Preparing for a Changing Climate: The Potential Consequences of Climate Variability and Change/Rocky Mountain/Great Basin Regional Climate Change Assessment.* A Report of the Rocky Mountain/Great Basin Regional Assessment Team for the U.S. Global Change Res. Prog. Utah State Univ. Press, Logan.

Romme, W. H., and M. G. Turner. 1991. Implications of global climate change for biogeographic patterns in the greater Yellowstone ecosystem. *Cons. Biol.* 5:373–386.

Sagarin, R. 2002. Historical studies of species' responses to climate change: Promises and pitfalls. Pp. 127–163 in S. H. Schneider and T. L. Root (eds.), *Wildlife Responses to Climate Change: North American Case Studies.* Island Press, Washington, D.C.

Sagarin, R. D., J. P Barry, S. E. Gilman, and C. H. Baxter. 1999. Climate-related change in an intertidal community over short and long time-scales. *Ecol. Monog.* 69:465–490.

Schneider, S. H., and T. L. Root (eds.). 2002. *Wildlife Responses to Climate Change: North American Case Studies.* Island Press, Washington, D.C.

Seager, R. 2007. An imminent transition to a more arid climate in southwestern North America. http://www.ldeo.columbia.edu/res/div/ocp/drought/science.shtml.

Siegenthaler, U., T. F. Stocker, E. Monnin, D. Luthi, J. Schwander, B. Stauffer, D. Raynaud, J.-M. Barnola, H. Fischer, V. Masson-Delmotte, and J. Jouzel. 2005. Stable carbon cycle-climate relationship during the late Pleistocene. *Science* 310:1313–1317.

Smith, S. D., K. J. Murray, F. H. Landau, and A. Sala. 1994. *The Woody Riparian Vegetation of Great Basin National Park.* Contrib. CPSU/UNLV 050/03, Coop. Nat. Park Res. Studies Unit, Nat. Biol. Surv., Univ. of Nevada, Las Vegas.

Sprigg, W. A., T. Hinkley, et al. 2000. *Preparing for a Changing Climate: The Potential Consequences of Climate Variability and Change/Southwest.* A Rept. of the Southwest Regional Assessment Group for the U.S. Global Change Res. Prog. Univ. of Arizona, Tucson.

Tebaldi, C., K. Hayhoe, J. M. Arblaster, and G. A. Meehl. 2006. Going to the extremes: An intercomparison of model-simulated historical and future changes in extreme events. *Clim. Change* (2006), doi:10.1007/s10584-006-9051-4.

Wagner, F. H., and C. K. Baldwin. 2003. Cultivated agriculture and ranching. Pp. 113–129 in F. H. Wagner (ed.), *Preparing for a Changing Climate: The Potential Consequences of Climate Variability and Change/Rocky Mountain/Great Basin Regional Climate-Change Assessment.* A Rept. for the Global Change Res. Prog. Utah State Univ. Press, Logan.

Wahl, E. R., and C. M. Ammann. 2005. Robustness of the Mann, Bradley, Hughes reconstruction of surface temperatures: Examination of criticisms based on the nature and processing of proxy climate evidence. *Clim. Change* (2006) http://www.cgd.ucar.edu/ammann/millennium/refs/WahlAmmann.

Contributors

Valerie A. Barber
Forest Sciences Department
School of Natural Resources and Agricultural Sciences
University of Alaska
Fairbanks

Marshall Bartlett
Department of Physical Sciences
BYU Hawaii
Manoa

Edward Berg
Kenai National Wildlife Refuge
Soldotna, Alaska

Brendan Buckley
Tree-Ring Laboratory
Lamont-Doherty Earth Observatory
Palisades, New York

Daniel R. Cayan
U.S. Geological Survey
Scripps Institution of Oceanography
La Jolla, California

David S. Chapman
Department of Geology and Geophysics
University of Utah
Salt Lake City

Rosanne D'Arrigo
Tree-Ring Laboratory
Lamont-Doherty Earth Observatory
Palisades, New York

Michael D. Dettinger
Scripps Institution of Oceanography
U.S. Geological Survey
La Jolla, California

Noah S. Diffenbaugh
Purdue Climate Change Research Center and
Department of Earth and Atmospheric Sciences
Purdue University
West Lafayette, Indiana

Robert N. Harris
College of Earth and Atmospheric Science
Oregon State University
Corvallis

Henry Huntington
Huntington Consulting
Anchorage, Alaska

Gregory V. Jones
Department of Environmental Studies
Southern Oregon University
Ashland

Torre Jorgensen
ABR, Inc. Environmental Research and Services
Fairbanks, Alaska

Glenn Patrick Juday
Forest Sciences Department
School of Natural Resources and Agricultural Sciences
University of Alaska
Fairbanks

Jesse A. Logan
USDA Forest Service
Rocky Mountain Research Station
Logan, Utah

Jessica D. Lundquist
University of Washington
Seattle

John H. Matthews
Section of Integrative Biology
University of Texas
Austin

David McGuire
U.S. Geological Survey
Alaska Cooperative Fish and Wildlife Research Unit
University of Alaska
Fairbanks

Kirk Miller
U.S. Geological Survey
Cheyenne, Wyoming

Philip W. Mote
Climate Impacts Group
Center for Science in the Earth System
University of Washington
Seattle

David L. Naftz
U.S. Geological Survey
Salt Lake City, Utah

Tom Osterkamp
Geophysical Institute
University of Alaska
Fairbanks

Liz Oswald
USDA Forest Service
Lander
Wyoming

Camille Parmesan
Section of Integrative Biology
University of Texas
Austin

James A. Powell
Department of Mathematics and Statistics
Utah State University
Logan

Brian Riordan
Forest Sciences Department
School of Natural Resources and Agricultural Sciences
University of Alaska
Fairbanks

Paul F. Schuster
U.S. Geological Survey
Boulder, Colorado

Iris T. Stewart
Santa Clara University
Santa Clara, California

Frederic H. Wagner
Department of Wildland Resources
Utah State University
Logan

Michael A. White
Department of Watershed Sciences
Utah State University
Logan

Alex Whiting
Native Village of Kotzebue
Alaska

Greg Wiles
Wooster College
Wooster, Ohio

Martin Wilmking
Grieswald University
Germany

Index